T0338583

NONLINEAR ANALYSIS AND SEMILINEAR ELLIPTIC PROBLEMS

Many problems in science and engineering are described by nonlinear differential equations, which can be notoriously difficult to solve. Through the interplay of topological and variational ideas, methods of nonlinear analysis are able to tackle such fundamental problems. This graduate text explains some of the key techniques in a way that will be appreciated by mathematicians, physicists and engineers. Starting from the elementary tools of bifurcation theory and analysis, the authors cover a number of more modern topics including critical point theory and elliptic partial differential equations. A series of appendices gives convenient accounts of a variety of advanced topics that will introduce the reader to areas of current research. The book is amply illustrated and many chapters are rounded off with a set of exercises.

Cambridge Studies in Advanced Mathematics

Editorial Board:

B. Bollobas, W. Fulton, A. Katok, F. Kirwan, P. Sarnak, B. Simon, B. Totaro

All the titles listed below can be obtained from good booksellers or from Cambridge University Press. For a complete series listing visit:
http://www.cambridge.org/mathematics/

NONLINEAR ANALYSIS
AND SEMILINEAR ELLIPTIC
PROBLEMS

ANTONIO AMBROSETTI

ANDREA MALCHIODI

CAMBRIDGE
UNIVERSITY PRESS

CAMBRIDGE
UNIVERSITY PRESS

University Printing House, Cambridge CB2 8BS, United Kingdom

One Liberty Plaza, 20th Floor, New York, NY 10006, USA

477 Williamstown Road, Port Melbourne, VIC 3207, Australia

314-321, 3rd Floor, Plot 3, Splendor Forum, Jasola District Centre, New Delhi - 110025, India

79 Anson Road, #06-04/06, Singapore 079906

Cambridge University Press is part of the University of Cambridge.

It furthers the University's mission by disseminating knowledge in the pursuit of education, learning and research at the highest international levels of excellence.

www.cambridge.org
Information on this title: www.cambridge.org/9780521863209

First published 2007

A catalogue record for this publication is available from the British Library

ISBN 978-0-521-86320-9 Hardback

Contents

Preface

The main purpose of nonlinear functional analysis is to develop abstract topological and variational methods to study nonlinear phenomena arising in applications. Although this is a rather recent field, initiated about one hundred years ago, remarkable advances have been made and there are now many results that are well established. The fundamental tools of the Leray–Schauder topological degree, local and global bifurcation and critical point theory, can be considered topics that any graduate student in mathematics and physics should know.

This book discusses a selection of the most basic results dealing with the aforementioned topics. The material is presented as simply as possible, in order to highlight the main ideas. In many cases we prefer to state results under slightly stronger assumptions, when this makes the exposition much more clear and avoids some unnecessary technicalities.

The abstract tools are discussed taking into account their applications to semilinear elliptic problems. In some sense, elliptic equations become like a guiding thread, along which the reader will recognize how one method is more suitable than another one, according to the specific feature of the nonlinearity. This is the reason why we discuss both topological methods and variational tools.

After a first chapter containing preliminary material, the book is divided into four parts. The first part is devoted to topological methods and bifurcation theory. Chapter 2 deals with the Lyapunov–Schmidt reduction method and the bifurcation from a simple eigenvalue and connects with the previous book *A Primer of Nonlinear Analysis* [20], of which the present book is a follow up. Chapter 3 deals with the topological degree. First, we define the degree in finite dimension using an analytical approach, which allows us to avoid several technical and cumbersome tools. Next, the Leray–Schauder degree is discussed together with some applications to elliptic boundary value problems.

Among the applications, we also prove the celebrated theorem by Krasnoselski dealing with the bifurcation from an odd eigenvalue for operators of the type *identity-compact*. In Chapter 4 global properties of the degree are discussed. In particular, the global bifurcation result due to Rabinowitz is proved. Special attention is also given to the existence of positive solutions of asymptotically linear boundary value problems.

Parts II and III are devoted to variational methods, namely to critical point theory. After some introductory material presented in Chapters 5 and 6, we discuss in Chapter 7 the main deformation lemmas and the Palais–Smale condition. Chapter 8 deals with the mountain pass and linking theorems. The Lusternik–Schnirelman theory and, in particular, the cases of even functionals on symmetric manifolds are discussed in Chapters 9 and 10, respectively.

Further results on elliptic boundary value problems are presented in Chapter 11, including the pioneering Brezis–Nirenberg result dealing with semilinear equations with critical nonlinearities.

An account of Morse theory is given in Chapter 12 which also contains applications to bifurcation for potential operators and to evaluation of the Morse index of a mountain pass critical point.

Part IV collects a number of appendices which deal with interesting problems that have been left out in the preceding parts because they are more specific in nature, or more complicated, or else because they are objects of current research and therefore are still in evolution. Here our main purpose is to bring the interested reader to the core of contemporary research. In many cases, we are somewhat sketchy, referring to original papers for more details.

Appendix 1 deals with the celebrated Gidas–Ni–Nirenberg symmetry result and with other qualitative results, such as the Liouville type theorem of Gidas and Spruck. Appendix 2 is concerned with the concentration-compactness method introduced by P. L. Lions and includes applications to problems with lack of compactness. Appendix 3 is related to bifurcation theory and deals with bifurcation problems in the absence of compactness, including bifurcation from the essential spectrum. Appendix 4, deals with the classical problem of vortex rings in an ideal fluid. In Appendix 5 we discuss some abstract perturbation methods in critical point theory with their applications to elliptic problems on $\mathbb{R}^n$, to nonlinear Schrödinger equations and to singular perturbation problems. Finally, in Appendix 6 we discuss some problems arising in differential geometry, from the classical Yamabe problem to more recent problems, dealing with fourth order invariants such as the Paneitz curvature.

The book is based on many sources. The first is the material taught in several courses given in past years at SISSA. Some of this material is based on previous lectures delivered by by Giovanni Prodi at the Scuola Normale of Pisa in the 1970s. Very special thanks are due to this great mathematician and friend.

The second source is the papers that we have written on nonlinear analysis. Most of them are works in collaboration with other people: we would like to thank all of them warmly (see the authors of joint papers with A. A. or A. M. listed in the references).

Another input has been discussions with many other friends, including V. Coti Zelati, I. Ekeland, M. Girardi, M. Matzeu and C. Stuart.

A. A. & A. M.

1

Preliminaries

In this chapter we will discuss some preliminary material we will use throughout the book.

1.1 Differential calculus

Let us begin with an outline, without proofs, of differential calculus in Banach spaces. For proofs and more details we refer to [20], Chapters 1 and 2.

The Fréchét derivative. Let X, Y be Banach spaces and let $L(X, Y)$ denote the space of linear continuous maps from X to Y. For $A \in L(X, Y)$ we will often write Ax or $A[x]$ instead of $A(x)$. Endowed with the norm

$$\|A\|_{L(X,Y)} = \sup\{\|Ax\|_Y : \|x\|_X \le 1\}, \quad A \in L(X, Y),$$

$L(X, Y)$ is a Banach space. If $U \subset X$ is an open set, $C(U, Y)$ denotes the space of continuous maps $f : U \to Y$.

Definition 1.1 *We say that $f : U \mapsto Y$ is (Fréchét) differentiable at $u \in U$ with derivative $\mathrm{d}f(u) \in L(X, Y)$ if*

$$f(u + h) = f(u) + \mathrm{d}f(u)[h] + o(\|h\|), \quad as \ h \to 0.$$

f is said differentiable on U if it is differentiable at every point $u \in U$.

From the definition it follows that if f is differentiable at $u \in U$ then f is continuous at u.

In order to find the derivative of a map f one can evaluate, for all $h \in X$, the limit

$$\lim_{\varepsilon \to 0} \frac{f(u + \varepsilon h) - f(u)}{\varepsilon} \overset{\mathrm{def}}{=} A_u h.$$

If $A_u \in L(X, Y)$ and if the map $u \mapsto A_u$ is continuous from U to $L(X, Y)$, then f is differentiable at u and $\mathrm{d}f(u) = A_u$. We will often use $f'(u)$ instead of $\mathrm{d}f(u)$.

Let $f : X \times Y \to Z$, and consider the map $f_v : u \mapsto f(u, v)$, respectively $f_u : v \mapsto f(u, v)$. The *partial derivative* of f with respect to u, respectively v, at $(u, v) \in X \times Y$ is defined by $\partial_u f(u, v) = df_v(u)$, respectively $\partial_v f(u, v) = df_u(v)$. In particular, $\partial_u f(u, v) \in L(X, Z)$ and $\partial_v f(u, v) \in L(Y, Z)$. It is easy to see that if $f : X \times Y \mapsto Z$ is differentiable at (u, v), then f is partially differentiable and $\partial_u f(u, v)[h] = df_v(u)[h] = df(u, v)[h, 0]$, respectively $\partial_v f(u, v)[k] = df_u(v)[k] = df(u, v)[0, k]$. Furthermore, the following result holds.

Proposition 1.2 *If f possesses the partial derivative with respect to u and v in a neighbourhood $\mathcal{N}$ of (u, v) and the maps $u \mapsto \partial_u f$ and $v \mapsto \partial_v f$ are continuous in $\mathcal{N}$, then f is differentiable at (u, v) and*

$$df(u, v)[h, k] = \partial_u f(u, v)[h] + \partial_v f(u, v)[k].$$

In the sequel, if no confusion arises, we will write f_u', f_v' instead of $\partial_u f$, $\partial_v f$, respectively.

Higher order derivatives. Let f be differentiable on U and the map $X \mapsto L(X, Y), u \mapsto df(u)$, be differentiable at $u \in U$. The derivative of such a map at u is a second derivative: $d^2 f(u) \in L(X, L(X, Y))$. From the canonical isomorphism between $L(X, L(X, Y))$ and $L_2(X, Y)$, the space of bilinear maps from X to Y, we can and will consider $d^2 f(u)$ belonging to $L_2(X, Y)$. By induction on k we can define the kth derivative $d^k f(u) \in L_k(X, Y)$, the space of k-linear maps from E into $\mathbb{R}$. If f is k times differentiable at every point of U, we say that f is k times differentiable on U.

We will use the following notation.

- C^k *maps.* If f is k times differentiable on U and the application
  $U \mapsto L_k(X, Y), u \mapsto d^k f(u)$, is continuous, we say that $f \in C^k(U, Y)$.
- $C^{0,\alpha}$ *maps.* If $f \in C(U, Y)$ satisfies

$$\sup \left[\frac{\|f(u) - f(v)\|_Y}{\|u - v\|_X^\alpha} : u, v \in U, \ u \neq v \right] < +\infty,$$

for some $\alpha \in (0, 1]$, we say that $f \in C^{0,\alpha}(U, Y)$. If $\alpha < 1$, respectively $\alpha = 1$, these maps are nothing but the *Hölder continuous* maps, respectively *Lipschitz* maps.

- $C^{k,\alpha}$ maps. If $f \in C^k(U, Y)$ and $d^k f(u) \in C^{0,\alpha}(U, L_k(X, Y))$, we say that $f \in C^{k,\alpha}(U, Y)$.

Let $f \in C^k(U, Y)$ and suppose that $u, h \in U$ be such that $u + th \in U$ for all $t \in [0, 1]$. Since one has

$$\frac{d^r}{dt^r} f(u + th) = d^r f(u + th)[h]^r, \qquad [h]^r = \underbrace{[h, \ldots, h]}_{r \text{ times}}, \qquad r = 1, \ldots, k,$$

the Taylor formula for $t \mapsto f(u + th)$ yields

$$f(u + h) = f(u) + df(u)[h] + \cdots + \frac{1}{k!} d^k f(u)[h]^k + o(\|h\|^k).$$

Local inversion and implicit function theorems. Let $f \in C(U, Y)$, $u^* \in U$ and $v^* = f(u^*) \in Y$. We say that f is *locally invertible* at u^* if there exist neighbourhoods U^* of u^* and V^* of v^* and a map $g \in C(V^*, U^*)$ such that

$$g(f(u)) = u, \quad \forall u \in U^*, \qquad f(g(v)) = v, \quad \forall v \in V^*.$$

The map g will be denoted by f^{-1}.

Theorem 1.3 (Local inversion theorem) *Suppose that $f \in C^1(U, Y)$ and that $df(u^*)$ is invertible (as a linear map in $L(X, Y)$). Then f is locally invertible at u^*, f^{-1} is of class C^1 and*

$$df^{-1}(v) = (df(u))^{-1}, \quad \forall v \in V^*, \quad \text{where } u = f^{-1}(v).$$

Furthermore, if $f \in C^k(U, Y)$ then f^{-1} is of class C^k, as well.

Let T, X be Banach spaces, $\Lambda \subset T$, $U \subset X$ be open subsets.

Theorem 1.4 (Implicit function theorem) *Let $f \in C^k(\Lambda \times U, Y)$, $k \geq 1$, and let $(\lambda^*, u^*) \in \Lambda \times U$ be such that $f(\lambda^*, u^*) = 0$. If $f'_u(\lambda^*, u^*) \in L(X, Y)$ is invertible, then there exist neighbourhoods Λ^* of λ^*, U^* of u^* and a map $g \in C^k(\Lambda^*, X)$ such that*

$$f(\lambda, u) = 0, \qquad (\lambda, u) \in \Lambda^* \times U^* \Leftrightarrow u = g(\lambda).$$

Moreover one has that

$$g'(\lambda) = -(f'_u(p))^{-1} \circ f'_\lambda(p), \quad \text{where } p = (\lambda, g(\lambda)), \; \lambda \in \Lambda^*.$$

1.2 Function spaces

We will deal with *bounded domains* Ω *contained in the Euclidean n-dimensional space* $\mathbb{R}^n$. We will mainly work in the following spaces of functions $u : \Omega \to \mathbb{R}$:

- $L^p(\Omega)$, Lebesgue spaces with norm $\| \cdot \|_{L^p}$;
- $H^{k,p}(\Omega)$, Sobolev spaces with norm $\| \cdot \|_{H^{k,p}}$;
- $C_0^\infty(\Omega)$, the space of functions $u \in C^\infty(\Omega)$ with compact support in Ω;
- $H_0^1(\Omega)$, the closure of $C_0^\infty(\Omega)$ in $H^{1,2}(\Omega)$.

For functions in $H_0^1(\Omega)$ the *Poincaré inequality* holds:

$$\int_\Omega |u|^2 dx \le c \int_\Omega |\nabla u|^2 dx,$$

where $c = c(\Omega)$ is a constant (possibly depending on Ω but independent of u).

As a consequence of the Poincaré inequality it follows that

$$\|u\| = \left(\int_\Omega |\nabla u|^2 \, dx \right)^{1/2}$$

is a norm equivalent to the standard one $\|u\|_{H_0^1}$.

Theorem 1.5 (Sobolev embedding theorem) *Let Ω be a bounded domain in $\mathbb{R}^n$ with Lipschitz boundary $\partial\Omega$ and let $k \ge 1$, $1 \le p \le \infty$.*

(i) *If $kp < n$, then $H^{k,p}(\Omega) \hookrightarrow L^q(\Omega)$ for all $1 \le q \le np/(n-kp)$; the embedding is compact provided $1 \le q < np/(n-kp)$.*

(ii) *If $kp = n$, then $H^{k,p}(\Omega) \hookrightarrow L^q(\Omega)$ for all $1 \le q < \infty$, and the embedding is compact.*

(iii) *If $kp > n$, then $H^{k,p}(\Omega) \hookrightarrow C^{0,\alpha}(\Omega)$, where*

$$\alpha = \begin{cases} k - n/p & \text{if } k - n/p < 1 \\ 1 & \text{if } k - n/p > 1. \end{cases}$$

If $k - n/p = 1$, the embedding holds for every $\alpha \in [0, 1)$.

When we deal with $H_0^1(\Omega)$ the requirement that $\partial\Omega$ is Lipschitz can be eliminated. For future references let us state explicitly what Theorem 1.5 becomes in such a case. We set

$$2^* = \begin{cases} 2n/(n-2) & \text{if } n > 2 \\ +\infty & \text{if } n = 2. \end{cases}$$

Theorem 1.6 *Let Ω be a bounded domain in $\mathbb{R}^n$. Then*

(i) *if $n > 2$, then $H_0^1(\Omega) \hookrightarrow L^q(\Omega)$ for all $1 \leq q \leq 2^*$; the embedding is compact provided $1 \leq q < 2^*$;*

(ii) *if $n = 2$, then $H_0^1(\Omega) \hookrightarrow L^q(\Omega)$ for all $1 \leq q < \infty$;*

(iii) *if $n < 2$, then $H_0^{1,p}(\Omega) \hookrightarrow C^{0,\alpha}(\Omega)$, where $\alpha = 1 - n/2$.*

1.3 Nemitski operators

Let $f : \Omega \times \mathbb{R} \to \mathbb{R}$. If $u : \Omega \to \mathbb{R}$ is a measurable real valued function, we can consider the map $u \mapsto f(u)$, where $f(u)$ is the real valued function defined on Ω by setting

$$f(u)(x) = f(x, u(x)).$$

Such a map is called the *Nemitski operator* associated to f and will be denoted with the same symbol f. For a discussion of the continuity and differentiability properties of Nemitski operators we refer to [20], Chapter 1, Section 2. Here we want to recall the following result.

Theorem 1.7 *Let $\alpha, \beta \geq 1$. Suppose that $f : \Omega \times \mathbb{R} \to \mathbb{R}$ satisfies*

(f.0) $\quad f(x, t)$ *is measurable with respect to $x \in \Omega$ for all $t \in \mathbb{R}$ and is continuous with respect $t \in \mathbb{R}$ for a.e. $x \in \Omega$,*

and that there exists $a_1 \in L^\beta(\Omega)$ and $a_2 > 0$ such that

$$|f(x, u)| \leq a_1(x) + a_2|u|^{\alpha/\beta} \quad \forall (x, u) \in \Omega \times \mathbb{R}, \quad (\alpha, \beta \geq 1). \tag{1.1}$$

Then the Nemitski operator f is continuous from $L^\alpha(\Omega)$ to $L^\beta(\Omega)$.

Condition (f.0) is also called the *Caratheodory condition* and a function $f(x, t)$ satisfying (f.0) is usually called a *Caratheodory function*. In most of the concrete applications we will deal with the functional space $H_0^1 = H_0^1(\Omega)$. In such a case, we shall suppose that f satisfies

(f.1) there exists $a_1 \in L^{2n/(n+2)}(\Omega)$ and $a_2 > 0$ such that

$$|f(x, u)| \leq a_1(x) + a_2|u|^p \quad \forall (x, u) \in \Omega \times \mathbb{R}, \tag{1.2}$$

where $p < 2^* - 1$.

In some cases we will weaken (f.1) requiring

(f.1$'$) f satisfies (1.2) with $p \leq 2^* - 1$.

The class of functions f which are locally Hölder continuous and satisfy $(f.1)$ or $(f.1')$, will be denoted by $\mathbb{F}_p$.

Let us point out that, in many cases, we could deal with functions that satisfy $(f.0)$, instead of being locally Hölder continuous; see, for example, Remark 1.9 below. The advantage of working with the class $\mathbb{F}_p$ is that we get classical solutions of elliptic equations we deal with, not merely weak solutions.

Moreover, in the sequel we also take $n > 2$. If $n = 1, 2$ one uses the stronger forms of the Sobolev embedding theorem and the arguments below require minor changes.

If $(f.1')$ holds then, according to Theorem 1.7, one has that

$$f \in C(L^{2^*}, L^{2n/(n+2)}). \tag{1.3}$$

Moreover, setting

$$F(x, u) = \int_0^u f(x, s) \, \mathrm{d}s,$$

it follows that

$$|F(x, u)| \le a_1 |u| + a_2 |u|^{p+1},$$

with $p + 1 \le 2^*$. Since $H_0^1 \hookrightarrow L^{2^*}$ then $F(\cdot, u(\cdot)) \in L^1$ provided $u \in H_0^1$ and it makes sense to consider the map $\Phi : H_0^1 \to \mathbb{R}$ defined by

$$\Phi(u) = \int_\Omega F(x, u) \, \mathrm{d}x. \tag{1.4}$$

One can show, see Theorem 2.9 of [20], that Φ is of class C^1 on H_0^1 and

$$\mathrm{d}\Phi(u)[v] = \int_\Omega f(x, u)v \, \mathrm{d}x.$$

Let us point out that, as remarked before, $f(x, u) \in L^{2n/(n+2)}$ while $v \in H_0^1 \subset L^{2^*}$; hence $f(x, u)v \in L^1$ so that the right hand side of the preceding formula makes sense.

Next, suppose that $f \in \mathbb{F}_p$ with $1 < p < (n+2)/(n-2)$, and let $u_n \to u$, weakly in H_0^1. Since $p + 1 < 2^*$, the embedding of $H_0^1 \hookrightarrow L^{p+1}$ is compact and thus, up to a subsequence, $u_n \to u$ strongly in L^{p+1}. This immediately implies that

$$\Phi(u_n) \to \Phi(u),$$

and shows that Φ is weakly continuous. Similarly, from Theorem 1.7 we infer that $f(u_n) \to f(u)$ in $L^\alpha(\Omega)$, with $\alpha = (p+1)/p$. Using the Hölder inequality we get

$$|\mathrm{d}\Phi(u_n)[v] - \mathrm{d}\Phi(u)[v]| \leq \left[\int_\Omega |f(x, u_n) - f(x, u)|^\alpha \mathrm{d}x\right]^{1/\alpha} \left[\int_\Omega |v|^{p+1} \mathrm{d}x\right]^{1/(p+1)}$$

$$= \|f(u_n) - f(u)\|_{L^\alpha} \|v\|_{L^{p+1}}.$$

Since $p + 1 < 2^*$ we deduce that $\|v\|_{L^{p+1}} \leq c \|v\|_{H_0^1}$, where $c > 0$ is a constant independent of v. In conclusion, we infer that

$$\|\mathrm{d}\Phi(u_n) - \mathrm{d}\Phi(u)\| \leq c \|f(u_n) - f(u)\|_{L^\alpha},$$

and this shows that $\mathrm{d}\Phi$ is a compact operator. Let us collect the above results in the following.

Theorem 1.8 *Suppose that $f \in \mathbb{F}_p$ with $1 < p \leq (n+2)/(n-2)$ and let Φ be defined on $H_0^1(\Omega)$ by (1.4). Then $\Phi \in C^1(H_0^1, \mathbb{R})$.*
Furthermore, if $f \in \mathbb{F}_p$ with $1 < p < (n+2)/(n-2)$, then Φ is weakly continuous and $\mathrm{d}\Phi$ is a compact operator.

Remark 1.9 The first (respectively second) statement holds true if we suppose that f satisfies $(f.0)$ and $(f.1')$, respectively $(f.0)$ and $(f.1)$. Furthermore, if f is Lipschitz, respectively of class C^k, with respect to u then it is easy to see that Φ is of class $C^{1,1}$, respectively C^{k+1}, on H_0^1. ∎

1.4 Elliptic equations

Consider the linear Dirichlet boundary value problem (BVP in short)

$$\begin{cases} -\Delta u(x) = h(x) & x \in \Omega \\ u(x) = 0 & x \in \partial\Omega \end{cases} \tag{1.5}$$

where h is a given function on Ω. If $h \in L^2(\Omega)$, a weak solution of (1.5) is a function $u \in H_0^1(\Omega)$ such that

$$\int_\Omega \nabla u \cdot \nabla v \, \mathrm{d}x = \int_\Omega hv \, \mathrm{d}x, \quad \forall v \in C_0^\infty(\Omega).$$

Hereafter c denotes a possibly different constant independent of u.

Theorem 1.10 *Let $\Omega \subset \mathbb{R}^n$ be a bounded domain.*

(i) If $h \in L^p(\Omega)$, $1 < p < +\infty$, then (1.5) has a unique weak solution $u \in H_0^1(\Omega) \cap H^{2,p}(\Omega)$ such that

$$\|u\|_{H^{2,p}} \leq c\|h\|_{L^p}.$$

(ii) *(Schauder estimates) If Ω is of class $C^{2,\alpha}$ and $h \in C^{0,\alpha}(\overline{\Omega})$ then $u \in C^{2,\alpha}(\overline{\Omega})$ is a classical solution of (1.5) and*

$$\|u\|_{C^{2,\alpha}} \leq c\|h\|_{C^{0,\alpha}}.$$

The statement (i) of the preceding theorem allows us to define a linear selfadjoint operator $K : L^2(\Omega) \to H_0^1(\Omega)$, $h \mapsto K(h) = u$, where u denotes the unique solution of (1.5). K is the *Green* operator of $-\Delta$ on $H_0^1(\Omega)$. Since the embedding of $H_0^1(\Omega)$ into $L^2(\Omega)$ is compact, it follows that K is compact as a map from $L^2(\Omega)$ in itself. Similarly, we can consider K as an operator in $X = C^{0,\alpha}(\Omega)$. From (ii) of Theorem 1.10 it follows that $K(X) \subset \{u \in C^{2,\alpha}(\Omega) : u_{|\partial\Omega} = 0\}$ and Ascoli's theorem implies that K is still compact.

Remark 1.11

(i) We point out that, here and always in the sequel, we can substitute $-\Delta$ with any uniformly elliptic second order operator with smooth coefficients and in divergence form.

(ii) The Schauder estimates stated before hold true when $-\Delta$ is replaced by any second order uniformly elliptic operator such as

$$-\mathcal{L}u = -\sum a_{ij}(x)\frac{\partial^2 u}{\partial x_i \partial x_j} + \sum b_i(x)\frac{\partial u}{\partial x_i} + c(x)u,$$

where a_{ij}, b_i, c are of class $C^1(\overline{\Omega})$, $c \leq 0$ in $\overline{\Omega}$ and $\exists \kappa > 0$ such that $\sum a_{ij}(x)\xi_i\xi_j \geq \kappa|\xi|^2, \forall x \in \overline{\Omega}, \xi \in \mathbb{R}^n$. ∎

1.4.1 Eigenvalues of linear Dirichlet boundary value problems

Consider the linear eigenvalue problem

$$\begin{cases} -\Delta u(x) = \lambda u(x) & x \in \Omega \\ u(x) = 0 & x \in \partial\Omega. \end{cases} \tag{1.6}$$

From the preceding discussion, it follows that (1.6) is equivalent to $u = \lambda K(u)$, $u \in L^2(\Omega)$ or $u \in C^{0,\alpha}(\Omega)$.

It is convenient to recall the main general properties of operators $A \in L(X)$ of the type *identity-compact*, according to the Riesz–Fredholm theory.

(**RF$_1$**) Ker(A) is finite dimensional, Range(A) is closed and has finite codimension;

(**RF$_2$**) Range(A) = [Ker(A^*)]$^\perp$ = $\{u \in X : \langle \psi, u \rangle = 0, \forall \psi \in \text{Ker}(A^*)\}$;

(**RF$_3$**) $\exists m \geq 1$ such that Ker(A^k) = Ker(A^{k+1}), $\forall k \geq m$. Moreover, $\forall k \geq m$ one has that Range(A^{k+1}) = Range(A^k), $X = \text{Ker}(A^m) \oplus \text{Range}(A^m)$

and the restriction of A to Range(A^m) is a linear homeomorphism of
Range(A^m) onto itself;

(RF₄) Ker(A) = {0} ⟺ Range(A) = X.

The preceding results apply to $A_\lambda = I - \lambda K$, where K is the Green operator of
$-\Delta$ with zero Dirichlet boundary conditions.

Definition 1.12 *A real number λ such that* Ker(A_λ) $\neq$ {0} *is an eigenvalue
of* (1.6). *The integer m such that* (RF₃) *holds (with $A_\lambda = A$) is called the
multiplicity of λ. When the multiplicity is equal to 1 we say that the eigenvalue
is simple.*

The following result holds.

Theorem 1.13
(i) *Equation* (1.6) *has a sequence of eigenvalues λ_k such that*

$$0 < \lambda_1 < \lambda_2 \leq \lambda_3 \leq \ldots, \quad \lambda_k \nearrow +\infty.$$

*(Here we use the convention that multiple eigenvalues are repeated
according to their multiplicity.) The first eigenvalue λ_1 is simple and the
corresponding eigenfunctions do not change sign in Ω. Moreover, λ_1 is
the only eigenvalue with this property.
We will denote by φ_1 the eigenfunction corresponding to λ_1, such that
$\varphi_1(x) > 0$ and $\|\varphi_1\|_{L^2} = 1$. We will also denote by φ_i the eigenfunctions
corresponding to λ_i such that*

$$\int_\Omega \varphi_h \varphi_k \mathrm{d}x = \begin{cases} 1 & \text{if } h = k \\ 0 & \text{if } h \neq k. \end{cases}$$

(ii) *There holds*

$$\lambda_1 = \min\left\{ \int_\Omega |\nabla u|^2 \mathrm{d}x : u \in H_0^1(\Omega), \int_\Omega u^2 \mathrm{d}x = 1 \right\}.$$

(iii) *Letting $W_k = \{u \in H_0^1(\Omega) : \int_\Omega \nabla u \cdot \nabla \varphi_h \mathrm{d}x = 0\}$ for $h = 1, \ldots, k - 1$,
one has that*

$$\lambda_k = \min\left\{ \int_\Omega |\nabla u|^2 \mathrm{d}x : u \in W_k, \int_\Omega u^2 \mathrm{d}x = 1 \right\}.$$

Properties (ii) and (iii) are the *variational characterizations* of eigenvalues, see
also Section 5.5. Let us also remark that from (ii) above we can deduce a more
precise form of the Poincaré inequality:

$$\lambda_1 \int_\Omega u^2 \mathrm{d}x \leq \int_\Omega |\nabla u|^2 \mathrm{d}x. \tag{1.7}$$

Concerning the nonhomogeneous problem

$$\begin{cases} -\Delta u(x) = \lambda u(x) + h(x) & x \in \Omega \\ u(x) = 0 & x \in \partial\Omega, \end{cases} \tag{1.8}$$

we can use the properties (RF_2)–(RF_4) to deduce the following.

Theorem 1.14

(i) *If $\lambda \neq \lambda_k$ for all integer $k \geq 1$, then (1.8) has a unique solution for any $h \in L^2(\Omega)$.*

(ii) *Let λ_k be an eigenvalue of (1.6) and let V_k denote the corresponding kernel. Then, given any $h \in L^2(\Omega)$, (1.8) has a unique solution if and only if $\int_\Omega hv \, dx = 0$, for all $v \in V_k$.*

Remark 1.15 If we work in Hölder spaces, the results stated in Theorems 1.13 and 1.14 hold with $L^2(\Omega)$ substituted by $C^{0,\alpha}(\Omega)$. ■

Let $a \in L^\infty(\Omega)$ be such that $a(x) \geq 0$ and $a(x) > 0$ in a set of positive measure in Ω. We will denote by $\lambda_k[a]$ the eigenvalues of

$$\begin{cases} -\Delta u(x) = \lambda a(x)u(x) & x \in \Omega \\ u(x) = 0 & x \in \partial\Omega. \end{cases}$$

There exist infinitely many eigenvalues $0 < \lambda_1[a] < \lambda_2[a] \leq \lambda_3[a] \leq \cdots$ satisfying properties similar to those listed in Theorem 1.13. In addition, the following properties hold.

(EP-1) *(Monotonicity property)* If $a \leq b$ then $\lambda_k[a] \geq \lambda_k[b]$, for all $k \geq 1$; moreover, if $a < b$ in a subset $\Omega' \subset \Omega$ with positive measure, then $\lambda_k[a] > \lambda_k[b]$, for all $k \geq 1$.

(EP-2) *(Continuity property)* If $a_m \to a$ in $L^{n/2}(\Omega)$, then $\lambda_k[a_m] \to \lambda_k[a]$, for all $k \geq 1$.

1.4.2 Regularity

It is a general fact that weak solutions of the Dirichlet BVP

$$\begin{cases} -\Delta u(x) = f(x, u(x)) & x \in \Omega \\ u(x) = 0 & x \in \partial\Omega \end{cases} \tag{1.9}$$

are indeed classical solutions provided $f \in \mathbb{F}_p$ with $1 < p < (n+2)/(n-2)$.

Theorem 1.16 *Let $\partial\Omega$ be smooth and suppose that $f \in \mathbb{F}_p$, with $1 < p < (n + 2)/(n - 2)$.*
Then every $u \in H_0^1(\Omega)$ which is a weak solution of (1.9) is a C^2-solution of (1.9).

Proof. We first use the so-called *bootstrap argument* to show that u is Hölder continuous on $\overline{\Omega}$. Let $n > 2$ and set $\kappa p = (n + 2)/(n - 2)$. Notice that $\kappa > 1$.

Step 1. By the Sobolev embedding theorem it follows that $u \in L^{2n/(n-2)}$.

Step 2. Using (1.3) with $q = 2^*$ we infer that $f(x, u) \in L^\beta$ with $\beta = 2^*/p = \kappa\, 2n/(n + 2)$.

Step 3. Since $-\Delta u = f(x, u)$, Theorem 1.10(i) yields $u \in H^{2,\beta}$.

If $2\beta > n$ then $u \in C^{0,\alpha}(\overline{\Omega})$. Otherwise, we can repeat steps 1–3:

(i) by the Sobolev embedding theorem 1.5(i) one has that $u \in L^{q'}$, with

$$q' = \frac{n\beta}{n - 2\beta} > \kappa\, \frac{2n}{n - 2};$$

(ii) from (1.3) it follows that $f(x, u) \in L^{\beta'}(\Omega)$ with $\beta' = q'/p > \kappa\beta$;
(iii) Theorem 1.10(i) implies that $u \in H^{2,\beta'}$.

In any case, after a finite number of times, one finds that $u \in H^{2,\gamma}$ with $\gamma > n/2$. Then the Sobolev embedding theorem 1.5(iii) yields $H^{2,\gamma} \subset C^{0,\alpha}(\overline{\Omega})$ with $\alpha \leq 1$.

At this point we can apply point (ii) of Theorem 1.10. Actually, letting $h(x) = f(x, u(x))$, u is a weak solution of $-\Delta u = h$ with $h \in C^{0,\nu}(\overline{\Omega})$ for some $\nu \leq 1$. Hence $u \in C^{2,\nu}(\overline{\Omega})$ and is a classical solution of (1.9). ■

Remark 1.17 From a result of Brezis and Kato [62] it follows that the same regularity is true if $p = (n + 2)/(n - 2)$. ■

In the sequel we will always deal with classical solutions.

1.4.3 Positive solutions

Dealing with equations like (1.9), one can use the maximum principle to obtain *positive solutions*.

Maximum Principle.   *Let $\Omega \subset \mathbb{R}^n$ be a bounded domain with smooth boundary and let $\lambda < \lambda_1$. Suppose that $u \in C^2(\Omega) \cup C(\overline{\Omega})$ satisfies*

$$\begin{cases} -\Delta u \geq \lambda u & in \ \Omega, \\ u \geq 0 & on \ \partial\Omega. \end{cases}$$

Then $u \geq 0$ in Ω. Moreover, either $u > 0$ in Ω or $u \equiv 0$ in Ω.

In order to apply the maximum principle to obtain a positive solution of (1.9) one can substitute f with its positive part $f^+(x, u) := \max\{f(x, u), 0\}$. If

$$\begin{cases} -\Delta u(x) &= f^+(x, u(x)), & x \in \Omega, \\ u(x) &= 0, & x \in \partial\Omega, \end{cases}$$

has a nontrivial solution u^*, namely such that $u^*(x) \not\equiv 0$, then $u^* > 0$ in Ω. Therefore $f^+(x, u^*(x)) = f(x, u^*(x))$ and u^* is a positive solution of (1.9).

PART I

Topological methods

2

A primer on bifurcation theory

A specific feature of many nonlinear problems is the existence of multiple solutions and often it is useful to introduce a parameter λ to detect when new solutions arise. From the mathematical point of view, one is led to consider a functional equation $S(\lambda, u) = 0$, depending on a parameter λ, and such that $S(\lambda, 0) \equiv 0$. Bifurcation theory deals with the existence of values λ^* at which *nontrivial solutions* branch off from the trivial one, $u = 0$. A very interesting survey on bifurcation theory is contained in the paper [145] by G. Prodi, which also contains applications to elasticity and fluid dynamics.

In this chapter we will address the simplest situation, the bifurcation from a simple eigenvalue. The material discussed in this chapter is closely related to that contained in [20], Chapter 5.

2.1 Bifurcation: definition and necessary conditions

Let X, Y be Banach spaces. We will deal with an equation like

$$S(\lambda, u) = 0 \tag{2.1}$$

where $S : \mathbb{R} \times X \to Y$ is such that

$$S(\lambda, 0) = 0, \qquad \forall \, \lambda \in \mathbb{R}.$$

The solution $u = 0$ will be called the *trivial solution* of (2.1). The set

$$\Sigma_S = \{(\lambda, u) \in \mathbb{R} \times X : u \neq 0, \ S(\lambda, u) = 0\}$$

will be called the set of *nontrivial solutions* of (2.1). When no confusion is possible, we will omit the subscript S.

Many problems arising in applications can be modelled in this way. For example, let us consider an elastic beam, with one hinged endpoint and length ℓ, which is compressed at the free edge by a force of intensity proportional to

$\lambda > 0$. The corresponding equation is given by

$$\begin{cases} u''(s) + \lambda \sin u(s) = 0 & s \in [0, \ell] \\ u'(0) = u'(\ell) = 0 \end{cases} \tag{2.2}$$

where s is the arc length, the *prime* denotes the derivative with respect to s and u is the angle between the horizontal line and the tangent to the beam. To frame such an equation in a form like (2.1), we set $X = \{u \in C^2(0, \ell) : u'(0) = u'(\ell) = 0\}$, $Y = C(0, \ell)$ and $S(\lambda, u) = u''(s) + \lambda \sin u(s)$. Let us point out that the choice of the function spaces is made taking into account the boundary conditions in (2.2). Clearly, $u(s) \equiv 0$ is a solution of (2.2) for any $\lambda \in \mathbb{R}$, corresponding to the unbended position of the beam. This is a stable equilibrium if λ remains small and corresponds to the *trivial solution* of our problem. When λ exceeds a certain critical threshold, the beam bends and this corresponds to a nontrivial solution. The trivial solution, which still exists, becomes unstable while the new solution is the stable one. Such a threshold is a bifurcation point of (2.2).

Coming back to (2.1), let us give the precise definition of bifurcation point.

Definition 2.1 *A bifurcation point for* (2.1) *is a number* $\lambda^* \in \mathbb{R}$ *such that* $(\lambda^*, 0)$ *belongs to the closure of* Σ. *In other words,* λ^* *is a bifurcation point if there exist sequences* $\lambda_n \in \mathbb{R}$, $u_n \in X \setminus \{0\}$ *such that*

(i) $S(\lambda_n, u_n) = 0$,
(ii) $(\lambda_n, u_n) \to (\lambda^*, 0)$.

The main purpose of the theory of bifurcation is to estabilish conditions for finding bifurcation points and, in general, to study the structure of Σ.

If $S \in C^1(\mathbb{R} \times X, Y)$ a necessary condition for λ^* to be a bifurcation point can be immediately deduced from the implicit function theorem.

Proposition 2.2 *If* λ^* *is a bifurcation point of* (2.1) *then* $S'_u(\lambda^*, 0) \in L(X, Y)$ *is not invertible.*
In particular, if $S(\lambda, u) = \lambda u - T(u)$, *then any bifurcation point of* (2.1) *belongs to the spectrum of* $T'(0)$.

Proof. If $S'_u(\lambda^*, 0)$ is invertible, the implicit function theorem implies that, locally near $(\lambda^*, 0)$, the unique solution of $S(\lambda, u) = 0$ is $u = 0$.

If $S(\lambda, u) = \lambda u - T(u)$, then $S'_u(\lambda^*, 0) : v \mapsto \lambda^* v - T'(0)v$ and hence $S'_u(\lambda^*, 0)$ is not invertible if and only if λ^* is in the spectrum of the linear map $T'(0)$. ∎

In the case of the beam equation (2.2), one has that

$$S'_u(\lambda, 0) : v \mapsto v'' + \lambda v, \quad v \in X.$$

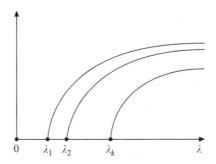

Figure 2.1 Bifurcation diagram of (2.2).

Thus the values of λ such that $S'_u(\lambda, 0)$ is not invertible are the eigenvalues of the linear problem

$$\begin{cases} u''(s) + \lambda u(s) = 0, & s \in [0, \ell] \\ u'(0) = u'(\ell) = 0. \end{cases}$$

which are given by $\lambda_k = k^2 \pi^2 / \ell^2$, $k = 1, 2, \dots$. Hence these λ_k are the only possible bifurcation points of (2.2). In this specific case, an elementary phase plane analysis shows that each λ_k is in fact a bifurcation point.

However, in general, one cannot expect that every value λ such that S'_u is not invertible, is a bifurcation point. Actually, the following example shows that, in general, the converse of Proposition 2.2 is not true.

Example 2.3 Let $X = Y = \mathbb{R}^2$ and let $S(\lambda, u) = \lambda u - T(u)$, where $u = (x_1, x_2)$ and $T : X \mapsto Y$ is defined by setting

$$T(x_1, x_2) = (x_1 + x_2^3, x_2 - x_1^3).$$

The solutions of $S(\lambda, u) = \lambda u - T(u) = 0$ are the pairs $u = (x_1, x_2) \in \mathbb{R}^2$ such that

$$\begin{cases} \lambda x_1 = x_1 + x_2^3 \\ \lambda x_2 = x_2 - x_1^3. \end{cases}$$

From this it follows that $x_1^4 + x_2^4 = 0$ and hence the only solution of $\lambda u = T(u)$ is $(x_1, x_2) = (0, 0)$. In other words, $f = 0$ has the trivial solution only and there is no bifurcation point. On the other hand, the derivative $T'(0)$ is the identity matrix in $\mathbb{R}^2$ and hence $\lambda^* = 1$ is an eigenvalue of $T'(0)$, indeed the only one. $\blacksquare$

2.2 The Lyapunov–Schmidt reduction

In this section we will discuss an abstract procedure which turns out to be very useful in many cases.

Let $S \in C^2(\mathbb{R} \times X, Y)$ and let $\lambda^* \in \mathbb{R}$ be such that

$$L = S_u'(\lambda^*, 0)$$

is not invertible. We will focus on the case in which this is due to the presence of a nontrivial kernel. Let $V = \mathrm{Ker}(L)$ and let R denote the range of L. We suppose

(V) V has a topological complement W in X.

(R) R is closed and has a topological complement Z in Y.

Remark 2.4 Any linear Fredholm operator L satisfies (V) and (R). Actually, in this case, V is finite dimensional, R is closed and the dimension of Z is finite. More precisely, if L is Fredholm with index zero, then $\dim(V) = \dim(Z)$. ∎

If (V) and (R) hold then (W and Z are closed and) $X = V \oplus W$, $Y = Z \oplus R$. In particular, for any $u \in X$ there exist unique $v \in V$ and $w \in W$ such that $u = v + w$. Similarly, we can define conjugate projections P, Q on Y onto Z and R, respectively.

Setting $u = v + w$ and applying P and Q to (2.1) we obtain the following equivalent system:

$$PS(\lambda, v + w) = 0, \tag{2.3}$$

$$QS(\lambda, v + w) = 0. \tag{2.4}$$

The latter is called the *auxiliary equation*.

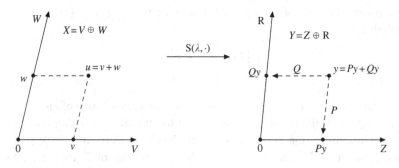

Figure 2.2

Lemma 2.5 *The auxiliary equation* (2.4) *is uniquely solvable in W, locally near* $(\lambda^*, 0)$. *Precisely, there exist neighbourhoods* Λ^* *of* λ^*, V_0 *of* $v = 0$ *in V,* W_0 *of* $w = 0$ *in W, and a map* $w = w(\lambda, v) \in C^2(\Lambda^* \times V_0, W)$, *such that*

$$QS(\lambda, v + w) = 0, \qquad (\lambda, v, w) \in \Lambda^* \times V_0 \times W_0 \quad \Longleftrightarrow \quad w = w(\lambda, v).$$

Furthermore, one has that

$$w(\lambda, 0) = 0, \quad \forall \lambda \in \Lambda^*, \tag{2.5}$$

$$w'_v(\lambda^*, 0) = 0. \tag{2.6}$$

Proof. Set $\phi(\lambda, v, w) = QS(\lambda, v + w)$. One has that $\phi \in C^1(\mathbb{R} \times V \times W, \mathbb{R})$ and $\partial_w \phi(\lambda^*, 0, 0)$ is the linear map from W to $\mathbb{R}$ given by

$$w \mapsto QS'_u(\lambda^*, 0)[w] = QLw = (\text{since } Lw \in \mathbb{R}) = Lw.$$

In other words, $\partial_w \phi(\lambda^*, 0, 0)$ is the restriction of L to W, and thus it is obviously injective and surjective (as a map from W to $\mathbb{R}$). Since $\mathbb{R}$ is closed, it follows that $\partial_w \phi(\lambda^*, 0, 0)$ is invertible and a straight application of the implicit function theorem yields the result. ∎

We can now substitute $w = w(\lambda, v)$ into (2.3) yielding the *bifurcation equation*

$$PS(\lambda, v + w(\lambda, v)) = 0. \tag{2.7}$$

Suppose that there exists a sequence of solutions of (2.7), $(\lambda_n, v_n) \rightarrow (\lambda^*, 0)$, with $v_n \neq 0$. From the preceding discussion it follows that, setting $u_n = v_n + w(\lambda_n, v_n)$, one has that $S(\lambda_n, u_n) = 0$. Moreover, according to (2.5), one has that $w(\lambda_n, v_n) \rightarrow 0$. Finally, if $v_n \neq 0$ then $u_n = v_n + w(\lambda_n, v_n) \neq 0$ and hence (λ_n, u_n) is a nontrivial solution of (2.1).

Summarizing, we have shown the following.

Theorem 2.6 *Let* $S \in C^1(\mathbb{R} \times X, Y)$ *satisfy* (V) *and* (R). *Suppose that the bifurcation equation* (2.7) *possesses a sequence of solutions* $(\lambda_n, v_n) \rightarrow (\lambda^*, 0)$, *with* $v_n \neq 0$. *Then, setting* $u_n = v_n + w(\lambda_n, v_n)$, *one has that* $(\lambda_n, u_n) \in \Sigma_S$, $u_n \rightarrow 0$ *and thus* λ^* *is a bifurcation point of* (2.1).

2.3 Bifurcation from the simple eigenvalue

According to Theorem 2.6, we need to impose conditions in such a way that the bifurcation equation (2.7) is solvable. The first case that we are going to discuss is when V is one dimensional and the codimension of $\mathbb{R}$ is also one. For this reason, this is called the *case of the simple eigenvalue*.

Precisely, we will assume that (V) and (R) hold and

(**V-1**) there exists $u^* \in X$, $u^* \neq 0$, such that $V = \text{span}\{u^*\}$;
(**R-1**) there exists $\psi \in Y^*$, $\psi \neq 0$, such that $R = \{y \in Y : \langle \psi, y \rangle = 0\}$.

Remark 2.7 In the specific case in which $X = Y$ and $S(\lambda, u) = u - \lambda Au - T(u)$, with T smooth and such that $T(0) = 0$, $T'(0) = 0$, and $A \in L(X, X)$ is compact, one has that $L = I - \lambda^* A$, where I denotes the identity in X. Suppose that λ^* is a simple characteristic value of A, in the sense that $\text{Ker}(I - \lambda^* A) \neq \{0\}$. Then

 (i) $\text{Ker}[I - \lambda^* A]$ is one dimensional,
(ii) the codimension of $\text{Range}[I - \lambda^* A]$ is one, and
     $\text{Ker}[I - \lambda^* A] \cap \text{Range}[I - \lambda^* A] = \{0\}$.

Then (i) and (ii) are obviously equivalent to (V-1) and (R-1), see also Remark 2.4.   ∎

Using the same notation as in the previous sections, we have that $v = tu^*$, $t \in \mathbb{R}$, and hence the solution of the auxiliary equation has the form $w(\lambda, v) = w(\lambda, tu^*)$. Moreover, one has that

$$PS(\lambda, v + w(\lambda, v)) = PS(\lambda, tu^* + w(\lambda, tu^*)).$$

Then, according to assumption (R-1), the bifurcation equation $PS = 0$ becomes

$$\beta(\lambda, t) \stackrel{\text{def}}{=} \langle \psi, S(\lambda, tu^* + w(\lambda, tu^*)) \rangle = 0. \qquad (2.8)$$

From property (2.5) of w we deduce:

$$\beta(\lambda, 0) = \langle \psi, S(\lambda, w(\lambda, 0)) \rangle = \langle \psi, S(\lambda, 0) \rangle.$$

Since $S(\lambda, 0) \equiv 0$ we infer

$$\beta(\lambda, 0) \equiv 0. \qquad (2.9)$$

Next, let us evaluate the partial derivative β'_t of β with respect to t:

$$\beta'_t(\lambda, t) = \langle \psi, S'_u(\lambda, tu^* + w(\lambda, tu^*))[u^* + w'_v(\lambda, tu^*)u^*] \rangle. \qquad (2.10)$$

For $t = 0$ we get

$$\beta'_t(\lambda, 0) = \langle \psi, S'_u(\lambda, 0)[u^* + w'_v(\lambda, 0)u^*] \rangle. \qquad (2.11)$$

In particular, for $\lambda = \lambda^*$, (2.6) yields

$$\beta'_t(\lambda^*, 0) = \langle \psi, S'_u(\lambda^*, 0)[u^*] \rangle = \langle \psi, Lu^* \rangle = 0. \qquad (2.12)$$

Furthermore, from (2.11) we deduce that the second mixed derivative $\beta''_{t,\lambda}$ satisfies

$$\beta''_{t,\lambda}(\lambda^*,0) = \langle \psi, S''_{u,\lambda}(\lambda^*,0)[u^* + w'_v(\lambda^*,0)]u^*\rangle$$
$$+ \langle \psi, S'_u(\lambda^*,0)[w''_{t,\lambda}(\lambda^*,0)u^*]\rangle$$
$$= \langle \psi, S''_{u,\lambda}(\lambda^*,0)[u^*]\rangle + \langle \psi, L[w''_{t,\lambda}(\lambda^*,0)u^*]\rangle.$$

Since $L[w''_{t,\lambda}(\lambda^*,0)u^*] \in R$, then $\langle \psi, L[w''_{t,\lambda}(\lambda^*,0)u^*]\rangle = 0$ and thus we find

$$\beta''_{t,\lambda}(\lambda^*,0) = \langle \psi, S''_{u,\lambda}(\lambda^*,0)[u^*]\rangle. \tag{2.13}$$

We are now in a position to state the main result of this section.

Theorem 2.8 *Let* (V-1) *and* (R-1) *hold and suppose that*

$$S''_{u,\lambda}(\lambda^*,0)[u^*] \notin R. \tag{2.14}$$

Then λ^* is a bifurcation point for S.

Proof. Define

$$h(\lambda,t) = \begin{cases} \dfrac{\beta(\lambda,t)}{t} & \text{if } t \neq 0 \\[2mm] \beta'_t(\lambda,0) & \text{if } t = 0. \end{cases}$$

Clearly, h is of class C^1 in a neighbourhood of $(\lambda^*,0) \in \mathbb{R}^2$. From (2.12) it follows that $h(\lambda^*,0) = 0$. Moreover, from (2.13) one infers that

$$h'_\lambda(\lambda^*,0) = \beta''_{t,\lambda}(\lambda^*,0) = \langle \psi, S''_{u,\lambda}(\lambda^*,0)[u^*]\rangle.$$

Then assumption (2.14) yields that $h'_\lambda(\lambda^*,0) \neq 0$. Applying the implicit function theorem to h we find $\lambda = \lambda(t)$, defined in an ε-neighbourhood of $t = 0$, such that

$$\lambda(0) = \lambda^*, \qquad h(\lambda(t),t) = 0, \quad \forall \, -\varepsilon \le t \le \varepsilon.$$

Now, if $h(\lambda(t),t) = 0, t \neq 0$, then $\beta(\lambda(t),t) = 0$ and thus the pair $(\lambda(t),u(t))$ with $u(t) = tu^* + w(\lambda(t),t)$ is a solution of the bifurcation equation $PS = 0$ such that $(\lambda(t),u(t)) \to (\lambda^*,0)$ as $t \to 0$. Let us point out that $u(t) \neq 0$ when $t \neq 0$. Then we deduce that λ^* is a bifurcation point for S. ∎

Remark 2.9
(a) The set of nontrivial solutions Σ_S is a smooth curve which has a Cartesian representation on the kernel V of L.

(b) We can describe Σ in a more precise fashion. Actually, we have:

$$\lambda'(0) = -\frac{h'_t(\lambda^*,0)}{h'_\lambda(\lambda^*,0)}.$$

We already know that

$$h'_\lambda(\lambda^*,0) = \langle \psi, S''_{u,\lambda}(\lambda^*,0)[u^*] \rangle \overset{\text{def}}{=} a.$$

In addition, one easily finds that

$$h'_t(\lambda^*,0) = \tfrac{1}{2}\beta''_{t,t}(\lambda^*,0) \overset{\text{def}}{=} b.$$

With straight calculations one deduces

$$b = \tfrac{1}{2}\langle \psi, S''_{u,u}(\lambda^*,0)[u^*]^2 \rangle.$$

Hence, if $b \neq 0$ we get

$$\lambda(t) = \lambda^* - \frac{b}{a}t + o(t), \qquad \text{as } t \to 0.$$

This implies that there are nontrivial solutions branching off from $(\lambda^*,0)$ both for $\lambda > \lambda^*$ and for $\lambda < \lambda^*$ (*transcritical bifurcation*). If $b = 0$ the structure of Σ depends on the higher order u-derivatives of S. For example, if S is odd with respect to u, one gets

$$\lambda''(0) = -\frac{1}{3a}\langle \psi, S'''_{u,u,u}(\lambda^*,0)[u^*]^3 \rangle,$$

and thus, if $\lambda''(0) \neq 0$ one finds

$$u = \pm \left(\frac{\lambda - \lambda^*}{2c} \right)^{1/2} u^* + O(\lambda - \lambda^*), \qquad c \overset{\text{def}}{=} \lambda''(0).$$

Then, if $c > 0$ the nontrivial solutions branch off on the right of λ^* (*supercritical bifurcation*), while if $c < 0$ the branching is on the left of λ^* (*subcritical bifurcation*). ∎

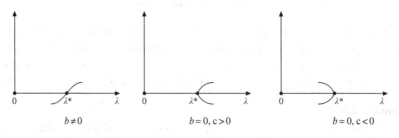

Figure 2.3

For future reference, it is convenient to state explicitly the result in the case
introduced in Remark 2.7. Let $X = Y$ and suppose that S has the form $S(\lambda, u) =
u - \lambda A u - T(u)$, where $A \in L(X, X)$. In this case Theorem 2.8 becomes the
following.

Theorem 2.10 *Let* $T \in C^2(X, X)$ *be such that* $T(0) = 0$ *and* $T'(0) = 0$.
Moreover, let A be compact. Then any simple characteristic value λ^* *of A is a
bifurcation point for* $S(\lambda, u) : u - \lambda A u - T(u) = 0$.

Proof. As anticipated in Remark 2.7, for any simple characteristic value of A the
assumptions (V-1) and (R-1) hold true. Moreover, one has that $V = \text{Ker}[I - \lambda^* A]$
verifies $V \cap R = \{0\}$. Since

$$S''_{u,\lambda}(\lambda^*, 0)[v] = Av,$$

assumption (2.14) follows immediately, too.   ∎

Applications to nonlinear eigenvalue problems

As a first application of the preceding results, let us consider the beam
equation (2.2). We have already seen that the problem can be handled by
means of bifurcation theory, setting $X = \{u \in C^2(0, \ell) : u(0) = u(\ell) = 0\}$,
$Y = C(0, \ell)$, $S(\lambda, u) = u'' + \lambda \sin u$, and $S'_u(\lambda, u) : v \mapsto v'' + \lambda v$. The eigenvalues $\lambda_k = k^2 \pi^2 / \ell^2$, $k = 1, 2, \ldots$ of $v'' + \lambda v = 0$, $v \in X$, are all simple and
thus, according to Theorem 2.8, are bifurcation points. Similar results hold true
for nonlinear Sturm–Liouville eigenvalue problems like

$$\begin{cases} -(p(x)u')' + q(x)u = \lambda u + f(x, u) & x \in [a, b] \\ u(a) = u(b) = 0, \end{cases}$$

where p, q and f are smooth, $p, q > 0$ in $[a, b]$ and $f(x, 0) \equiv 0$. For example, if
$f'_u(x, 0) = 0$ then all the eigenvalues of

$$\begin{cases} -(p(x)u')' + q(x)u = \lambda u & x \in [a, b] \\ u(a) = u(b) = 0, \end{cases}$$

are bifurcation points.

Next, we will study the bifurcation for semilinear elliptic eigenvalue
problems like

$$\begin{cases} -\Delta u = \lambda u + f(x, u) & x \in \Omega \\ u = 0 & x \in \partial\Omega. \end{cases} \qquad (D_\lambda)$$

Here Ω is a smooth bounded domain in $\mathbb{R}^N$ and $f \in C^2(\Omega \times \mathbb{R})$. We introduce
the Hölder spaces $X = \{u \in C^{2,\alpha}(\Omega) : u_{|\partial\Omega} = 0\}$ and $Y = C^{0,\alpha}(\Omega)$, and define

$$S(\lambda, u) = \Delta u + \lambda u + f(x, u).$$

If $(\lambda, u) \in \mathbb{R} \times X$ is a solution of $S(\lambda, u) = 0$ then u solves $(\boldsymbol{D_\lambda})$.
Suppose that $f(x, 0) = f'_u(x, 0) = 0$. Then $S(\lambda, 0) \equiv 0$ and

$$S'_u(\lambda, 0) : v \mapsto \Delta v + \lambda v.$$

Let us take $\lambda^* = \lambda_k$, an eigenvalue of the linear boundary value problem

$$\begin{cases} -\Delta u = \lambda u & x \in \Omega \\ u = 0 & x \in \partial\Omega. \end{cases} \tag{2.15}$$

Then $L = L_k : v \mapsto \Delta u + \lambda_k u$ has a nontrivial kernel $V = V_k$. Assumption
(V-1) requires that V_k is one dimensional, namely that λ_k is simple. Let φ_k, with
$\int_\Omega \varphi_k^2 \, dx = 1$, be an eigenfunction corresponding to λ_k. Then $u^* = \varphi_k$ and
$V_k = \text{span}\{\varphi_k\}$. Moreover, according to Theorem 1.14, see also Remark 1.15,
one has that the range R_k of L_k has codimension one and is given by

$$R_k = \left\{ u \in Y : \int_\Omega u\varphi_k \, dx = 0 \right\}.$$

In other words, here $R_k = \text{Ker}(\psi)$ where ψ is defined by setting $\langle \psi, u \rangle = \int u\varphi_k \, dx$. Finally, $S''_{u,\lambda}(\lambda_k, 0) : v \mapsto v$ and hence $S''_{u,\lambda}(\lambda_k, 0)[\varphi_k] = \varphi_k \notin R_k$. In
conclusion, Theorem 2.8 applies and ensures that every simple eigenvalue of
(2.15) is a bifurcation point for S, namely for the semilinear Dirichlet boundary
value problem $(\boldsymbol{D_\lambda})$.

Remark 2.11 We could also use a slightly different approach that leads to an
equation in the form considered in Theorem 2.10. Let us set $X = C_0^{0,\alpha}(\Omega) = \{u \in C^{0,\alpha}(\Omega) : u(x) = 0, \ \forall x \in \partial\Omega\}$, let K denote the Green operator of
$-\Delta$ on X, see the discussion after Theorem 1.10, and let f be the Nemitski
operator associated to $f(x, u)$. Let us recall that $K \in L(X)$ is compact. With this
notation, $(\boldsymbol{D_\lambda})$ is equivalent to the equation $u = \lambda K(u) - K(f(u))$, which is in
the form addressed in Theorem 2.10, with $A = K$ and $T = K \circ f$. Of course,
the characteristic values of K are nothing but the eigenvalues of (2.15). ∎

Completing the above discussion we can establish whether the bifurcation is
transcritical, supercritical or subcritical. One has:

$$a = \langle \psi, \varphi_k \rangle = \int_\Omega \varphi_k^2 \, dx = 1,$$

$$b = \tfrac{1}{2}\langle \psi, S''_{u,u}(\lambda_k, 0)[\varphi_k]^2 \rangle = \tfrac{1}{2} \int_\Omega f''_{u,u}(x, 0)\varphi_k^3 \, dx.$$

Furthermore, if $b = 0$ and f is odd with respect to u, then

$$c = -\tfrac{1}{6} \int_\Omega f'''_{u,u,u}(x, 0)\varphi_k^4 \, dx.$$

Example 2.12 Consider the problem

$$\begin{cases} -\Delta u = \lambda u \pm |u|^{p-1}u & x \in \Omega \\ u = 0 & x \in \partial\Omega, \end{cases}$$

where $p \geq 2$. Let us consider the case $\lambda^* = \lambda_1$, the first eigenvalue of (2.15). We can take $\varphi_1 > 0$.

(i) If $p = 2$ then $b = \int_\Omega \varphi_1^3 \, dx > 0$ and the bifurcation is transcritical.
(ii) If $p = 3$ then $b = 0$, while the sign of c depends on whether the nonlinearity is $+u^3$ or $-u^3$. In the former case we have that

$$c = -\int_\Omega \varphi_1^4 dx < 0,$$

while in the latter

$$c = \int_\Omega \varphi_1^4 \, dx > 0.$$

Therefore the bifurcation is subcritical if the nonlinearity is $+u^3$, and supercritical if the nonlinearity is $-u^3$. ∎

Remark 2.13 Let us point out that the solutions branching off from λ_1 are small in $C^{2,\alpha}$ and hence the behaviour of $f(x, u)$ as $|u| \to \infty$ does not play any role. In other words, bifurcation takes place whatever the exponent is and the restriction $|f(x, u)| \leq c_1 + c_2|u|^p$, with $1 < p \leq (n+2)/(n-2)$ $(n > 2)$ is not required here. ∎

3

Topological degree, I

The topological degree (in short, degree) of a map is a classical tool which is very useful for solving functional equations. It was introduced by L. Brouwer for finite dimension and extended by J. Leray and J. Schauder to infinite dimension. There is a very broad literature dealing with degree. We limit ourselves to citing the books [4, 68, 87, 95, 148, 160] which are most closely related to the topics discussed here.

We divide the material into Chapters 3 and 4. The former is organized as follows. First we give an account of the degree in a somewhat axiomatic way, listing its main properties. The Brouwer fixed point theorem is discussed in Section 3.2. In Section 3.3 we carry out the construction of the degree. The Leray–Schauder degree is discussed in Section 3.4. Applications include the Leray–Schauder fixed point theorem (Section 3.5), the Krasnoselski bifurcation theorem (Section 3.7) as well as examples in which the degree is used to solve elliptic differential equations (Section 3.6). Further theoretical tools, global in nature, and their applications to bifurcation theory and elliptic BVP are discussed in Chapter 4.

3.1 Brouwer degree and its properties

Let us assume that:

(a) Ω is an open bounded set in $\mathbb{R}^n$, with boundary $\partial\Omega$;
(b) f is a continuous map from $\overline{\Omega}$ to $\mathbb{R}^n$; the components of f will be denoted by f_i;
(c) p is a point in $\mathbb{R}^n$ such that $p \notin f(\partial D)$.

To each triple (f, Ω, p) satisfying (a)–(c), one can associate an integer $\deg(f, \Omega, p)$, called the degree of f (with respect to Ω and p), with the following basic properties.

(P.1) Normalization: if $I_{\mathbb{R}^n}$ denotes the identity map in $\mathbb{R}^n$, then

$$\deg(I_{\mathbb{R}^n}, \Omega, p) = \begin{cases} 1 & \text{if } p \in \Omega \\ 0 & \text{if } p \notin \Omega. \end{cases}$$

(P.2) Solution property: if $\deg(f, \Omega, p) \neq 0$ then there exists $z \in \Omega$ such that $f(z) = p$.

(P.3) $\deg(f, \Omega, p) = \deg(f - p, \Omega, 0)$.

(P.4) Decomposition: if $\Omega_1 \cap \Omega_2 = \emptyset$ then

$$\deg(f, \Omega_1 \cup \Omega_2, p) = \deg(f, \Omega_1, p) + \deg(f, \Omega_2, p).$$

Below we will give an outline of the procedure usually followed to define the degree, omitting the consistency of the definition and the verification of (P.1–P.4). The complete construction, with proofs, will be carried out in Section 3.3. For details about the *standard* construction, see the references given at the beginning of the chapter. First one considers a C^1 map f and a regular value p. Let us recall that, by definition, p is said to be a regular value for f, if the Jacobian $J_f(x)$ is different from zero for every $x \in f^{-1}(p)$. The Jacobian is the determinant of the matrix $f'(x)$ with entries

$$a_{ij} = \frac{\partial f_i}{\partial x_j}.$$

If p is a regular value then the set $f^{-1}(p)$ is finite and one can define the degree by setting

$$\deg(f, \Omega, p) = \sum_{x \in f^{-1}(p)} \text{sgn}[J_f(x)], \tag{3.1}$$

where, for $b \in \mathbb{R} \setminus \{0\}$, we set

$$\text{sgn}[b] = \begin{cases} 1 & \text{if } b > 0 \\ -1 & \text{if } b < 0. \end{cases}$$

It is immediate to verify that the degree defined above satisfies the properties (P.1)–(P.5).

In order to extend the preceding definition to any continuous function f and any point p, one uses an approximation procedure. First, in order to approximate p with regular values p_k one applies the Sard theorem.

Theorem 3.1 (Sard theorem) *Let* $f \in C^1(\Omega, \mathbb{R}^n)$ *and set* $\mathcal{S}_f = \{x \in \Omega : J_f(x) = 0\}$. *Then* $f(\mathcal{S}_f)$ *is a set of zero measure.*

For a proof, see [95, Lemma 1.4]. The set $\mathcal{S}_f$ is called the set of singular points of f. Any u such that $f(u) = p$ is called a nonsingular solution of the equation $f = p$, provided $u \notin \mathcal{S}_f$.

According to the Sard theorem, there exists a sequence $p_k \notin S_f$, such that $p_k \to p$. When p_k is sufficiently close to p, p_k verifies (c) and hence it makes sense to consider the $\deg(f, \Omega, p_k)$, given by (3.1). Moreover, one can show that, for $k \gg 1$, $\deg(f, \Omega, p_k)$ is a constant which is independent of the approximating sequence p_k. Hence one can define the degree of $f \in C^1(\Omega, \mathbb{R}^n) \cap C(\overline{\Omega}, \mathbb{R}^n)$ at any p by setting $\deg(f, \Omega, p) = \lim_k \deg(f, \Omega, p_k)$. Similarly, given $f \in C(\overline{\Omega}, \mathbb{R}^n)$, let $f_k \in C^1(\Omega, \mathbb{R}^n) \cap C(\overline{\Omega}, \mathbb{R}^n)$ be such that $f_k \to f$ uniformly on $\overline{\Omega}$. If $k \gg 1$, then any (f_k, Ω, p) satisfies (a)–(c) and one can consider the degree $\deg(f_k, \Omega, p_k)$. Once more, one can show that $\lim \deg(f_k, \Omega, p)$ does not depend upon the choice of the sequence f_k and thus one can define the degree of f by setting $\deg(f, \Omega, p) = \lim_k \deg(f_k, \Omega, p)$.

An important property of the degree defined above is the invariance by homotopy. An homotopy is a map $h = h(\lambda, x)$ such that $h \in C([0, 1] \times \overline{\Omega}, \mathbb{R}^n)$. An homotopy is *admissible* (with respect to Ω and p), if $h(\lambda, x) \neq p$ for all $(\lambda, x) \in [0, 1] \times \partial\Omega$.

One can prove that the degree defined before satisfies

(P.5) Homotopy invariance: if h is an admissible homotopy, then $\deg(h(\lambda, \cdot), \Omega, p)$ is constant with respect to $\lambda \in [0, 1]$. In particular, if $f(x) = h(0, x)$ and $g(x) = h(1, x)$ then $\deg(f, \Omega, p) = \deg(g, \Omega, p)$.

As an immediate consequence of the homotopy invariance, we can deduce the following.

Theorem 3.2 (Dependence on the boundary values) *Let $f, g \in C(\Omega, \mathbb{R}^n)$ be such that $f(x) = g(x)$ for all $x \in \partial\Omega$ and let $p \notin f(\partial\Omega) = g(\partial\Omega)$. Then $\deg(f, \Omega, p) = \deg(g, \Omega, p)$.*

Proof. Consider the homotopy

$$h(\lambda, x) = \lambda g(x) + (1 - \lambda)f(x).$$

For all $x \in \partial\Omega$ one has that $f(x) = g(x)$ and hence $h(\lambda, x) = f(x) \neq p$. Thus h is admissible and the homotopy invariance yields:

$$\deg(f, \Omega, p) = \deg(h(\cdot, 0), \Omega, p) = \deg(h(\cdot, 1), \Omega, p) = \deg(g, \Omega, p),$$

proving the result. ∎

Let us list below (without proofs) some further properties of the degree.

(P.6) Continuity: if $f_k \to f$ uniformly in $\overline{\Omega}$, then $\deg(f_k, \Omega, p) \to \deg(f, \Omega, p)$. Moreover, $\deg(f, \Omega, p)$ is continuous with respect to p.

(P.7) Excision property: let $\Omega_0 \subset \Omega$ be an open set such that $f(x) \neq p$, for all $x \in \Omega \setminus \Omega_0$. Then $\deg(f, \Omega, p) = \deg(f, \Omega_0, p)$.

The excision property allows us to define the index of an isolated solution of $f(x) = p$. Let $x_0 \in \Omega$ be such that $f(x_0) = p$ and suppose that there exists $r > 0$ such that $f(x) \neq p$ for all $x \in \overline{B_r(x_0)} \setminus \{x_0\}$. Using the excision property, with $\Omega = B_r(x_0)$ and $\Omega_0 = B_\rho(x_0)$, $\rho \in (0, r)$, we deduce that

$$\deg(f, B_\rho(x_0), p) = \deg(f, B_r(x_0), p), \quad \forall \rho \in (0, r).$$

This common value is, by definition, the *index* of f with respect to x_0:

$$i(f, x_0) = \lim_{\rho \to 0} \deg(f, B_r(x_0), p), \quad p = f(x_0).$$

Moreover, if $f^{-1}(p) = \{x_1, \ldots, x_k\}$, $x_j \in \Omega$, then

(P.8)
$$\deg(f, \Omega, p) = \sum_1^k i(f, x_j).$$

To see this, it suffices to take $\rho > 0$ such that $B_\rho(x_i) \cap B_\rho(x_j)$ for all $i \neq j$. Letting $\Omega_0 = B_\rho(x_1) \cup \cdots \cup B_\rho(x_k)$, using the excision property (P.7) and the decomposition property (P.4), we find

$$\deg(f, \Omega, p) = \deg(f, \Omega_0, p) = \sum_1^k \deg(f, B_\rho(x_j), p) = \sum_1^k i(f, x_j),$$

proving (P.8).

Let $f \in C^1$ and let p be a regular value of f (i.e. $J_f(x_0) \neq 0$ for all $x_0 \in f^{-1}(p)$). As already pointed out before, if p is a regular value of f then the set $f^{-1}(p)$ is discrete. In particular, any solution x_0 of $f(x) = p$ is isolated and it makes sense to consider the index $i(f, x_0)$.

Lemma 3.3 *Suppose that $f \in C^1(\Omega, \mathbb{R}^n) \cap C(\overline{\Omega}, \mathbb{R}^n)$ and let $x_0 \in \Omega$ be such that $p = f(x_0)$ is a regular value of f. Then*

$$i(f, x_0) = (-1)^\beta,$$

where β is the sum of the algebraic multiplicities of all the negative eigenvalues of $f'(x_0)$.

Proof. Let $r > 0$ be such that the only solution of $f(x) = p$ in $B_r = B_r(x_0)$ is x_0. Then $i(f, x_0) = \deg(f, B_r, p)$ and (3.1) yields $i(f, x_0) = \text{sgn}[J_f(x_0)]$. Using the Jordan normal form, we know that the Jacobian determinant $J_f(x_0)$ is given by

$$J_f(x_0) = \lambda_1 \cdots \lambda_n,$$

where λ_j are the eigenvalues of $f'(x_0)$ repeated according to their algebraic multiplicity. Now, let us remark that:

- each λ_j is different from zero, because x_0 is a regular value;
- if an eigenvalue is complex, say equal to $a + ib$, then its complex conjugate $a - ib$ is also an eigenvalue of $f'(x_0)$, and their product is $a^2 + b^2 > 0$.

From this it follows that $\mathrm{sgn}[J_f(x_0)] = (-1)^\beta$, completing the proof. $\blacksquare$

Remark 3.4 It has been shown in [7] that the topological degree $\deg(f, \Omega, p) \in \mathbb{Z}$ is uniquely determined by properties (P.1), (P.4) and (P.5). $\blacksquare$

3.2 Application: the Brouwer fixed point theorem

In this section we will assume that a degree $deg(f, \Omega, p)$ satisfying the properties listed in the previous section has been defined, and we will show how it can be used to obtain the classical Brouwer fixed point theorem. We start with a preliminary result.

Let $B_1 = \{x \in \mathbb{R}^n : |x| < 1\}$ denote the unit ball in $\mathbb{R}^n$.

Theorem 3.5 *The unit sphere ∂B_1 is not a 'retract' of the unit ball B_1. Namely, there is no continuous map $f : B_1 \mapsto \partial B_1$ such that $f(x) \equiv x$, for all $x \in \partial B_1$.*

Proof. Assuming the contrary, Theorem 3.2, with $g(x) = x$, implies

$$\deg(f, \Omega, 0) = \deg(g, \Omega, 0) = 1.$$

Using the *solution property* we infer that there exists $x \in B_1$ such that $f(x) = 0$ and this is in contradiction with the assumption that $f(B_1) \subseteq \partial B_1$. $\blacksquare$

Remark 3.6 More in general, it is easy to see that, if Ω is any bounded open *convex* set, or else if Ω is a bounded domain homeomorphic to a convex set, then it is not possible to retract Ω onto its boundary $\partial \Omega$. $\blacksquare$

We are now ready to prove the Brouwer fixed point theorem.

Theorem 3.7 *If f is a continuous map from the a bounded closed convex set $C \subset \mathbb{R}^n$ into itself, then there exists $z \in C$ such that $f(z) = z$.*

Proof. First, let C be the closure of the unit ball B_1 in $\mathbb{R}^n$. If $f(x) \neq x$ for all $x \in \overline{B}_1$, we can define a map $\widetilde{f} : \overline{B}_1 \mapsto \partial B_1$ by letting $\widetilde{f}(x)$ be the intersection of ∂B_1 with the half-line starting from $f(x)$ and crossing x, see Figure 3.1.

It is easy to see that $\widetilde{f}$ is continuous. Moreover, $\widetilde{f}(x) = x$ for all $x \in \partial B_1$, and hence ∂B_1 is a deformation retract of B_1, a contradiction to Theorem 3.5.

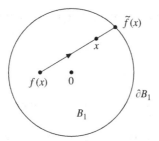

Figure 3.1

The case in which C is a bounded closed convex set, follows similarly, by using Remark 3.6. ∎

Remark 3.8

(a) According to Remark 3.6, one can extend the Brouwer fixed point theorem proving that if Ω is a bounded domain homeomorphic to a convex set then any continuous map from $\overline{\Omega}$ into itself has at least a fixed point $z \in \overline{\Omega}$.

(b) Another proof of Theorem 3.7 can be carried out using the homotopy invariance of the degree. Roughly, one shows that the homotopy $(\lambda, x) \mapsto x - \lambda f(x)$ is admissible and thus $\deg(I - f, B_1, 0) = \deg(I, B_1, 0) = 1$. Using the solution property, one deduces that there exists zero of $x - f(x)$, namely a fixed point of f. For details, see the proof of Theorem 3.21 in the sequel. ∎

It is worth mentioning that there are several further topological results that can be proved using the finite dimensional degree, see [95, Chapter 3]. Here we limit ourselves to stating a result that we will need in the sequel, see Theorem 10.5.

Theorem 3.9 (Borsuk–Ulam theorem) *Let $\Omega \subset \mathbb{R}^n$ be bounded, open, symmetric (namely $x \in \Omega \Leftrightarrow -x \in \Omega$) and such that $0 \notin \Omega$. Suppose that $f \in C(\overline{\Omega}, \mathbb{R}^n)$ is an odd map such that $0 \notin f(\partial\Omega)$. Then $\deg(f, \Omega, 0) = 1$ (mod 2).*

3.3 An analytic definition of the degree

In this section we give a complete account of the topological degree and of its properties. The definition we will give below is not the one sketched in

Section 3.1. Instead, we prefer to follow a more analytical approach due to E. Heinz [105] which is slightly simpler from the technical point of view.

3.3.1 Degree for C^2 maps

We begin with the case of C^2 maps. It is always understood that the preceding conditions (a)–(c) are verified. Let $f \in C^2(\Omega, \mathbb{R}^n)$ and let $J_f(x)$ denote the Jacobian of f. From (c) one has that $\min\{|f(x) - p| : x \in \partial\Omega\} > 0$ and we can choose $\alpha > 0$ such that

$$\alpha < \min_{x \in \partial\Omega} |f(x) - p|.$$

Consider a real valued continuous function φ defined on $[0, \infty)$ and such that

(i) $\text{supp}[\varphi] \subset\,]0, \alpha[\, ,$

(ii) $\int_{\mathbb{R}^n} \varphi(|x|)\, dx = 1.$

Definition 3.10 *For $f \in C^1(\Omega, \mathbb{R}^n)$ we set*

$$\deg(f, \Omega, p) = \int_\Omega \varphi(|f(x) - p|) J_f(x)\, dx.$$

Remark 3.11 Obviously, one has that $\deg(f, \Omega, p) = \deg(f - p, \Omega, 0)$, namely property (P.3) holds. ∎

We shall justify the definition by showing that it is independent of the choice of α and φ satisfying (i) and (ii) above. Precisely, let us show that if $\alpha_1, \varphi_1, \alpha_2, \varphi_2$ satisfy (i) and (ii), then

$$\int_\Omega \varphi_1(|f(x) - p|) J_f(x)\, dx = \int_\Omega \varphi_2(|f(x) - p|) J_f(x)\, dx. \qquad (3.2)$$

We can take $p = 0$. Setting $\widetilde{\varphi} = \varphi_1 - \varphi_2$, (3.2) becomes

$$\int_\Omega \widetilde{\varphi}(|f(x)|) J_f(x)\, dx = 0.$$

From (ii) we infer

$$\int_0^\infty r^{n-1} \widetilde{\varphi}(r)\, dr = 0. \qquad (3.3)$$

Moreover, if $\text{supp}[\varphi_i] \subset\,]0, \alpha_i[$, $(i = 1, 2)$, then

$$\text{supp}[\widetilde{\varphi}] \subset\,]0, \alpha[, \qquad (3.4)$$

where $\alpha = \max\{\alpha_1, \alpha_2\}$. We set

$$\psi(r) = \begin{cases} r^{-n} \int_0^r s^{n-1} \widetilde{\varphi}(s)\, ds & \text{if } r > 0 \\ 0 & \text{if } r = 0. \end{cases}$$

From (3.3) and (3.4) it follows that for $r > \alpha$ one has that

$$\psi(r) = r^{-n} \int_0^r s^{n-1} \widetilde{\varphi}(s) \, ds = r^{-n} \int_0^\infty s^{n-1} \widetilde{\varphi}(s) \, ds = 0.$$

Thus $\mathrm{supp}[\psi] \subset \,]0, \alpha[$. Moreover, ψ is of class C^1 and

$$\widetilde{\varphi}(r) = r\psi'(r) + n\psi(r).$$

Let A_{ij} denote the cofactor of $a_{ij} = \partial f_i / \partial x_j$ in the Jacobian J_f and consider the vector field $V \in C^1(\Omega, \mathbb{R}^n)$ with components

$$V_i(x) = \sum_{j=1}^n A_{ji}(x) \, \psi(|f(x)|) f_j(x).$$

Taking into account the following property of the cofactors A_{ij}

$$\sum_{i=1}^n \frac{\partial A_{ji}}{\partial x_i} = 0, \quad \forall j = 1, \dots, n,$$

a straight calculation yields

$$\mathrm{div}(V(x)) = \sum_{i=1}^n \frac{\partial V_i}{\partial x_i}$$

$$= J_f(x)(|f(x)| \psi'(|f(x)|) + n\psi(|f(x)|)) = \widetilde{\varphi}(|f(x)|)J_f(x).$$

Integrating on Ω we find

$$\int_\Omega \widetilde{\varphi}(|f(x)|)J_f(x) \, dx = \int_{\partial\Omega} V(x) \cdot \nu \, d\sigma, \tag{3.5}$$

where ν denotes the unit outer normal at $\partial\Omega$. Now, let us remark that for $x \in \partial\Omega$ one has that $|f(x)| \geq \min\{|f(x)| : x \in \partial\Omega\} > \alpha$ and thus

$$V(x) = 0, \quad \forall x \in \partial\Omega. \tag{3.6}$$

Using (3.6), we infer that the last integral in (3.5) is zero and hence

$$\int_\Omega \widetilde{\varphi}(|f(x)|)J_f(x) \, dx = 0.$$

This proves that (3.2) holds true.

We end this subsection by evaluating the degree in some specific cases.

Example 3.12 Let $f = A \in L(\mathbb{R}^n)$ with A nonsingular. Then

$$\deg(A, \Omega, p) = \int_\Omega \varphi(|Ax - p|)J_A(x) \, dx$$

$$= \int_\Omega \varphi(|Ax - p|) \det(A) \, dx = \int_{A(\Omega)} \varphi(|y - p|) \, \mathrm{sgn}[\det(A)] \, dy.$$

We can take $\alpha < \min_{\partial\Omega} |Ax - p|$ such that $B_\alpha(p) \subset A(\Omega)$ if $p \in A(\Omega)$, respectively $B_\alpha(p) \cap \overline{\Omega}$ if $p \notin A(\Omega)$. Since the support of $\varphi(\cdot - p)$ is contained in the ball $B_\alpha(p)$, then one gets

$$\deg(A, \Omega, p) = \begin{cases} \text{sgn}[\det(A)] & \text{if } p \in A(\Omega) \\ 0 & \text{if } p \notin A(\Omega). \end{cases}$$

In particular, we find

$$\deg(I_{\mathbb{R}^n}, \Omega, p) = \begin{cases} 1 & \text{if } p \in \Omega \\ 0 & \text{if } p \notin \Omega, \end{cases}$$

namely (P.1) holds.   ∎

As a second example, we consider the case in which p is a regular value of f.

Example 3.13 Let D be an open subset of $\mathbb{R}^n$, consider $f \in C^2(D, \mathbb{R}^n)$ and let $x_0 \in D$ be such that $p = f(x_0)$ is a regular value of f. We know that f induces a diffeomorphism between a ball $B_\varepsilon(x_0)$, $\varepsilon > 0$ small enough, and its image $U_\varepsilon = f(B_\varepsilon(x_0))$. We can also assume that J_f has constant sign in $B_\varepsilon(x_0)$. Let us evaluate the degree $\deg(f, B_\varepsilon(x_0), p)$. With arguments similar to those used in the preceding example, we get

$$\begin{aligned} \deg(f, B_\varepsilon(x_0), p) &= \int_{B_\varepsilon(x_0)} \varphi(|f(x) - p|) J_f(x) \, dx \\ &= \text{sgn}[J_f(x_0)] \int_{B_\varepsilon(x_0)} \varphi(|f(x) - p|) |J_f(x)| \, dx \\ &= \text{sgn}[J_f(x_0)] \int_{U_\varepsilon} \varphi(|y - p|) \, dy. \end{aligned}$$

In the definition of the degree we can take α such that $B_\alpha(p) \subset U_\varepsilon$. Since $\varphi(|y - p|) = 0$ for $|y - p| > \alpha$, then

$$\int_{U_\varepsilon} \varphi(|y - p|) \, dy = \int_{\mathbb{R}^n} \varphi(|y - p|) \, dy = 1,$$

and hence

$$\deg(f, B_\varepsilon(x_0), p) = \text{sgn}[J_f(x_0)].$$ ∎

In the above examples the degree is an integer. In general, we will show that the degree, as defined in Definition 3.10, is always an integer. Moreover, at the end of the section we will recover formula (3.1) for C^1 maps and regular values, see Corollary 3.15.

3.3.2 Degree for continuous maps

We are now going to define the degree for any continuous f verifying the conditions (a)–(c) stated at the beginning of Section 3.1. For this purpose, we need the following lemma.

Lemma 3.14 *For $i = 1, 2$, let $f_i \in C^2(\Omega, \mathbb{R}^n)$ be such that*

$$|f_i(x) - p| > \alpha > 0, \quad \forall x \in \partial\Omega.$$

Given $\varepsilon \in]0, \alpha/6[$, we suppose that

$$|f_2(x) - f_1(x)| < \varepsilon, \quad \forall x \in \overline{\Omega}.$$

Then $\deg(f_1, \Omega, p) = \deg(f_2, \Omega, p)$.

Proof. According to Remark 3.11 we can take $p = 0$. Let $\chi \in C^1(0, \infty)$ be a nondecreasing function such that

$$\chi(r) = \begin{cases} 1 & \text{if } 0 \le r \le 2\varepsilon \\ 0 & \text{if } r \ge 3\varepsilon, \end{cases}$$

and define $f_3 \in C^1(\Omega, \mathbb{R}^n) \cap C(\overline{\Omega}, \mathbb{R}^n)$ by setting

$$f_3(x) = (1 - \chi(|f_1(x)|))f_1(x) + \chi(|f_1(x)|)f_2(x).$$

From the definition of χ we infer:

$$f_3(x) = f_1(x), \quad \forall x \in \overline{\Omega} : |f_1(x)| > 3\varepsilon, \tag{3.7}$$

$$f_3(x) = f_2(x), \quad \forall x \in \overline{\Omega} : |f_1(x)| < 2\varepsilon. \tag{3.8}$$

In particular, since $|f_1(x)| > \alpha$, for all $x \in \partial\Omega$, and $\alpha > 6\varepsilon$, then $|f_3(x)| = |f_1(x)| > \alpha$ for all $x \in \partial\Omega$. Furthermore, for all $x \in \overline{\Omega}$ one has that

$$f_3(x) - f_1(x) = \chi(|f_1(x)|) \cdot (f_2(x) - f_1(x)),$$

$$f_3(x) - f_2(x) = (1 - \chi(|f_1(x)|)) \cdot (f_1(x) - f_2(x)),$$

and therefore, for $i = 1, 2$

$$|f_3(x) - f_i(x)| < \varepsilon, \quad \forall x \in \overline{\Omega}. \tag{3.9}$$

Choose two functions $\varphi_i \in C(0, \infty)$ with the following properties:

$$\int_{\mathbb{R}^n} \varphi_i(|x|) \, dx = 1, \quad i = 1, 2,$$

$$\text{supp}[\varphi_1] \subset]4\varepsilon, 5\varepsilon[,$$

$$\text{supp}[\varphi_2] \subset]0, \varepsilon[.$$

According to Definition 3.10, the functions φ_1 and φ_2 can be used to evaluate the degree of f_i, $i = 1, 2, 3$. In particular, one has

$$\begin{aligned} \deg(f_3, \Omega, 0) &= \int_\Omega \varphi_1(|f_3(x)|) J_{f_3}(x) \, dx, \\ \deg(f_1, \Omega, 0) &= \int_\Omega \varphi_1(|f_1(x)|) J_{f_1}(x) \, dx. \end{aligned} \tag{3.10}$$

Now, since supp$[\varphi_1] \subset \,]4\varepsilon, 5\varepsilon[$ we get:

$$\varphi_1(|f_3(x)|) \neq 0 \iff 4\varepsilon < |f_3(x)| < 5\varepsilon.$$

Using (3.9), we deduce that $3\varepsilon < |f_1(x)| < 6\varepsilon$ provided $4\varepsilon < |f_3(x)| < 5\varepsilon$ and then (3.7) yields $f_3(x) = f_1(x)$ for all $x \in \overline{\Omega}$ such that $4\varepsilon < |f_3(x)| < 5\varepsilon$. In other words,

$$\varphi_1(|f_3(x)|) J_{f_3}(x) = \varphi_1(|f_1(x)|) J_{f_1}(x), \quad \forall x \in \overline{\Omega}$$

and this, jointly with (3.10), implies that $\deg(f_3, \Omega, 0) = \deg(f_1, \Omega, 0)$. Similarly, we see that

$$\varphi_2(|f_3(x)|) J_{f_3}(x) = \varphi_2(|f_2(x)|) J_{f_2}(x), \quad \forall x \in \overline{\Omega}$$

and we deduce that $\deg(f_3, \Omega, 0) = \deg(f_2, \Omega, 0)$. In conclusion, one has $\deg(f_1, \Omega, 0) = \deg(f_3, \Omega, 0) = \deg(f_2, \Omega, 0)$, proving the lemma. ∎

We are now in a position to define the degree for any continuous map $f : \overline{\Omega} \mapsto \mathbb{R}^n$ such that $f(x) \neq p$ for all $x \in \partial\Omega$. By density, there exists a sequence of functions $f_k \in C^2(\Omega, \mathbb{R}^n) \cap C(\overline{\Omega}, \mathbb{R}^n)$ converging to f uniformly on $\overline{\Omega}$. Clearly, for $k \gg 1$ we have that $f_k(x) \neq p$ on $\partial\Omega$ and hence the degree $\deg(f_k, \Omega, p)$ is well defined. Moreover, Lemma 3.14 implies that $\deg(f_k, \Omega, p)$ is constant for k sufficiently large. This allows us to define the degree of f with respect to Ω and p, by setting

$$\deg(f, \Omega, p) = \lim_{k \to \infty} \deg(f_k, \Omega, p). \tag{3.11}$$

3.3.3 Properties of the degree

In this subsection we will show that properties (P.1)–(P.7) hold. We have already checked that (P.1) and (P.3) are verified. According to the definition and to Lemma 3.14, it suffices to carry out the proofs under the additional assumption that $f \in C^2$. Let us point out that $p \notin f(\partial\Omega)$ implies that $p \notin f_k(\partial\Omega)$ for $k \gg 1$, as well as $q \notin f(\partial\Omega)$ for all q near p, so that it makes sense to consider $\deg(f_k, \Omega, p)$ and $\deg(f, \Omega, q)$.

Proof of (P.2). If $f(x) \neq p$ for all $x \in \Omega$, then $f(x) \neq p$ on all the compact set $\overline{\Omega}$ and hence $\exists \delta > 0$, $\delta \leq \alpha$, such that $|f(x) - p| > \delta$, for all $x \in \overline{\Omega}$.

Choose φ such that $\int_{\mathbb{R}^n} \varphi(|x|)\, dx = 1$ and supp$[\varphi] \subset]0, \delta[$. From this and the fact that $|f(x) - p| > \delta$ in Ω, we infer that $\varphi(|f(x) - p|) \equiv 0$ in Ω and one finds

$$\deg(f, \Omega, p) = \int_\Omega \varphi(|f(x) - p|)J_f(x)\, dx = 0.$$

This is in contradiction with our assumptions. ∎

Proof of (P.4). Since $\Omega_1 \cap \Omega_2 = \emptyset$ one has

$$\int_{\Omega_1 \cup \Omega_2} \varphi(|f(x) - p|)J_f(x)\, dx = \int_{\Omega_1} \varphi(|f(x) - p|)J_f(x)\, dx$$
$$+ \int_{\Omega_2} \varphi(|f(x) - p|)J_f(x)\, dx.$$

Then (P.4) immediately follows from Definition 3.10. ∎

Proof of (P.5). For any $\varepsilon > 0$ small, we can find $\delta(\varepsilon) > 0$ such that $|h(x, \lambda_1) - h(x, \lambda_2)| < \varepsilon$ for all $x \in \overline{\Omega}$ provided $|\lambda_1 - \lambda_2| < \delta$. Using Lemma 3.14 we infer

$$\deg(h(\cdot, \lambda_1), \Omega, p) = \deg(h(\cdot, \lambda_2), \Omega, p).$$

Covering the interval $[0, 1]$ with a finite number of subintervals with length smaller than δ, and applying the preceding equation, the result follows. ∎

Proof of (P.6). This follows immediately from Lemma 3.14. Since $\deg(f, \Omega, p) = \deg(f - p, \Omega, 0)$ we also deduce the continuity of the degree with respect to p. ∎

Proof of (P.7). Since $f(x) \neq p, \forall x \in \Omega \setminus \Omega_0$, then $f(x) \neq p$, on the compact set $\overline{\Omega} \setminus \Omega_0$, and there exists $\alpha_1 > 0$ such that $|f(x) - p| > \alpha_1, \forall x \in \overline{\Omega} \setminus \Omega_0$. In the definition of the degree, let us choose φ in such a way that supp$[\varphi] \subset]0, \alpha_1[$. Then $\varphi(|f(x) - p|) \equiv 0$ on $\Omega \setminus \Omega_0$ and this yields

$$\int_\Omega \varphi(|f(x) - p|)J_f(x)\, dx = \int_{\Omega_0} \varphi(|f(x) - p|)J_f(x)\, dx.$$

Since, by definition, the former integral equals $\deg(f, \Omega, p)$ while the latter equals $\deg(f, \Omega_0, p)$, we conclude that $\deg(f, \Omega, p) = \deg(f, \Omega_0, p)$. ∎

Let us point out that the definition and properties of the index depend on (P.1)–(P.7) only. In particular, (P.8) holds. Moreover, arguing as in Example 3.13, we immediately deduce the following corollary, which is nothing but the definition of the degree for regular values and C^1 functions, given in (3.1).

Corollary 3.15 *If p is a regular value of* $f \in C^1(\Omega, \mathbb{R}^n) \cap C(\overline{\Omega}, \mathbb{R}^n)$, *then*

$$\deg(f, \Omega, p) = \sum_{x \in f^{-1}(p)} \operatorname{sgn}[J_f(x)].$$

We end this section by showing that $\deg(f, \Omega, p)$ defined in Definition 3.10, is always an integer. As usual, it suffices to consider C^1 maps. If p is a regular value, the claim follows from the corollary above. Otherwise, let $\mathcal{S}_f$ denote the set of points $x \in \Omega$ such that $J_f(x) = 0$. The Sard theorem ensures that $f(\mathcal{S}_f)$ has zero Lebesgue measure. Hence, there exists a sequence of regular values p_k with $p_k \to p$. Since $\deg(f, \Omega, p_k)$ is an integer then, by continuity, we infer that $\deg(f, \Omega, p)$ is an integer too.

3.4 The Leray–Schauder degree

In this section we will define the Leray–Schauder degree, namely the degree for maps $f \in C(X, X)$, where X is a Banach space and f is a compact perturbation of the identity $I = I_X$. This extension is particularly important for applications to differential equations.

3.4.1 Defining the Leray–Schauder degree

Let D be an open bounded subset of the Banach space X. We will deal with compact perturbations of the identity, namely with operators $S \in C(\overline{D}, X)$ such that $S = I - T$, where T is compact.

Let $p \notin S(\partial D)$. It is easy to check that $S(\partial D)$ is closed and hence

$$r := \operatorname{dist}(p, S(\partial D)) > 0.$$

It is known, see [60, VI.1], that there exists a sequence $T_k \in C(\overline{D}, X)$ such that $T_k \to T$ uniformly in $\overline{D}$ and

$$T_k(\overline{D}) \subset E_k \subset X, \qquad \text{with } \dim(E_k) < \infty. \tag{3.12}$$

We shall define the degree of $I - T$ as the limit of the degrees of $I - T_k$, which we are going to introduce. First, some preliminaries are in order.

Let us consider a map $\phi \in C(\overline{\Omega}, \mathbb{R}^m)$, where $\Omega \subset \mathbb{R}^n$ and $m \leq n$. We will identify $\mathbb{R}^m$ as the subset of $\mathbb{R}^n$ whose points have the last $n - m$ components equal to zero: $\mathbb{R}^m = \{x \in \mathbb{R}^n : x_{m+1} = \cdots = x_n = 0\}$. The above function ϕ can be considered as a map with values on $\mathbb{R}^n$ by understanding that the last $n - m$ components are zero: $\phi_{m+1} = \cdots = \phi_n = 0$. Let $g(x) = x - \phi(x)$ and let $g_m \in C(\overline{\Omega} \cap \mathbb{R}^m, \mathbb{R}^m)$ denote the restriction of g to $\overline{\Omega} \cap \mathbb{R}^m$. Let us show that if $p \in \mathbb{R}^m \setminus g(\partial\Omega)$ then

$$\deg(g, \Omega, p) = \deg(g_m, \Omega \cap \mathbb{R}^m, p). \tag{3.13}$$

Let $x \in \Omega$ be such that $g(x) = p$. This means that $x = \phi(x) + p$. Thus $x \in \Omega \cap \mathbb{R}^m$ and so $g_m(x) = g(x) = p$. This shows that $g^{-1}(p) \subset g_m^{-1}(p)$. Since the converse is trivially true, it follows that

$$g^{-1}(p) = g_m^{-1}(p). \tag{3.14}$$

We can suppose that $\Omega \cap \mathbb{R}^m \neq \emptyset$, otherwise, $g_m^{-1}(p) = \emptyset$ and by (3.14), $g^{-1}(p) = \emptyset$. As usual, we can suppose that ϕ is of class C^1 and, moreover, that p is a regular value of g_m. Then, according to (3.1), we get

$$\deg(g, \Omega, p) = \sum_{x \in g^{-1}(p)} \operatorname{sgn}[J_g(x)].$$

Now the Jacobian matrix $g'(x)$ is in triangular form

$$\begin{pmatrix} g_m'(x) & \cdot \\ 0 & I_{\mathbb{R}^{n-m}} \end{pmatrix},$$

and hence $\operatorname{sgn}[J_g(x)] = \operatorname{sgn}[J_{g_m}(x)]$. From this and (3.14) we infer

$$\deg(g, \Omega, p) = \sum_{x \in g^{-1}(p)} \operatorname{sgn}[J_g(x)] = \sum_{x \in g_m^{-1}(p)} \operatorname{sgn}[J_{g_m}(x)] = \deg(g_m, \Omega \cap \mathbb{R}^m, p),$$

proving (3.13), provided p is a regular value. In the general case, we use the Sard Lemma and argue as at the end of the previous section. This completes the proof of (3.13).

The preceding discussion allows us to define the degree for a map g such that $g(x) = x - \phi(x)$, where $\phi(\overline{D})$ is contained in a *finite dimensional* subspace E of X. Let $p \in X, p \notin g(\overline{D})$. Let E_1 be a subspace of X containing E and p. We set $g_1 = g_{|\overline{D} \cap E_1}$ and define

$$\deg(g, D, p) = \deg(g_1, D \cap E_1, p). \tag{3.15}$$

Let us show that the preceding definition is independent of E_1. Let E_2 be another subspace of X such that $E \subset E_2$ and $p \in E_2$. Then $E \subset E_1 \cap E_2$ and $p \in E_1 \cap E_2$. Applying (3.13) we infer:

$$\deg(g_i, D \cap E_i, p) = \deg(g_{|\overline{D} \cap E_1 \cap E_2}, D \cap E_1 \cap E_2, p), \quad i = 1, 2.$$

This justifies the defintion given in (3.15).

Now, let us come back to the map $S = I - T$, with T compact. Let $T_k \to T$ satisfy (3.12) and set $S_k = I - T_k$. Taking k such that

$$\sup_{x \in \overline{D}} \|T(x) - T_k(x)\| \leq r/2, \tag{3.16}$$

we infer that $p \notin S_k(\overline{D})$ and hence it makes sense to consider the degree $\deg(S_k, D, p)$ defined as in (3.15).

Definition 3.16 *Let $p \notin S(\partial D)$, where $S = I - T$ with T compact. We set*

$$\deg(S, D, p) = \deg(I - T_k, D, p),$$

for any T_k satisfying (3.12) and (3.16).

Once more, we have to justify the definition, by showing that the degree does not depend on the approximation T_k. To prove this claim, let T_i, $i = 1, 2$, be such that (3.12)–(3.16) hold. Let E_i be finite dimensional spaces such that $T_i(\overline{D}) \subset E_i$. If E is the space spanned by E_1 and E_2, we use the definition (3.15) to get

$$\deg(S_i, D, p) = \deg((S_i)_{|\overline{D} \cap E}, D \cap E, p), \qquad i = 1, 2. \tag{3.17}$$

Consider the homotopy

$$h(\lambda, \cdot) = \lambda (S_1)_{|\overline{D} \cap E} + (1 - \lambda)(S_2)_{|\overline{D} \cap E}.$$

It is easy to check that h is admissible on $D \cap E$ and thus

$$\deg((S_1)_{|\overline{D} \cap E}, D \cap E, p) = \deg((S_2)_{|\overline{D} \cap E}, D \cap E, p).$$

This together with (3.17) proves that Definition 3.16 is justified.

The Leray–Schauder degree has the same properties (P.1)–(P.8) as the finite dimensional degree (with Ω substitued by D). Precisely, as far as the homotopy invariance (P.5) is concerned, one has to deal with homotopies $h(\lambda, x) \in C([0, 1] \times \overline{D}, X)$ such that, for every $\lambda \in [0, 1]$, $h(\lambda, \cdot)$ is a compact perturbation of the identity.

Of course, one can also extend the notion of index of an isolated solution x_0 of $S(x) = x - T(x) = p$ by setting

$$i(S, x_0) = \lim_{r \to 0} \deg(S, B_r(x_0), p), \qquad p = S(x_0),$$

$$B_r(x_0) = \{x \in X : \|x - x_0\| < r\}.$$

We end this subsection by proving the counterpart of Lemma 3.3. First some preliminaries are in order. Recall that $\mu \neq 0$ is a characteristic value of a linear map A if and only if μ^{-1} is an eigenvalue of A. Moreover, if 1 is not a characteristic value of $T'(x_0)$, then $S'(x_0)$ is invertible. We will refer to these solutions as nonsingular solutions of $S = 0$. In particular, the local inversion theorem applies and hence x_0 is an isolated solution of $x - T(x) = p$.

Lemma 3.17 *Let* $T \in C(X,X)$ *be compact and differentiable at* x_0. *Then* $T'(x_0)$ *is a linear compact operator, hence there are only a finite number of characteristic values of* $T'(x_0)$ *contained in* $]0,1[$ *and each has finite multiplicity.*

Proof. Setting for $0 < t < 1$

$$R_t(x) = \frac{T(x_0 + tx) - T(x_0)}{t},$$

one has that R_t is compact and $T'(x_0)[x] = \lim_{t \to 0} R_t(x)$. It follows that the set $T'(x_0)[B_r(x_0)]$ is precompact, for all $r > 0$. ∎

Lemma 3.18 *Let* $T \in C^1(\overline{D}, X)$ *be compact and suppose that* 1 *is not a characteristic value of* $T'(0)$. *Set* $S(x) = x - T(x)$ *and, let* $x_0 \in X$ *be such that* $S(x_0) = p$. *Then one has*

$$i(S, x_0) = \deg(S'(x_0), B_r(x_0), p), \qquad r \ll 1.$$

Proof. To simplify the notation we take $x_0 = 0$ and $p = 0$. We have $S(x) = S'(0)[x] + R(x) = x - T'(0)[x] + R(x)$, where $R(x) = o(\|x\|)$ as $\|x\| \to 0$. Consider the homotopy

$$h(\lambda, x) = x - T'(0)[x] + \lambda R(x).$$

From Lemma 3.17 it follows immediately that $h(\lambda, \cdot)$ is a compact perturbation of the identity. Let us check that there exists $r > 0$ small enough such that h is admissible on $D = B_r(0)$. Otherwise, there exist $x_i \to 0$ and $\lambda_i \in [0, 1]$ such that $h(x_i, \lambda_i) = 0$, namely $x_i - T'(0)[x_i] + \lambda_i R(x_i) = 0$. Setting $z_i = \|x_i\|^{-1} x_i$, z_i satisfies

$$z_i = T'(0)[z_i] - \lambda_i \frac{R(x_i)}{\|x_i\|}.$$

Without relabelling the indices, we can assume that $z_i \to z^*$, weakly in X, and $\lambda_i \to \lambda^* \in [0, 1]$. Since $R(x) = o(\|x\|)$ and using the fact that $T'(0)$ and R are compact, we infer that $z_i \to z^*$ strongly and there holds $\|z^*\| = 1$ as well as $z^* = T'(0)[z^*]$. This is in contradiction with the assumption that $\mu = 1$ is not a characteristic value of $T'(0)$.

Using the homotopy invariance we get

$$\deg(S, D, 0) = \deg(h(1, \cdot), D, 0) = \deg(h(0, \cdot), D, 0) = \deg(S'(0), D, 0),$$

and the lemma follows. ∎

Lemma 3.19 *Let L be a linear, compact map in X and suppose that 1 is not a characteristic value of L, Then*

$$\deg(I - L, B_r(0), 0) = (-1)^\beta, \qquad r > 0,$$

where β is the sum of the algebraic multiplicities of all the characteristic values of L contained in $]0, 1[$.

Proof. For each characteristic value μ_i of L, let us set

$$N_i = \bigcup_{m=1}^{\infty} \text{Ker}[(I - \mu_i L)^m].$$

We also denote by $q_i = \dim[N_i]$ the algebraic multiplicity of μ_i. Letting μ_i, $1 \le i \le k$ denote the characteristic values of L contained in $]0, 1[$, and setting $N = \oplus_{k=1}^{\infty} N_i$, we have that $\dim[N] = q_1 + \cdots + q_k = \beta$ (if there are no characteristic values in $]0, 1[$, we take $N = \emptyset$; in such a case $\beta = 0$). Let W be such that $X = N \oplus W$ and let P, Q denote the projections on N and W, respectively. Consider now the homotopy

$$h(\lambda, x) = x - L[Px] - \lambda L[Qx].$$

Clearly, for each λ, h is a compact perturbation of the identity. Moreover, if there exist $r > 0$ and (λ, x) such that

$$x - L[Px] - \lambda L[Qx] = 0, \qquad \lambda \in [0, 1], \quad \|x\| = r,$$

one finds $Px - L[Px] = \lambda L[Qx] - Qx$. Then $Px - L[Px] \in N, \lambda L[Qx] - Qx \in W$ and $N \cap W = \emptyset$, imply

$$Px = L[Px]$$

$$Qx = \lambda L[Qx].$$

Since 1 is not a characteristic value of L, from $Px = L[Px]$ it follows that $Px = 0$. Then $x = Qx$ and the second equation yields $x = \lambda L[x]$. Obviously, λ cannot be 0 nor 1, hence $0 < \lambda < 1$. But, in such a case, λ must coincide with one of the $\mu_i \in]0, 1[$ and thus $x \in N$, a contradiction to the fact that $x = Qx \in W$.

Using the homotopy invariance, we infer that

$$\deg(I - L, B_r(0), 0) = \deg(I - LP, B_r(0), 0).$$

The latter is a finite dimensional perturbation of the identity and we can use Lemma 3.3 to deduce that $\deg(I - LP, B_r(0), 0) = (-1)^\beta$. ∎

We are now ready to prove the following.

Theorem 3.20 *Let* $T \in C^1(\overline{D}, X)$ *be compact and such that 1 is not a characteristic value of* $T'(x_0)$, *for some* $x_0 \in D$. *Then, setting* $S(x) = x - T(x)$ *and* $S(x_0) = p$, *one has that* x_0 *is an isolated solution of* $S(x) = p$ *and there holds*

$$i(S, x_0) = (-1)^\beta,$$

where β *is the sum of the algebraic multiplicities of all the characteristic values of* $T'(x_0)$ *contained in* $]0, 1[$.

Proof. It suffices to use Lemmas 3.18 and 3.19. ∎

3.5 The Schauder fixed point theorem

Before giving applications to differential equations, let us show how the degree allows us to obtain a classical result on the existence of fixed points of a compact map.

Theorem 3.21 *Let D be a bounded, open convex subset of the Banach space* X *such that* $0 \in D$ *and let* $T \in C(\overline{D}, X)$ *be compact and such that* $T(\overline{D}) \subset \overline{D}$. *Then T has a fixed point in* $\overline{D}$, *namely there exists* $x \in \overline{D}$ *such that* $T(x) = x$.

Proof. Without loss of generality, we can assume that

$$T(x) \neq x, \qquad \forall\, x \in \partial D, \tag{3.18}$$

otherwise we are done. Thus, we can define the degree $\deg(I - T, D, 0)$ and we will prove the theorem by showing that $\deg(I - T, D, 0) \neq 0$. Define

$$h(\lambda, x) = x - \lambda T(x), \qquad \lambda \in [0, 1], \quad x \in \overline{D}.$$

For each $\lambda \in [0, 1]$ the map $x \mapsto h(\lambda, x)$ is a compact perturbation of the identity in X. We claim that

$$h(\lambda, x) \neq 0, \qquad \forall\, (\lambda, x) \in [0, 1] \times \partial D. \tag{3.19}$$

Otherwise, there exist $x^* \in \partial D$ and $\lambda^* \in [0, 1]$ such that $h(\lambda^*, x^*) = 0$, namely $x^* = \lambda^* T(x^*)$. From (3.18) it follows that $\lambda^* < 1$. Since $T(\overline{D}) \subset \overline{D}$, one has that $T(x^*) \in \overline{D}$. Then $\lambda^* < 1$ and the convexity of D imply that $\lambda^* T(x^*) \in D$, a contradiction with the fact that $\lambda^* T(x^*) = x^* \in \partial D$. Since (3.19) holds, we can use the homotopy invariance of the degree to find

$$\deg(I - T, D, 0) = \deg(I, D, 0) = 1,$$

because $0 \in D$. Using the solution property, we infer that there exists $x \in D$ such that $x - T(x) = 0$. ∎

Of course, if $X = \mathbb{R}^n$ Theorem 3.21 is nothing but the Brouwer fixed point theorem.

3.6 Some applications of the Leray–Schauder degree to elliptic equations

In this section we will discuss some first applications of the Leray–Schauder degree to nonlinear elliptic boundary value problems such as

$$\begin{cases} -\Delta u(x) = f(x, u(x)) & x \in \Omega \\ u(x) = 0 & x \in \partial\Omega. \end{cases} \tag{3.20}$$

The general strategy will be the following:

(1) to choose a Banach space X and convert the boundary value problem into a functional equation like $u = T(u)$, $u \in X$ in such a way that T is compact;
(2) to use the homotopy invariance of the degree to show that $u = T(u)$ has a solution. Usually, one takes the homotopy $h(\lambda, u) = u - \lambda T(u)$ and shows that there exists $R > 0$ such that $u \neq \lambda T(u)$ for all $(\lambda, u) \in [0, 1] \times X$ with $\|u\| = R$.

Dealing with (3.20), we can take $X = L^2(\Omega)$ and denote by K the inverse of $-\Delta$ on $H_0^1(\Omega)$. According to the discussion after Theorem 1.10 in Section 1.4, K, as a map from X into itself, is compact.

Remark 3.22 The Laplace operator $-\Delta$ could be substituted by any second order uniformly elliptic operator such as

$$-\mathcal{L}u = -\sum a_{ij}(x)\frac{\partial^2 u}{\partial x_i \partial x_j} + \sum b_i(x)\frac{\partial u}{\partial x_i} + c(x)u,$$

with smooth coefficients satisfying the same assumptions listed in Remark 1.11(ii). If $\mathcal{L}$ is not variational, it is convenient to work in the Hölder space $C^{0,\alpha}(\Omega)$. The arguments used below require minor changes and make use of the Schauder estimates, see Theorem 1.10(ii) and Remark 1.11(ii). ∎

The results one can obtain mainly depend upon the asymptotic behaviour of f. We first consider the easy case in which f is sublinear at infinity.

3.6.1 Sublinear problems

Theorem 3.23 *Suppose that* $f : \Omega \times \mathbb{R} \mapsto \mathbb{R}$ *is locally Hölder continuous and satisfies*

$$\lim_{|s| \to \infty} \frac{f(x, s)}{s} = 0, \qquad (3.21)$$

uniformly with respect to $x \in \Omega$. *Then* (3.20) *has a (classical) solution.*

Proof. From (3.21) it follows that, for all $\varepsilon > 0$, there exists $C_\varepsilon > 0$ such that

$$|f(x, s)| \le C_\varepsilon + \varepsilon |s|. \qquad (3.22)$$

Then, according to Theorem 1.7, f induces a Nemitski operator on $X = L^2(\Omega)$, still denoted by f. Setting $T(u) = Kf(u)$, $T \in C(X, X)$ is compact and (3.20) can be written in the form $u = T(u)$, $u \in X$. Let us show that there exists $R > 0$ such that the homotopy $h(t, u) = u - tT(u)$ is admissible in $B_R = \{u \in X : \|u\| < R\}$. Otherwise, there exist $u_j \in X$, with $\|u_j\| \to \infty$, and $t_j \in [0, 1]$ such that $u_j = t_j T(u_j)$. This is equivalent to $-\Delta u_j = t_j f(x, u_j)$, with $u_j \in H_0^1(\Omega)$. Taking u_j as test function, using (3.22) and the fact that $t_j \le 1$, we get

$$\int_\Omega |\nabla u_j|^2 \, dx \le t_j \int_\Omega |f(x, u_j) u_j| \, dx \le C_\varepsilon \int_\Omega |u_j| \, dx + \varepsilon \int_\Omega |u_j|^2 \, dx.$$

Then, using the Hölder and Poincaré inequality (1.7), we deduce

$$\lambda_1 \|u_j\|_{L^2}^2 \le \int_\Omega |\nabla u_j|^2 \, dx \le C_\varepsilon \|u_j\|_{L^2} + \varepsilon \|u_j\|_{L^2}^2.$$

If we take ε such that $\varepsilon < \lambda_1$, this equation implies that $\|u_j\|_{L^2} \le C$, for some $C > 0$, a contradiction. Thus the homotopy $h(t, u) = u - tT(u)$ is admissible on the ball B_R. Using the homotopy invariance (P.5), it follows that $\deg(I - T, B_R, 0) = \deg(I, B_R, 0) = 1$ and hence, by the solution property (P.2), there exists $u \in B_R$ such that $u = T(u)$, giving rise to a solution of (3.20). ∎

Remark 3.24 It is clear that the same existence result holds when the equation in (3.20) is replaced by $-\Delta u = \beta u + f(x, u)$, where $\beta < \lambda_1$, the first eigenvalue of $-\Delta$ on Ω with zero Dirichlet boundary conditions. We first note that the linear elliptic operator $-\Delta - \beta$ is invertible on $H_0^1(\Omega)$, with inverse K_β which is compact in X. Next, one repeats the preceding proof, with K substituted by K_β. ∎

It is worth recalling that, dealing with sublinear problems, one can also find a solution by using variational methods or else by sub- and super-solutions. Regarding the former, we refer to Section 5.4, see, in particular Theorem 5.9 and Remark 5.10. For the latter, we will outline the method below.

Consider the equation

$$\begin{cases} -\Delta u(x) = f(x, u(x)) & x \in \Omega \\ u(x) = 0 & x \in \partial\Omega. \end{cases} \tag{3.23}$$

We say that $\underline{u} \in C^2(\Omega) \cap C^1(\overline{\Omega})$ is a sub-solution of (3.23) if

$$\begin{cases} -\Delta\underline{u}(x) \leq f(x, \underline{u}(x)) & x \in \Omega \\ \underline{u}(x) \leq 0 & x \in \partial\Omega. \end{cases}$$

A super-solution $\overline{u}$ is defined by reversing the above inequalities[1]. It is well known that if (3.23) has a sub-solution $\underline{u}$ and a super-solution $\overline{u}$ such that $\underline{u}(x) \leq \overline{u}(x)$ in Ω, then it has a solution u with $\underline{u}(x) \leq u(x) \leq \overline{u}(x)$. This solution can be found by using a monotone iteration scheme, see [154], Theorem 2.3.1.

Let us show how one argues in the specific case of the problem

$$-\Delta u = u^q, \ x \in \Omega, \quad u = 0, \ x \in \partial\Omega, \quad 0 < q < 1. \tag{3.24}$$

The general case is left to the reader as an exercise. Taking $\underline{u}_\varepsilon := \varepsilon\varphi_1$, one finds that $-\Delta\underline{u}_\varepsilon = \varepsilon\lambda_1\varphi_1$ while $\underline{u}_\varepsilon^q = \varepsilon^q\varphi_1^q$. For all $\varepsilon < \lambda_1^{1/(q-1)}\|\varphi_1\|_\infty^{-1}$ one has $\varepsilon\lambda_1\varphi_1 < \varepsilon^q\varphi_1^q$ and thus $\underline{u}_\varepsilon$ is a sub-solution. On the other hand, let ψ be such that $-\Delta\psi = 1$ in Ω and $\psi = 0$ on $\partial\Omega$. If $M^{1-q} > \|\psi\|_\infty^q$ then $\overline{u}_M := M\psi$ satisfies $-\Delta\overline{u}_M > \overline{u}_M^q$ and thus is a super-solution of (3.24). Choosing ε possibly smaller, one has that $\underline{u}_\varepsilon < \overline{u}_M$ in Ω and therefore (3.24) has a (positive) solution u such that $\underline{u}_\varepsilon \leq u \leq \overline{u}_M$. It is also possible to prove that such a u is unique. For other results on sublinear problems, see Section 11.4 in Chapter 11.

3.6.2 Problems at resonance

In this subsection we will deal with a class of elliptic *problems at resonance*. By a problem at resonance we mean an equation like

$$\begin{cases} -\Delta u(x) = \lambda^* u(x) + f(x, u(x)) - h(x) & x \in \Omega \\ u(x) = 0 & x \in \partial\Omega, \end{cases} \tag{3.25}$$

where f is a bounded function and λ^* is an eigenvalue of $-\Delta$ with zero Dirichlet boundary conditions. In contrast with the cases discussed in the preceding subsection, problem (3.25) might have no solution at all. Actually, if u is a

---

[1] We consider smooth sub- and super-solutions, but it would be possible to deal with weak (say H^1) sub- and super-solutions.

(classical) solution of (3.25), letting $V^* = \{v \in H_0^1(\Omega) : -\Delta v = \lambda^* v\}$ (the eigenspace associated to λ^*), one has that

$$-\int_\Omega \Delta u\, v\, dx = \lambda^* \int_\Omega uv\, dx + \int_\Omega f(x, u)v\, dx - \int_\Omega hv\, dx, \quad \forall v \in V^*.$$

Integrating by parts, we find a necessary condition for the existence of solutions of (3.25):

$$\int_\Omega f(x, u)v\, dx = \int_\Omega hv\, dx, \quad \forall v \in V^*.$$

For example, if $A < f < B$, if $\lambda^* = \lambda_k$ is simple and if φ_k denotes a corresponding eigenfunction, the above condition becomes

$$A\int_{\varphi_k > 0} \varphi_k\, dx + B\int_{\varphi_k < 0} \varphi_k\, dx < \int_\Omega h\varphi_k\, dx < B\int_{\varphi_k > 0} \varphi_k\, dx + A\int_{\varphi_k < 0} \varphi_k\, dx.$$

$$(3.26)$$

Problems at resonance have been broadly studied, beginning with the pioneering paper by E. A. Landesman and A. C. Lazer [112].

We will show the following result.

Theorem 3.25 *Suppose that f is bounded, locally Hölder continuous and $\exists M > 0$ such that $|f_u(x, u)| \leq M$. Moreover, suppose that f possesses limits at $\pm\infty$ and there holds*

$$f_-(x) = \lim_{u \to -\infty} f(x, u), \quad f_+(x) = \lim_{u \to +\infty} f(x, u).$$

Then (3.25) has a solution provided

$$\int_\Omega hv\, dx < \int_{v > 0} f_+ v\, dx + \int_{v < 0} f_- v\, dx, \quad \forall v \in V^*, \|v\| = 1. \qquad (3.27)$$

Proof. The proof will be carried out in two steps. First we use a Lyapunov–Schmidt reduction and solve the auxiliary equation by means of the global inversion theorem. Afterwards, we use the degree to solve the bifurcation equation.

As usual, we denote by $\{\lambda_i\}_{i \geq 1}$ the eigenvalues of $-\Delta$ with zero Dirichlet boundary conditions and by φ_i a corresponding orthonormal set of eigenfunctions. For the sake of simplicity, we will further assume that

$$\begin{cases} \text{(i)} & \exists c_1, c_2 > 0 : c_1 \leq \lambda^* + f_u(x, u) \leq c_2, \quad \forall (x, u) \in \Omega \times \mathbb{R}, \\ \text{(ii)} & \text{the interval } [c_1, c_2] \text{ does not contain any } \lambda_j \neq \lambda^*. \end{cases} \qquad (3.28)$$

At the end of the proof we will indicate how one can handle the general case. Let $V = \text{Ker}[-\Delta - \lambda^*]$, let W denote the L^2-orthogonal complement of V in $E = H_0^1(\Omega)$ and let P, Q denote the corresponding orthogonal projections.

Setting $Pu = v$ and $Qu = w$, we have that $u = v + w$ and (3.25) is equivalent to the Lyapunov–Schmidt system

$$\begin{cases} -\Delta w = \lambda^* w + Qf(v + w) - Qh \\ 0 = Pf(v + w) - Ph. \end{cases}$$

Using the global inversion theorem (see [20], Chapter 3, Theorem 1.8) it is easy to check that the auxiliary equation has a unique solution $w = w(v)$ and $\|w\| \leq$ constant. Actually, the map $F(w) = -\Delta w - \lambda^* w - Qf(v+w)$ is proper and (3.28) implies that $F'(w) : \omega \mapsto -\Delta\omega - \lambda_k\omega - Qf_u(v + w)\omega$ is invertible. In order to solve the bifurcation equation we set $\Phi(v) = Pf(v + w(v))$. Let us show that if (3.27) holds, then $\exists R > 0$ such that

$$\deg(\Phi - Ph, B_R, 0) \neq 0, \quad \text{where } B_R = \{v \in V : \|v\| < R\}. \tag{3.29}$$

Clearly, if (3.29) holds, then the solution property of the degree implies that the bifurcation equation $\Phi(v) = Ph$ has a solution $\bar{v}$, yielding a solution $\bar{v} + w(\bar{v})$ of (3.25).

In order to prove (3.29), we consider the homotopy

$$h(v, s) = s(\Phi(v) - Ph) + (1 - s)v, \quad v \in V, \ s \in [0, 1].$$

If there is $R > 0$ such that h is admissible on $B_R = \{v \in V : \|v\| < R\}$, then $\deg(\Phi - Ph, B_R, 0) = \deg(I_V, B_R, 0) = 1$ and (3.29) follows. Arguing by contradiction, let $v_j \in V$, with $\|v_j\| \to +\infty$, and $s_j \to \bar{s} \in [0, 1]$, be such that

$$s_j(\Phi(v_j) - Ph) + (1 - s_j)v_j = 0.$$

Remark that $s_j \neq 0$. Setting $z_j = \|v_j\|^{-1}v_j$, we can assume (recall that V is finite dimensional) that $z_j \to z$, where $z \in V$ and $\|z\| = 1$. From the preceding equation it follows that

$$s_j \left[\int_\Omega f(v_j + w_j)z_j \, dx - \int_\Omega hz_j \, dx \right] = -(1 - s_j) \int_\Omega v_j z_j \, dx \leq 0, \tag{3.30}$$

where $w_j = w(v_j)$. Since $\|w_j\| \leq c$, we can assume that, up to a subsequence, $w_j \to w$ a.e. in Ω. Then, for almost every $x \in \{z > 0\}$, respectively $x \in \{z < 0\}$, one has that $v_j(x) + w_j(x) \to +\infty$, respectively $v_j(x) + w_j(x) \to -\infty$. Then, using Fatou's lemma we get

$$\liminf \int_\Omega f(v_j + w_j)z_j \, dx$$

$$\geq \liminf \int_{z>0} f(v_j + w_j)z_j \, dx + \liminf \int_{z<0} f(v_j + w_j)z_j \, dx$$

$$\geq \int_{z>0} f_+ z \, dx + \int_{z<0} f_- z \, dx.$$

This and (3.30) yield

$$\int_\Omega hz\,dx \geq \int_{z>0} f_+ z\,dx + \int_{z<0} f_- z\,dx,$$

a contradiction to the assumption (3.27).

In the general case that (3.28) does not hold, we argue as follows. Let $m \geq 1$ be such that

$$\lambda^* + f_u(x, u) \leq c_2 < \lambda_{m+1}, \qquad \forall\, (x, u) \in \Omega \times \mathbb{R},$$

and set $V^* = \langle \varphi_1, \ldots, \varphi_m \rangle$. Letting P^* denote the projection of V^* on V and letting $L : V^* \to V^*$ be defined by setting $Lv = -\Delta v - \lambda^* v$, one still performs a Lyapunov–Schmidt reduction and solves uniquely the auxiliary equation. To prove that the bifurcation equation has a solution, one makes an homotopy between $\Phi - Ph$ and $L - P^*$. The details are left to the reader. ∎

Remark 3.26 If $f_-(x) < f(x, u) < f_+(x)$ for all $(x, u) \in \Omega \times \mathbb{R}$, then (3.26) shows that (3.27) is also a necessary condition for the existence of a solution. ∎

3.6.3 Exact multiplicity results

In our next application we will show how the degree can be used to find precise multiplicity results. We consider the problem

$$\begin{cases} -\Delta u(x) = f(u(x)) & x \in \Omega \\ u(x) = 0 & x \in \partial\Omega, \end{cases} \qquad (3.31)$$

under the following assumptions on $f \in C^2(\mathbb{R})$:

(a) $f(0) = 0, \quad uf''(u) > 0, \quad \forall u \neq 0,$

(b) $\lim_{u \to \pm\infty} f(u) = f_\pm,$ (3.32)

(c) $\lambda_{k-1} < f'(0) < \lambda_k < f_\pm < \lambda_{k+1}.$

Theorem 3.27 *If* (3.32) *holds, then* (3.31) *has exactly three solutions: the trivial solution* $u \equiv 0$ *and two nontrivial ones.*

Using the same notation as in the preceding theorem, we will look for solutions of $S(u) = 0$, $u \in X$, with $S = I - T$ and $T = Kf$. We need first some lemmas.

Lemma 3.28 *Any $u^* \in X$ such that $S(u^*) = 0$ is a nonsingular solution, namely $S'(u^*)$ is invertible. Moreover there holds*

$$i(S, u^*) = \begin{cases} (-1)^{k-1} & \text{if } u^* = 0 \\ (-1)^k & \text{if } u^* \neq 0. \end{cases}$$

Proof. The equation $S'(u^*)[v] = 0$ is equivalent to the linearized problem

$$-\Delta v = f'(u^*)v, \qquad v_{|\partial\Omega} = 0.$$

Denote by $\mu_j[a]$ the jth eigenvalue of $\Delta a = \mu a v$ with zero Dirichlet boundary conditions. By the assumptions it follows that $f'(u) < \lambda_{k+1}$ and thus the monotonicity property of the eigenvalues (EP-1), see Section 1.4.1, yields

$$\mu_{k+1}[f'(u^*)] > \mu_{k+1}[\lambda_{k+1}] = 1. \tag{3.33}$$

Let us set $f(u) = u\psi(u)$. From (3.32) it follows that ψ is bounded and satisfies

$$\begin{cases} \text{(a)} & \psi(u) < f'(u), \quad \forall u \neq 0, \\ \text{(b)} & \lambda_{k-1} < \psi(u) < \lambda_{k+1}. \end{cases} \tag{3.34}$$

If $u^* \neq 0$, from $-\Delta u^* = \psi(u^*)u^*$ we infer that there exists an integer $j \geq 0$ such that $\mu_j[\psi(u^*)] = 1$. Then (3.34)(b) implies

$$\mu_{k-1}[\psi(u^*)] < \mu_{k-1}[\lambda_{k-1}] = 1, \qquad \mu_{k+1}[\psi(u^*)] > \mu_{k+1}[\lambda_{k+1}] = 1,$$

and thus $\mu_j[\psi(u^*)] = \mu_k[\psi(u^*)] = 1$. Furthermore, from (3.34)(a) it follows that $\mu_k[f'(u^*)] < \mu_k[\psi(u^*)] = 1$. This and (3.33) yield $\mu_k[f'(u^*)] < 1 < \mu_{k+1}[f'(u^*)]$, which proves that u^* is nonsingular and $i(S, u^*) = (-1)^k$. If $u^* = 0$, the linearized equation $S'(0)[v] = 0$ becomes $-\Delta v = f'(0)v = 0$. Repeating the previous arguments and using the assumption that $\lambda_{k-1} < f'(0) < \lambda_k$, see (3.32)(c), we immediately deduce that $u^* = 0$ is also nonsingular and $i(S, 0) = (-1)^{k-1}$. ∎

Lemma 3.29 *There exists $R > 0$ such that $\deg(S, B_R, 0) = (-1)^k$.*

Proof. Let us first show that $\exists R > 0$ such that

$$S(u) \neq 0, \qquad \forall u \in X, \|u\| = R. \tag{3.35}$$

Arguing by contradiction, there exist $\{u_i\}_i$ with $\|u_i\| \to \infty$, such that $S(u_i) = 0$, namely

$$-\Delta u_i = \psi(u_i)u_i, \text{ in } \Omega, \qquad u_i = 0, \text{ on } \partial\Omega.$$

Setting $z_i = \|u_i\|^{-1}u_i$, one finds

$$-\Delta z_i = \psi(u_i)z_i. \tag{3.36}$$

Since $\|z_i\| = 1$, then $z_i \to z^*$ weakly (up to a subsequence). Using (3.36) and the fact that ψ is bounded, elliptic regularity implies that $z_i \to z^*$ strongly in X and uniformly in Ω. Taken any $a \in]\lambda_k, \lambda_{k+1}[$, let us set

$$\psi^*(x) = \begin{cases} f_+ & \text{if } z^*(x) > 0 \\ f_- & \text{if } z^*(x) < 0 \\ a & \text{if } z^*(x) = 0. \end{cases}$$

Taking into account that $u_i(x) = z_i(x)\|u_i\|$ and that $\|u_i\| \to \infty$, it follows that $\psi(u_i(x))z_i(x)$ converges to $\psi^*(x)z^*(x)$, a.e. in Ω. From (3.36) we get $\int_\Omega \nabla z_i \cdot \nabla \varphi \, dx = \int_\Omega \psi(u_i)z_i\varphi \, dx$ for all test function $\varphi \in C_0^\infty(\Omega)$ and, passing to the limit, we find

$$\int_\Omega \nabla z^* \cdot \nabla \varphi \, dx = \int_\Omega \psi^* z^* \varphi \, dx, \qquad \forall \, \varphi \in C_0^\infty(\Omega).$$

This implies that z^* is a solution of $-\Delta z^* = \psi^* z^*$ in Ω, $z^* = 0$ on $\partial\Omega$ and thus $\lambda_j[\psi^*] = 1$ for some integer $j \geq 1$. On the other hand, by the definition of ψ^* and from assumption (3.32)(c), it follows that $\lambda_k < \psi^* < \lambda_{k+1}$. Using the monotonicity property of the eigenvalues, see (EP-1) in Section 1.4.1, this implies that $\lambda_k[\psi^*], 1 < \lambda_{k+1}[\psi^*]$, a contradiction. This proves (3.35).

We can now use the homotopy invariance of the degree. Let us set

$$h(\lambda, u) = (1 - \lambda)S(u) + \lambda(u - K(\psi^*u)),$$

where $v = K(\psi^*u) \iff -\Delta v = \psi^*u$. Arguing as in the previous proof of (3.35), one readily shows that h is admissible in B_R and thus $\deg(S, B_R, 0) = \deg(I - K\psi^*, B_R, 0) = (-1)^k$. ∎

We are now ready to prove Theorem 3.27.

Proof of Theorem 3.27. S is a compact perturbation of the identity and thus it is a proper map. Since every solution of $S = 0$ is nonsingular by Lemma 3.28, the solution set $S^{-1}(0)$ is finite. Let m denote the number of nontrivial solutions of $S = 0$. Using Lemmas 3.28 and 3.29 jointly with the property (P.8) of the degree, we infer

$$(-1)^k = \deg(S, B_R, 0) = \sum_{u^* \in S^{-1}(0)} i(S, u^*) = m(-1)^k + (-1)^{k-1},$$

and this implies $m = 2$. ∎

The same arguments carried out before can be used to study the problem

$$\begin{cases} -\Delta u(x) = \lambda u - g(u) & x \in \Omega \\ u(x) = 0 & x \in \partial\Omega. \end{cases} \tag{3.37}$$

Theorem 3.30 *Suppose that* $g \in C^2(\mathbb{R})$ *is such that* $g(0) = 0$, $g'(0) = 0$, $g'(u) \to +\infty$, *respectively* $-\infty$, *if* $u \to +\infty$, *respectively* $-\infty$, *and* $ug''(u) > 0$ *for all* $u \neq 0$. *Then for all* $\lambda \in]\lambda_1, \lambda_2[$, *problem* (3.37) *has precisely two nontrivial solutions.*

For some further results on (3.37), see also the exercises at the end of the chapter.

3.7 The Krasnoselski bifurcation theorem

In this section we will prove a remarkable bifurcation result due to M. A. Krasnoselski [110].

Theorem 3.31 *Let* X *be a Banach space and let* $T \in C^1(X,X)$ *be a compact operator such that* $T(0) = 0$ *and* $T'(0) = 0$. *Moreover, let* $A \in L(X)$ *also be compact. Then every characteristic value* λ^* *of* A *with odd (algebraic) multiplicity is a bifurcation point for* $u = \lambda Au + T(u)$.

Proof. Setting $S_\lambda(u) = u - \lambda Au - T(u)$, let us suppose by contradiction that λ^* is not a bifurcation point. Then there exist $\varepsilon_0 > 0$ such that for all $r \in (0, \varepsilon_0)$ and $\varepsilon \in (0, \varepsilon_0)$ one has

$$S_\lambda(u) \neq 0, \qquad \forall \lambda \in [\lambda^* - \varepsilon, \lambda^* + \varepsilon], \quad \forall \|u\| = r. \qquad (3.38)$$

We can also choose ε_0 in such a way that the interval $[\lambda^* - \varepsilon_0, \lambda^* + \varepsilon_0]$ does not contain other characteristic values of A but λ^*. Notice that S_λ is a compact

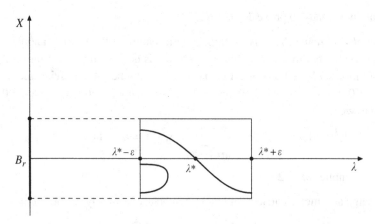

Figure 3.2 The line in bold is the solution set $\{(\lambda, u) \in [\lambda^* - \varepsilon, \lambda^* + \varepsilon] \times B_r : S_\lambda(u) = 0\}$.

perturbation of the identity. Then, taking into account also that (3.38) holds, it follows that it makes sense to consider the Leray–Schauder topological degree $\deg(S_\lambda, B_r, 0)$ of S_λ on the ball $B_r = \{u \in X : \|u\| < r\}$. By the property of the invariance by homotopy, one has that

$$\deg(S_{\lambda^*-\varepsilon}, B_r, 0) = \deg(S_{\lambda^*+\varepsilon}, B_r, 0). \tag{3.39}$$

On the other hand, taking ε_0 possibly smaller, $\deg(S_\lambda, B_r, 0)$ equals the Leray–Schauder index $i(S_\lambda, 0)$, which can be evaluated using Theorem 3.20. If β denotes the sum of the algebraic multiplicities of the characteristic values μ of A such that $\mu > \lambda^* - \varepsilon$, then Theorem 3.20 yields

$$\deg(S_{\lambda^*-\varepsilon}, B_r, 0) = i(S_{\lambda^*-\varepsilon}, 0) = (-1)^\beta.$$

Similarly, if β' denotes the sum of the algebraic multiplicities of the characteristic values μ of A such that $\mu > \lambda^* + \varepsilon$, we have

$$\deg(S_{\lambda^*+\varepsilon}, B_r, 0) = i(S_{\lambda^*+\varepsilon}, 0) = (-1)^{\beta'}.$$

Since $[\lambda^*-\varepsilon, \lambda^*+\varepsilon]$ contains only the eigenvalue λ^*, it follows that $\beta = \beta'+\nu^*$, where ν^* denotes the algebraic multiplicity of λ^*. As a consequence we get

$$\deg(S_{\lambda^*-\varepsilon}, B_r, 0) = (-1)^{\beta'+\nu^*} = (-1)^{\nu^*}\deg(S_{\lambda^*+\varepsilon}, B_r, 0).$$

Since ν^* is an odd integer we infer

$$\deg(S_{\lambda^*-\varepsilon}, B_r, 0) = -\deg(S_{\lambda^*+\varepsilon}, B_r, 0),$$

in contradiction to (3.39). ∎

Theorem 3.31 applies to the nonlinear eigenvalue problem

$$\begin{cases} -\mathcal{L}u = \lambda u + f(x, u), & x \in \Omega \\ u = 0 & x \in \partial\Omega, \end{cases} \tag{D_λ}$$

where $\mathcal{L}$ is the elliptic operator introduced in Remark 3.22. Using the notation introduced in Remark 2.11, we set

- $X = C_0^{0,\alpha}(\Omega)$,
- $K = (-\mathcal{L})^{-1}$ on X,
- f, the Nemitski operator associated to $f(x, u)$,
- $T = K \circ f$,
- $S_\lambda(u) = u - \lambda Ku - T(u)$.

Then (D_λ) is equivalent to the equation $S_\lambda(u) = 0$. Since K is compact, then also T is. A straight application of Theorem 3.31 with $A = K$ yields that every characteristic value of A with odd (algebraic) multiplicity is a bifurcation point for S_λ. Finally, since the characteristic values of A coincide with the eigenvalues of $-\mathcal{L}u = \lambda u$, $u \in X$, we have the following.

Corollary 3.32 *Every eigenvalue with odd multiplicity of* $-\mathcal{L}$ *with Dirichlet boundary conditions on* Ω *is a bifurcation point for* (D_λ).

We anticipate that, using variational tools, it is possible to show that when $\mathcal{L}$ is variational, say $\mathcal{L} = \Delta$, every eigenvalue of $-\mathcal{L}$ with Dirichlet boundary condition, is indeed a bifurcation point for (D_λ). See Theorem 12.20 in Chapter 12.

3.8 Exercises

(i) Let $f \in C(\mathbb{R}^n, \mathbb{R}^n)$ and suppose that $\exists R > 0$ such that $(f(x) \mid x) > 0$ for all $x \in \mathbb{R}^n$, $|x| = R$. Show that $\deg(f, B_R, 0) = 1$.

(ii) As in the preceding exercise, but suppose that $\exists R > 0$ such that $(f(x) \mid x) < 0$ for all $x \in \mathbb{R}^n$, $|x| = R$. Show that $\deg(f, B_R, 0) = (-1)^n$.

(iii) Prove the Shaeffer fixed point theorem. Let T be a compact operator in X and suppose that $\exists r > 0$ such that $u = \lambda T(u)$, $u \in X$, $\lambda \in [0, 1] \Rightarrow \|u\| < r$. Then T has a fixed point in $B_r(0)$.

(iv) Let T be a compact operator in X such that $T(\partial B_r(0)) \subset B_r(0)$, for some $r > 0$. Then T has a fixed point in $\overline{B_r(0)}$.

(v) Consider the problem at resonance (3.25) and assume that f and f_u are bounded, and that (3.28) holds. Furthermore, let $f_- = f_+ = 0$ and suppose that $\lim_{|u| \to \infty} uf(x, u) = \mu > 0$. Show that there exists $\varepsilon > 0$ such that

    (a) (3.25) has at least one solution, provided $\|Ph\| < \varepsilon$,

    (b) if $0 < \|Ph\| < \varepsilon$, then (3.25) has at least two solutions.

(vi) Give an alternative proof of (3.29) showing that $(\Phi(v) \mid v) > (Ph \mid v)$ on $\|v\| = R \gg 1$.

(vii) Prove Theorem 3.30.

(viii) Consider (3.37) with g as in Theorem 3.30.

    (a) Show that it has only the trivial solution for $\lambda \leq \lambda_1$.

    (b) Extend Theorem 3.30 by proving that (3.37) has exactly two nontrivial solutions for $\lambda = \lambda_2$, too.

    (c) Suppose that λ_2 is simple and prove that there is $\varepsilon > 0$ such that (3.37) has exactly four nontrivial solutions for all $\lambda \in]\lambda_2, \lambda_2 + \varepsilon[$. [Hint: use the behaviour of the branch bifurcating from λ_2.]

4

Topological degree, II: global properties

In this chapter we will exploit the global properties of the Leray–Schauder topological degree to discuss the nonlocal structure of the solutions set of several classes of elliptic equations and to prove the Rabinowitz global bifurcation theorem.

As recalled before, another important bifurcation result dealing with the case of variational operators will be discussed in Section 12.3.

4.1 Improving the homotopy invariance

The main purpose of this section is to prove a more general version of the homotopy invariance property (P.5). This will be useful in several applications later on.

Let X be a Banach space, $U \subset [a,b] \times X$ be open and bounded. We set $U_\lambda = \{x \in X : (\lambda, x) \in U\}$, whose boundary is denoted by ∂U_λ.
Let us remark that one has to distinguish ∂U_λ from $(\partial U)_\lambda$: in general, one has that $\partial U_\lambda \subset (\partial U)_\lambda$, see Figure 4.1.

Consider a map $h(\lambda, x) = x - k(\lambda, x)$ such that $k(\lambda, \cdot)$ is compact and $0 \notin h(\partial U)$. Such a map h will also be called an *admissible homotopy* on U. If h is an admissible homotopy, for every $\lambda \in [a,b]$ and every $x \in \partial U_\lambda$, one has that $h_\lambda(x) := h(\lambda, x) \neq 0$ and it makes sense to evaluate $\deg(h_\lambda, U_\lambda, 0)$.

Theorem 4.1 *If h is an admissible homotopy on $U \subset [a,b] \times X$, then* $\deg(h_\lambda, U_\lambda, 0)$ *is constant for all $\lambda \in [a,b]$.*

Proof. For fixed $\lambda \in]a,b[$, let H_λ denote the set $\{x \in U_\lambda : h(\lambda, x) = 0\}$. H_λ is compact and $H_\lambda \cap \partial(U_\lambda) = \emptyset$ and thus there exists an open neighbourhood $\mathcal{O}_\lambda$ of H_λ and $\varepsilon > 0$ such that

$$[\lambda - \varepsilon, \lambda + \varepsilon] \times \mathcal{O}_\lambda \subset U.$$

55

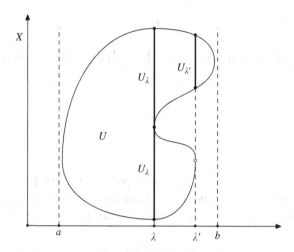

Figure 4.1 Points marked with bold circles are points of ∂U_λ. The point marked with an open circle belongs to $(\partial U)_{\lambda'}$ but not to $\partial U_{\lambda'}$.

Let us show that, if ε is sufficiently small, there holds

$$\{(\ell, x) : h(\ell, x) = 0, \ \lambda - \varepsilon \le \ell \le \lambda + \varepsilon\} \subset [\lambda - \varepsilon, \lambda + \varepsilon] \times \mathcal{O}_\lambda. \qquad (4.1)$$

Otherwise, there exist $\varepsilon_i \downarrow 0$, $\ell_i \in [a, b]$, $x_i \in \mathbb{R}^n$ such that

$$|\ell_i - \lambda| \le \varepsilon_i, \quad h(\ell_i, x_i) = 0, \quad (\ell_i, x_i) \notin [\lambda - \varepsilon_i, \lambda + \varepsilon_i] \times \mathcal{O}_\lambda.$$

By compactness, we can also assume that, up to a subsequence, $(\ell_i, x_i) \to (\lambda, x^*)$. By continuity, $h(\ell, x^*) = \lim h(\ell_i, x_i) = 0$ and hence $x^* \in H_\lambda$. On the other hand, this is not possible because one also has that $x^* \notin \mathcal{O}_\lambda$, and $H_\lambda \subset \mathcal{O}_\lambda$. Equation (4.1) shows that h is an admissible homotopy on $[\lambda - \varepsilon, \lambda + \varepsilon] \times \mathcal{O}_\lambda$, and hence we can use the (standard) homotopy invariance (P.5) to deduce

$$\deg(h_\ell, \mathcal{O}_\lambda, 0) = \text{constant}, \quad \forall \ell \in [\lambda - \varepsilon, \lambda + \varepsilon].$$

Finally, since $H_\ell \subset \mathcal{O}_\lambda$, the excision property implies

$$\deg(h_\ell, U_\ell, 0) = \deg(h_\ell, \mathcal{O}_\lambda, 0), \quad \forall \ell \in [\lambda - \varepsilon, \lambda + \varepsilon].$$

Small modifications allow us to handle the cases $\lambda = a$ and $\lambda = b$. The preceding arguments show that $\deg(h_\lambda, U_\lambda, 0)$ is locally constant on $[a, b]$. Since the degree is an integer, the conclusion follows.   ∎

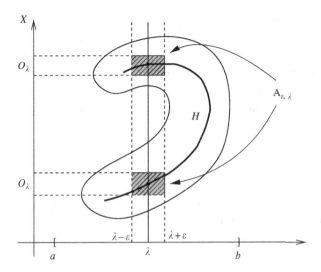

Figure 4.2 $H = \{(\lambda, x) \in U : h_\lambda(x) = 0\}; A_{\varepsilon, \lambda} = [\lambda - \varepsilon, \lambda + \varepsilon] \times \mathcal{O}_\lambda.$

4.2 An application to a boundary value problem with sub- and super-solutions

As a first application of Theorem 4.1 we will study a problem with sub- and super-solutions. The definition of sub- and super-solution has been given in Section 3.6.1 in the preceding chapter, where we also recalled that if

$$\begin{cases} -\Delta u(x) = f(x, u(x)) & x \in \Omega \\ u(x) = 0 & x \in \partial\Omega, \end{cases} \tag{4.2}$$

has a sub-solution $\underline{u}$ and a super-solution $\overline{u}$ such that $\underline{u}(x) \leq \overline{u}(x)$ in Ω, then it has a solution u with $\underline{u}(x) \leq u(x) \leq \overline{u}(x)$. We will show that when $f = \lambda_1 u + g$ with g bounded, the assumption that $\underline{u}(x) \leq \overline{u}(x)$ is no longer necessary.

Theorem 4.2 *Suppose that* $f(x, u) = \lambda_1 u + g(x, u)$ *where* λ_1 *is the first eigenvalue of* $-\Delta$ *on* Ω *with zero boundary conditions and* g *is bounded and locally Hölder continuous. Then* (4.2) *has a solution provided it possesses a sub-solution and a super-solution.*

Proof. Setting, as above, $X = L^2(\Omega)$, equation (4.2) with $f = \lambda_1 u + g$, can be written in the form $Lu = g(u)$, where L is the densely defined operator given by $Lu = -\Delta u - \lambda_1 u$. Let V denote the kernel of L: V is nothing but the one-dimensional space spanned by the function $\varphi_1 > 0$ satisfying $-\Delta\varphi_1 = \lambda_1\varphi_1$ in Ω, $\varphi_1 = 0$ on $\partial\Omega$.

We will solve this equation by using the Lyapunov–Schmidt reduction, see Section 2.2. Let W denote the orthogonal complement of V in X, so that $X = V \oplus W$. Letting P, Q denote the orthogonal projections on V and W, respectively, one has that any $u \in X$ can be written in the form $u = v + w$, with $v = Pu \in V$ and $w = Qu \in W$. With this notation, our equation becomes $Lw = g(v + w)$. Now, L is invertible on W, with compact inverse L^{-1} and, setting $T = L^{-1}g$, we get $w = T(v + w)$ which is equivalent to the Lyapunov–Schmidt system

$$\begin{cases} w = QT(v + w), \\ 0 = PT(v + w). \end{cases}$$

We will use the degree to solve the auxiliary equation, finding a connected set of solutions. Putting $v = t\varphi_1$, let us consider the following set of solutions of the auxiliary equation

$$\Sigma = \{(t, w) \in \mathbb{R} \times W : w = QT(t\varphi_1 + w)\}.$$

We need a preliminary lemma. If $A \subset \mathbb{R} \times W$, we say that $t \in \mathrm{proj}_{\mathbb{R}} A$ if and only if $(t, w) \in A$ for some $w \in W$. ∎

Lemma 4.3 *For all $\alpha > 0$ there exists $\Sigma_\alpha \subset \Sigma$ which is connected and such that $\mathrm{proj}_{\mathbb{R}} \Sigma_\alpha \supseteq [-\alpha, \alpha]$.*

Proof. Since g is bounded, then there exists $r > 0$ such that $\|QT(t\varphi_1 + w)\| < r$ and thus $\Sigma \subset \mathbb{R} \times B_r$. We introduce the following notation: $K = \Sigma \cap \big([-\alpha, \alpha] \times \overline{B}_r\big)$ and $K_\pm = \Sigma \cap \big(\{\pm\alpha\} \times \overline{B}_r\big)$. Suppose, by contradiction, that there is no connected subset of K which joins K_- and K_+. Then there are two closed sets $C_\pm \supseteq K_\pm$ such that

$$C_- \cap C_+ = \emptyset, \qquad K = C_- \cup C_+.$$

In addition, we can find an open set $U \subset \big([-\alpha, \alpha] \times \overline{B}_r\big)$ such that $C_- \subset U$ while $C_+ \cap \overline{U} = \emptyset$. Since $C_- \cup C_+ = K = \Sigma \cap \big([-\alpha, \alpha] \times \overline{B}_r\big)$ and $\Sigma \subset \mathbb{R} \times B_r$, then $w \neq QT(t\varphi_1 + w)$ on $(\partial U)_t$ for all $t \in [-\alpha, \alpha]$. Thus we can use the general invariance property of homotopy, see Theorem 4.1, to get (here the homotopy parameter is $t \in [-\alpha, \alpha]$)

$$\deg(I - QT(-\alpha\varphi_1 + \cdot), U_{-\alpha}, 0) = \deg(I - QT(\alpha\varphi_1 + \cdot), U_\alpha, 0).$$

Since $U_\alpha = \emptyset$ we deduce

$$\deg(I - QT(-\alpha\varphi_1 + \cdot), U_{-\alpha}, 0) = 0. \tag{4.3}$$

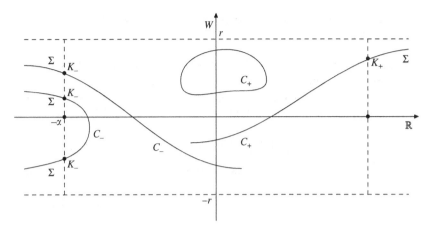

Figure 4.3 $K = C_- \cup C_+, C_- \cap C_+ = \emptyset$.

On the other hand, using the homotopy $(\mu, w) \mapsto I - \mu QT(-\alpha\varphi_1 + w)$ on B_r one easily finds that

$$\deg(I - QT(-\alpha\varphi_1 + \cdot), B_r, 0) = 1.$$

Then, by the excision property, we find

$$\deg(I - QT(-\alpha\varphi_1 + \cdot), U_{-\alpha}, 0) = \deg(I - QT(-\alpha\varphi_1 + \cdot), B_r, 0) = 1,$$

a contradiction to (4.3). ∎

Setting

$$G(t, w) = \int_\Omega g(t\varphi_1 + w)\varphi_1 \, dx,$$

the bifurcation equation $QT(t\varphi_1 + w) = 0$ can be written in the form $G(t, w) = 0$.

Claim 4.4 If $(t, w) \in \Sigma$ and $G(t, w) \geq 0$, (respectively ≤ 0), then $t\varphi_1 + w$ is a sub-solution (respectively super-solution) of (4.2).

. To prove this claim it suffices to remark that $t\varphi_1 + w = 0$ on $\partial\Omega$ and

$$-\Delta(t\varphi_1 + w) = t\lambda_1\varphi_1 - \Delta w = t\lambda_1\varphi_1 + \lambda_1 w + Qg(t\varphi_1 + w)$$
$$= \lambda_1(t\varphi_1 + w) + g(t\varphi_1 + w) - G(t, w)\varphi_1.$$

If $G(t, w) \geq 0$, then the preceding equation implies that

$$-\Delta(t\varphi_1 + w) \leq \lambda_1(t\varphi_1 + w) + g(t\varphi_1 + w), \qquad \text{in } \Omega,$$

and thus $t\varphi_1 + w$ is a sub-solution. Similarly, if $G(t, w) \geq 0$ then $t\varphi_1 + w$ is a super-solution.

Proof of Theorem 4.2 completed. We have already pointed out that the solutions of the auxiliary equation are bounded: $\|w\| < r$. Moreover, by regularity, w is also bounded in the C^1 norm. Then there is $\tau > 0$ such that for every solution w of the auxiliary equation there holds

$$-\tau\varphi_1(x) < w(x) < \tau\varphi_1(x), \qquad \forall x \in \Omega.$$

Let $\underline{u}$ and $\overline{u}$ denote a pair of sub- and super-solution, respectively, of (4.2). Take $\alpha > \tau$ such that

$$\begin{cases} (\alpha - \tau)\varphi_1(x) \geq \underline{u}(x) & \forall x \in \Omega \\ -(\alpha - \tau)\varphi_1(x) \leq \overline{u}(x) & \forall x \in \Omega. \end{cases}$$

Then one has (for every w solving the auxiliary equation)

$$-\alpha\varphi_1(x) + w(x) < -(\alpha - \tau)\varphi_1(x) \leq \overline{u}(x), \quad \forall x \in \Omega, \tag{4.4}$$

$$\alpha\varphi_1(x) + w(x) > \quad (\alpha - \tau)\varphi_1(x) \geq \underline{u}(x), \quad \forall x \in \Omega. \tag{4.5}$$

For fixed α as above, let Σ_α be the connected set given by Lemma 4.3. Consider G restricted on Σ_α. If $G(t, w) > 0$ for all $(t, w) \in \Sigma_\alpha$ then, according to Claim 4.4, all $t\varphi_1 + w$ are sub-solutions. In particular, for all $w \in W$ such that $(-\alpha, w) \in \Sigma_\alpha$, the function $-\alpha\varphi_1 + w$ is a sub-solution of (4.2). By (4.4), $-\alpha\varphi_1 + w < \overline{u}$ and hence there exists a solution u such that $-\alpha\varphi_1 + w \leq u \leq \overline{u}$. Similarly, if $G(t, w) < 0$ for all $(t, w) \in \Sigma_\alpha$ then $\alpha\varphi_1 + w$ is a super-solution such that $\underline{u} < \alpha\varphi_1 + w$ and there exists a solution u such that $\underline{u} \leq u \leq \alpha\varphi_1 + w$. It remains to consider the case in which G changes sign on Σ_α. Since Σ_α is connected and G is continuous, then there exists $(t_0, w_0) \in \Sigma_\alpha$ such that $G(t_0, w_0) = 0$. Then $t_0\varphi_1 + w_0$ is a solution of (4.2). ■

Remark 4.5 An example shows that, in general, (4.2) can have no solution at all if we do not assume that $\underline{u} \leq \overline{u}$. See [5]. ■

4.3 The Rabinowitz global bifurcation theorem

Throughout this section, X is a Banach space, $A \in L(X)$ is compact and $T \in C^1(X, X)$ is compact and such that $T(0) = 0$ and $T'(0) = 0$. We also set $S_\lambda(u) = u - \lambda Au - T(u)$ and denote by Σ the set

$$\Sigma = \{(\lambda, u) \in \mathbb{R} \times X, u \neq 0 : S_\lambda(u) = 0\}.$$

If $(\lambda^*, 0) \in \overline{\Sigma}$ then λ^* is a bifurcation point for $S_\lambda = 0$. A connected component of $\overline{\Sigma}$ is a closed connected set $C \subset \overline{\Sigma}$ which is maximal with respect to the inclusion. According to the Krasnoselski bifurcation theorem 3.31, if λ^* is an odd characteristic value of A, then λ^* is a bifurcation point, namely $(\lambda^*, 0) \in \overline{\Sigma}$. Let C be the connected component of $\overline{\Sigma}$ containing $(\lambda^*, 0)$. We are going to discuss a celebrated paper [147] by P. Rabinowitz, which improves the Krasnoselski result by showing that C is either unbounded in $\mathbb{R} \times X$ or meets another bifurcation point of $S_\lambda = 0$. The set of characteristic values of A will be denoted by $r(A)$.

We first need a topological lemma. See Figure 4.4.

Lemma 4.6 *Let C be the connected component of $\overline{\Sigma}$ containing $(\lambda^*, 0)$ and suppose that C is bounded and does not contain any point $(\hat{\lambda}, 0)$ with $\hat{\lambda} \in r(A)$, $\hat{\lambda} \neq \lambda^*$. Then there exists an open bounded set $\mathcal{O} \subset \mathbb{R} \times X$ such that*

(i) $C \subset \mathcal{O}$,
(ii) $\partial \mathcal{O} \cap \Sigma = \emptyset$,
(iii) $\mathcal{O} \cap (\mathbb{R} \times \{0\}) =]\lambda^* - \varepsilon, \lambda^* + \varepsilon[$, *with $\varepsilon > 0$ and smaller than δ, the distance from C and $(r(A) \setminus \{\lambda^*\}) \times \{0\}$,*
(iv) $\exists\, \alpha > 0$ *such that if $(\lambda, u) \in \mathcal{O}$ with $|\lambda - \lambda^*| \geq \varepsilon$, then $\|u\| \geq \alpha$.*

The proof of the lemma will make use of the following result in point set topology. For a proof see for example [87], Lemma 29.1.

Lemma 4.7 *Let C_1, C_2 be closed disjoint subsets of the metric compact space $\mathcal{K}$. If there are no connected components of $\mathcal{K}$ with nonempty intersection with C_1, C_2, then $\mathcal{K} = K_1 \cup K_2$ with K_1, K_2 closed, $K_1 \cap K_2 = \emptyset$, $C_1 \subset K_1$, $C_2 \subset K_2$.*

Proof of Lemma 4.6. Let $\varepsilon < \delta$, and let U_ε be an ε-neighbourhood of C. Such a U_ε satisfies (i) and (iii), but (ii) could possibly fail. To overcome this problem, we will use the preceding lemma. We set $C_1 = C$, $C_2 = \partial U_\varepsilon \cap \overline{\Sigma}$ and $\mathcal{K} = \overline{U}_\varepsilon \cap \overline{\Sigma}$. Since C is bounded and A, T are compact, then C_1, C_2 and $\mathcal{K}$ are compact and the preceding lemma applies. Let $d = \min\{\mathrm{dist}[K_1, K_2], \mathrm{dist}[K_1, \partial U_\varepsilon]\}$ and let U'_ε be an ε-neighbourhood of K_1, with $\varepsilon < d/2$. Let us check that U'_ε satisfies (ii). This follows from the fact that $C_2 = \partial U_\varepsilon \cap \overline{\Sigma} \subset K_2$, $C_1 = C \subset K_1$ and $K_2 \cap K_1 = \emptyset$. Taking ε possibly smaller, (i) and (iii) hold true as well. As for the property (iv), we remark that the distance d' between $\mathcal{K}$ and the set $(\mathbb{R} \setminus]\lambda^* - \varepsilon, \lambda^* + \varepsilon[) \times \{0\}$ is positive because U'_ε does not contain any characteristic value of A, but λ^*. Then, taking $\alpha < d'$ and setting $\mathcal{O} = U'_\varepsilon \setminus [(\mathbb{R} \setminus]\lambda^* - \varepsilon, \lambda^* + \varepsilon[) \times \overline{B}_\alpha]$, one readily verifies that $\mathcal{O}$ satisfies (iv), too. ∎

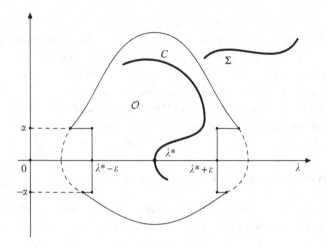

Figure 4.4

We are now in a position to prove the Rabinowitz global bifurcation theorem.

Theorem 4.8 *Let $A \in L(X)$ be compact and let $T \in C^1(X, X)$ be compact and such that $T(0) = 0$ and $T'(0) = 0$. Suppose that λ^* is a characteristic value of A with odd multiplicity. Let C be the connected component of $\overline{\Sigma}$ containing $(\lambda^*, 0)$. Then either*

(a) C is unbounded in $\mathbb{R} \times X$, or
(b) $\exists \hat{\lambda} \in \rho(A) \setminus \{\lambda^*\}$ such that $(\hat{\lambda}, 0) \in C$.

Proof. The proof will be carried out by contradiction. If neither (a) nor (b) holds, then Lemma 4.6 yields a bounded open set $\mathcal{O}$ satisfying (i–iv). We are going to use the general homotopy invariance (see Theorem 4.1), applied to the homotopy S_λ. We will split the arguments into several steps.

Step 1. Let us take β such that $\mathcal{O}_\beta = \emptyset$ and consider the interval $J = [\lambda^* + 2\varepsilon, \beta]$. Taking ε possibly smaller, we can assume that no points of $r(A)$ belong to $]\lambda^*, \lambda^* + 2\varepsilon]$. Using property (iv) of Lemma 4.6 one has that $\mathcal{O}_\lambda \cap B_\alpha = \emptyset$, for all $\lambda \in J$. This and the other properties of $\mathcal{O}$ stated in Lemma 4.6 imply that $S_\lambda(u) \neq 0$ for all $u \in \partial \mathcal{O}_\lambda$ and all $\lambda \in J$. Hence by the homotopy invariance we deduce that $\deg(S_\lambda, \mathcal{O}_\lambda, 0)$ is constant for all $\lambda \in J$. In particular, since $\mathcal{O}_\beta = \emptyset$ we find

$$\deg(S_{\lambda + 2\varepsilon}, \mathcal{O}_{\lambda + 2\varepsilon}, 0) = 0. \tag{4.6}$$

Step 2. Take $\varepsilon' \in]0, \varepsilon[$ and set $J' = [\lambda^* + \varepsilon', \lambda^* + 2\varepsilon]$. Since J' does not contain any point in $r(A)$, and $C \cap (J' \times X)$ is compact, there exists $\rho_0 > 0$ (with $\rho_0 \leq \alpha$)

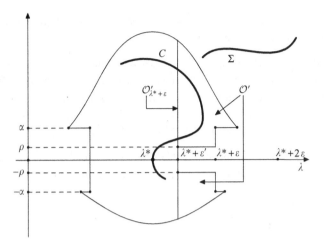

Figure 4.5

such that $\overline{\Sigma} \cap (J' \times \overline{B}_\rho) = \emptyset$ for all $0 < \rho \leq \rho_0$. Thus S_λ is admissible on the set $\mathcal{O}' = \mathcal{O} \cap (J' \times (X \setminus \overline{B}_\rho))$ and there holds

$$\deg(S_{\lambda+\varepsilon'}, \mathcal{O}'_{\lambda^*+\varepsilon'}, 0) = \deg(S_{\lambda^*+2\varepsilon}, \mathcal{O}'_{\lambda^*+2\varepsilon}, 0).$$

This, the fact that $\mathcal{O}'_{\lambda^*+\varepsilon'} = \mathcal{O}_{\lambda^*+\varepsilon'} \setminus \overline{B}_\rho$ and $\mathcal{O}'_{\lambda^*+2\varepsilon} = \mathcal{O}_{\lambda^*+2\varepsilon}$, jointly with (4.6) imply

$$\deg(S_{\lambda^*+\varepsilon'}, \mathcal{O}_{\lambda^*+\varepsilon'} \setminus \overline{B}_\rho, 0) = 0, \qquad \forall \rho \in (0, \rho_0],$$

and thus

$$\deg(S_{\lambda^*+\varepsilon'}, \mathcal{O}_{\lambda^*+\varepsilon'}, 0) = \deg(S_{\lambda^*+\varepsilon'}, B_\rho, 0), \qquad \forall \rho \in (0, \rho_0]. \qquad (4.7)$$

By a quite similar argument we find

$$\deg(S_{\lambda^*-\varepsilon'}, \mathcal{O}_{\lambda^*-\varepsilon'}, 0) = \deg(S_{\lambda^*-\varepsilon'}, B_\rho, 0), \qquad \forall \rho \in (0, \rho_0]. \qquad (4.8)$$

Step 3. We use once more the homotopy invariance of S_λ on the set $\mathcal{O} \cap ([\lambda^* - \varepsilon', \lambda^* + \varepsilon'] \times X)$ to infer

$$\deg(S_{\lambda^*-\varepsilon'}, \mathcal{O}_{\lambda^*-\varepsilon'}, 0) = \deg(S_{\lambda^*+\varepsilon'}, \mathcal{O}_{\lambda^*+\varepsilon'}, 0), \qquad \forall \rho \in (0, \rho_0]. \qquad (4.9)$$

Step 4. Putting together (4.7), (4.8) and (4.9) we deduce

$$\deg(S_{\lambda^*+\varepsilon'}, B_\rho, 0) = \deg(S_{\lambda^*-\varepsilon'}, B_\rho, 0), \qquad \forall \rho \in (0, \rho_0].$$

This means that the index $i(S_{\lambda^*+\varepsilon'}, 0)$ is equal to the index $i(S_{\lambda^*-\varepsilon'}, 0)$, and, as we have seen in the proof of the Krasnoselski theorem, this cannot be true, because $[\lambda^* - \varepsilon', \lambda^* + \varepsilon'] \cap r(A) = \{\lambda^*\}$ and λ^* has odd multiplicity. The contradiction proves the theorem. ∎

Remark 4.9 When applying Theorem 4.8 to differential equations it is important to rule out one of the two alternatives (a) and (b). Although often the connected component containing $(\lambda^*, 0)$ turns out to be unbounded, elementary examples show that the alternative (b) of Theorem 4.8 can arise as well. See Exercise 4.5(i). See also [79] for an elliptic problem where case (b) arises. ∎

Theorem 4.8 has many applications. Here we limit ourselves to citing a paper by M. Crandall and P. Rabinowitz [83] dealing with nonlinear Sturm–Liouville problems, of the form

$$\begin{cases} -u'' = \lambda u + f(x, u, u'), & x \in (0, \pi), \\ u(0) = u(\pi) = 0, \end{cases} \tag{4.10}$$

where f is Lipschitz and $f(x, u, \xi) = o(\sqrt{u^2 + |\xi|^2})$ as $(u, \xi) \to (0, 0)$, uniformly with respect to $x \in [0, \pi]$. We already know (see the example in Section 2.3 after Theorem 2.10) that the numbers k^2, $k \in \mathbb{N}$, are simple eigenvalues of the linearized problem $-u'' = \lambda u$, $u(0) = u(\pi) = 0$ and hence are bifurcation points for (4.10). As an application of Theorem 4.8 we want to show the following result, which is a particular case of a more general one in [83].

Theorem 4.10 *From each k^2, $k \in \mathbb{N}$, bifurcates an unbounded connected component $\mathcal{C}_k \subset \Sigma$ of nontrivial solutions of (4.10). Moreover $\mathcal{C}_k \cap \mathcal{C}_j = \emptyset$ if $k \neq j$.*

Proof. Under the preceding simplifying assumptions, the proof is not too complicated and we will give an outline of the arguments, leaving the details as an exercise. One works on $E = \{u \in C^1(0, \pi) : u(0) = u(\pi) = 0\}$ endowed with the standard norm. First one shows that there exists a neighbourhood U_k of $(k^2, 0) \in \mathbb{R} \times E$ such that if $(\lambda, u) \in \Sigma \cap U_k$, then u has exactly $k - 1$ simple zeros in $(0, \pi)$. Moreover, by the uniqueness of the Cauchy problem it follows that the nontrivial solutions of (4.10) have only simple zeros in $(0, \pi)$. These two properties, together with the fact that the branch $\mathcal{C}_k \subset \Sigma$ emanating from $(k^2, 0)$ is connected, allow us to rule out the alternative (b) of Theorem 4.8 and to show that $\mathcal{C}_k \cap \mathcal{C}_j = \emptyset$ if $k \neq j$, proving the theorem. ∎

Finally, it is worth mentioning that the arguments carried out in Sections 4.2 and 4.3 can be adapted to prove a classical global result by Leray and Schauder [115].

Theorem 4.11 *Consider the equation $u = \lambda T(u)$, where $T \in C(X,X)$ is compact, let $\Sigma = \{(\lambda, u) \in \mathbb{R} \times X : u = \lambda T(u)\}$ and let C denote the connected component of $\overline{\Sigma}$ containing $(0,0)$. Then $C = C^+ \cup C^-$ where $C^\pm \subset \mathbb{R}^\pm \times X$ and $C^+ \cap C^- = \{(0,0)\}$. Moreover, $C^\pm$ are unbounded in $\mathbb{R}^\pm \times X$.*

4.4 Bifurcation from infinity and positive solutions of asymptotically linear elliptic problems

In this section we will discuss, following [16], the existence of positive solutions of a class of asymptotically linear elliptic boundary value problems like

$$
\begin{cases}
-\Delta u = \lambda f(u) & x \in \Omega \\
u = 0 & x \in \partial\Omega,
\end{cases}
\tag{4.11}
$$

where $f \in C(\mathbb{R}^+, \mathbb{R})$ is asymptotically linear. We will see that it is possible to study in a rather precise way the global behaviour of the set of solutions of (4.11) and this will allow us to obtain existence and multiplicity results. In carrying out this analysis, a fundamental role will be played by suitable applications of Theorem 4.8.

Let us start with an abstract setting. Let X be a Banach space and consider a map $S(\lambda, u) = u - \lambda T(u)$, with $T \in C(X,X)$ compact. We set $\Sigma = \{(\lambda, u) \in \mathbb{R} \times (X \setminus \{0\}) : S(\lambda, u) = 0\}$. To investigate the asymptotic behaviour of Σ, it is convenient to give the following definition.

Definition 4.12 *We say that $\lambda_\infty \in \mathbb{R}$ is a bifurcation from infinity for $S = 0$ if there exist $\lambda_j \to \lambda_\infty$ and $u_j \in X$, such that $\|u_j\| \to \infty$ and $(\lambda_j, u_j) \in \Sigma$.*

Let us now assume that $T = A + G$, with A linear and G bounded on bounded sets. Let us set $z = \|u\|^{-1}u$, and

$$
\Psi(\lambda, z) =
\begin{cases}
z - \lambda \|z\|^2 T\left(\dfrac{z}{\|z\|^2}\right) & \text{if } z \neq 0 \\
0 & \text{if } z = 0.
\end{cases}
\tag{4.12}
$$

For $z \neq 0$ one has that

$$
\Psi(\lambda, z) = z - \lambda A z - \lambda \|z\|^2 G\left(\frac{z}{\|z\|^2}\right),
$$

and hence Ψ is continuous at $z = 0$. Moreover, setting

$$
\Gamma = \{(\lambda, z) : z \neq 0, \ \Psi(\lambda, z) = 0\},
$$

there holds

$$(\lambda, u) \in \Sigma \quad \Longleftrightarrow \quad (\lambda, z) \in \Gamma. \tag{4.13}$$

In addition, $\|u_j\| \to \infty$ if and only if $\|z_j\| = \|u_j\|^{-1} \to 0$. This and (4.13) immediately imply the following.

Lemma 4.13 λ_∞ *is a bifurcation from infinity for* $S = 0$ *if and only if* λ_∞ *is a bifurcation from the trivial solution for* $\Psi = 0$. *In such a case we will say that* Σ *bifurcates from* (λ_∞, ∞).

Dealing with the bifurcation from the trivial solution for Ψ, one has to be careful because $\Psi(\lambda, \cdot)$ is not differentiable at $z = 0$. See Remark 4.17 later on.

We will suppose that

$$f(u) = mu + g(u), \quad m > 0, \quad g \in C^{0,\alpha}(\mathbb{R}^+, \mathbb{R}), \quad |g(u)| \le \text{constant}, \quad g(0) \ge 0. \tag{4.14}$$

It is convenient to consider a function $\widetilde{f} \in C^{0,\alpha}(\mathbb{R})$ defined by setting $\widetilde{f}(u) = f(u)$ for $u \ge 0$, such that $0 < \widetilde{f}(u) \le C_1$ for all $u < 0$. We will also write $\widetilde{f}(u) = mu + \widetilde{g}(u)$. Of course, $\widetilde{g}(u) = g(u)$ for all $u \ge 0$. Consider the problem

$$-\Delta u = \lambda \widetilde{f}(u), \quad \text{in } \Omega \quad u = 0, \text{in } \partial\Omega. \tag{4.15}$$

Letting $X = L^2(\Omega)$, (4.15) be equivalent to the functional equation $S(\lambda, u) = u - \lambda(Au + G(u))$, where $Au = K(mu)$, $G = K\widetilde{g}(u)$, and K is the inverse of $-\Delta$ with zero Dirichlet boundary conditions. For some $\lambda > 0$, let u be a non-trivial solution of (4.15) and let $x^* \in \Omega$ be such that $u(x^*) = \min_\Omega u(x)$. If $u(x^*) < 0$, then, since $\widetilde{f}(u) > 0$ for $u < 0$, we have $-\Delta u(x^*) = \lambda \widetilde{f}(u(x^*)) > 0$, a contradiction. This shows that $u \ge 0$ in Ω, and hence is a solution of (4.11). Furthermore, from (4.14) it follows that there exists $\delta > 0$ such that $f(u) + \delta u > 0$ for all $u > 0$. Since $-\Delta u + \lambda \delta u = \lambda(f(u) + \delta u)$, the maximum principle implies that $u > 0$ in Ω.

As usual, we let λ_1 denote the first eigenvalue of $-\Delta$ with zero Dirichlet boundary conditions. The corresponding positive (normalized) eigenfunction will be indicated by φ_1.

Theorem 4.14 *Let (4.14) hold. Then* $\lambda_\infty := \lambda_1/m$ *is a bifurcation from infinity for S, and the only one. More precisely, there exists a connected component* Σ_∞ *of* Σ *bifurcating from* (λ_∞, ∞) *which corresponds to an unbounded connected component* $\Gamma_\infty \subset \Gamma$ *bifurcating from the trivial solution of* $\Psi_\lambda(u) = 0$ *at* $(\lambda_\infty, 0)$.

According to the preceding discussion, we will show that λ_∞ is a bifurcation from $z = 0$ for $\Psi(\lambda, z) = 0$, and the only one. In the sequel we use the notation $\Psi_\lambda(u)$ to indicate the map $\Psi(\lambda, u)$ defined in (4.12). We need some preliminary lemmas.

Lemma 4.15 *Let $J \subset \mathbb{R}^+$ be any compact interval such that $\lambda_\infty \notin J$. Then*

(a) $\exists\, r > 0$ such that $S_\lambda(u) \neq 0$, $\forall \lambda \in J$, $\forall \|u\| \geq r$,
(b) λ_∞ is the only possible bifurcation from infinity,
(c) $i(\Psi_\lambda, 0) = 1$ for all $\lambda < \lambda_\infty$.

Proof. (a) Assuming the contrary, there are sequences $\lambda_j \to \bar{\lambda} \in J$ and $\|u_j\| \to \infty$, such that $u_j = \lambda_j(Au_j + G(u_j))$. Setting $v_j = \|u_j\|^{-1}u_j$, we find $v_j = \lambda_j(Av_j + G(v_j))$, by compactness, $v_j \to \bar{v}$ strongly in X, and by elliptic regularity $v_j \to \bar{v}$ in C^2. Since v_j satisfy

$$-\Delta v_j = \lambda_j \frac{\widetilde{f}(u_j)}{\|u_j\|},$$

and $\widetilde{f}(u) < 0$ for $u < 0$, it follows that $v_j > 0$. Thus, passing to the limit in the preceding equation, one finds

$$-\Delta \bar{v} = \bar{\lambda} m \bar{v}, \qquad \bar{v} \geq 0, \quad \|\bar{v}\| = 1.$$

This implies that $\bar{v} = \varphi_1$ and $\bar{\lambda} m = \lambda_1$, namely $\bar{\lambda} = \lambda_\infty$. This is a contradiction, because $\bar{\lambda} \in J$, while J does not contain λ_∞. This proves (a). Statement (b) follows immediately from (a). Regarding (c), fix any $\lambda < \lambda_\infty$ and take $J = [0, \lambda]$. For $t \in [0, 1]$, the parameter $t\lambda$ belongs to J and from (a) it follows that $u \neq t\lambda T(u)$ for all $\|u\| \geq r$. This implies that $\Psi(t\lambda, z) \neq 0$ for all $\|z\| \leq 1/r$. Consider the homotopy $h(t, u) = \Psi(t\lambda, u)$. Using the homotopy invariance, we get $\deg(h(1, \cdot), B_{1/r}, 0) = \deg(h(0, \cdot), B_{1/r}, 0)$, namely

$$\deg(\Psi_\lambda, B_{1/r}, 0) = \deg(I, B_{1/r}, 0) = 1,$$

proving (c). ∎

Lemma 4.16 *Let $\lambda > \lambda_\infty$. Then*

(a) $\exists\, r > 0$ such that $S_\lambda(u) \neq \tau\varphi_1$, $\forall \tau \geq 0$, $\forall \|u\| \geq r$,
(b) $i(\Psi_\lambda, 0) = 0$ for all $\lambda > \lambda_\infty$.

Proof. (a) By contradiction, there exist $\tau_j \geq 0$ and $\|u_j\| \to \infty$ such that $S_\lambda(u_j) = \tau_j\varphi_1$, namely

$$-\Delta u_j = \lambda\widetilde{f}(u_j) + \tau_j\lambda_1\varphi_1.$$

Since $\tau_j \lambda_1 \varphi_1 \geq 0$, and $\widetilde{f}(u) > 0$ for $u < 0$, it follows as before that $u_j \geq 0$ and thus $\widetilde{f}(u_j) = f(u_j)$. Let us write $u_j = s_j \varphi_1 + w_j$, with $s_j = \int_\Omega u_j \varphi_1 \, dx \geq 0$ and $\int_\Omega w_j \varphi_1 \, dx = 0$. The function w_j satisfies

$$-\Delta w_j + s_j \lambda_1 \varphi_1 = \lambda f(u_j) + \tau_j \lambda_1 \varphi_1. \tag{4.16}$$

We first claim that $s_j \to +\infty$. Otherwise, $\|w_j\| \to \infty$ and setting $v_j = \|w_j\|^{-1} w_j$, from (4.16) we get

$$-\Delta v_j + \frac{s_j \lambda_1 \varphi_1}{\|w_j\|} = \lambda \frac{f(u_j)}{\|w_j\|} + \tau_j \lambda_1 \frac{\varphi_1}{\|w_j\|}.$$

From this equation we easily deduce that $v_j \to v^*$ strongly in X (and in C^2). In particular, $\|v^*\| = 1$. Moreover, from $\int_\Omega v_j \varphi_1 dx = 0$ it follows that $\int_\Omega v^* \varphi_1 \, dx = 0$. On the other hand, $u_j \geq 0$ implies that $v_j \geq \|w_j\|^{-1} s_j \varphi_1$ and, passing to the limit, $v^* \geq 0$, which, jointly with $\|v^*\| = 1$, gives a contradiction that proves the claim.

Next, we use (4.16) again to infer

$$s_j \lambda_1 = \lambda \int_\Omega f(u_j) \varphi_1 dx + \tau_j \lambda_1 \geq \lambda \int_\Omega f(u_j) \varphi_1 dx.$$

Since $f(u) = mu + g(u)$, we get

$$s_j \lambda_1 \geq \lambda \left[\int_\Omega mu_j \varphi_1 \, dx + \int_\Omega g(u_j) \varphi_1 \, dx \right] = \lambda m s_j + \int_\Omega g(u_j) \varphi_1 \, dx.$$

Thus

$$\lambda_1 \geq \lambda m + \int_\Omega s_j^{-1} g(u_j) \varphi_1 dx.$$

Since g is bounded and $s_j \to \infty$ (by the claim), passing to the limit we find $\lambda_1 \geq \lambda m$, while λ has been chosen strictly greater than $\lambda_\infty = \lambda_1/m$. This contradiction proves (a).

(b) Take $\tau = t\|u\|^2$, with $t \in [0, 1]$. By (a) it follows that $S_\lambda(u) \neq t\|u\|^2 \varphi_1$ for all $\|u\| \geq r$. This implies

$$\Psi_\lambda(z) \neq t\varphi_1, \quad \forall \, 0 < \|z\| \leq \frac{1}{r}, \quad \forall \, t \in [0, 1]. \tag{4.17}$$

Using the homotopy $h(t, z) = \Psi_\lambda(z) - t\varphi_1$ on the ball $B_{1/r}$ we find

$$\deg(\Psi_\lambda, B_{1/r}, 0) = \deg(\Psi_\lambda - \varphi_1, B_{1/r}, 0).$$

The latter degree is zero because (4.17), with $t = 1$, implies that $\Psi_\lambda(z) = \varphi_1$ has no solution on $B_{1/r}$. This proves (b). ∎

Remark 4.17 The fact that the index $i(\Psi_\lambda, 0)$ is zero for $\lambda > \lambda_\infty$, makes it clear that Ψ_λ is not differentiable at $z = 0$. Compare with Lemma 3.19. ∎

Proof of Theorem 4.14. We cannot apply directly the Rabinowitz Theorem 4.8 with $S_\lambda = \Psi_\lambda$ and $\lambda^* = \lambda_\infty$, because Ψ_λ is not differentiable at $z = 0$. However, an inspection of the proof shows that such an assumption is used only to evaluate the index of the map S_λ when λ crosses λ^* proving that $i(S_{\lambda-\varepsilon}, 0) \neq i(S_{\lambda+\varepsilon}, 0)$. In our case, the index has been evaluated in Lemmas 4.15 and 4.16 and it has been shown that $i(\Psi_\lambda, 0) = 1$ for $\lambda < \lambda_\infty$, while $i(\Psi_\lambda, 0) = 0$ for $\lambda > \lambda_\infty$. Since the two indices are different, one can repeat the arguments carried out in the proof of Theorem 4.8, yielding the existence of a connected component $\Gamma_\infty \subset \Gamma$, bifurcating from $(\lambda_\infty, 0)$. As a consequence of Lemma 4.15(b), there are no other bifurcation points (from the trivial solution) for Ψ_λ, but $(\lambda_\infty, 0)$ and thus Γ_∞ is unbounded. This Γ_∞ corresponds to a connected component $\Sigma_\infty \subset \Sigma$ emanating from (λ_∞, ∞). ∎

Using similar arguments one can study the bifurcation from the trivial solution for $S_\lambda = 0$, to obtain the following theorem.

Theorem 4.18 *Let* (4.14) *hold.*

(a) *If $f(0) > 0$ there exists an unbounded connected component $\Sigma_0 \subset \Sigma$, with $\Sigma_0 \subset]0, \infty) \times X$, such that $(0, 0) \in \overline{\Sigma}_0$. Moreover, $(\lambda, 0) \in \overline{\Sigma}_0 \Rightarrow \lambda = 0$.*

(b) *If $f(0) = 0$ and the right-derivative $f'_+(0)$ exists and is positive, then letting*

$$\lambda_0 := \frac{\lambda_1}{f'_+(0)},$$

there exists an unbounded connected component $\Sigma_0 \subset \Sigma$ such that $(\lambda_0, 0) \in \overline{\Sigma}_0$ and $(\lambda, 0) \in \overline{\Sigma}_0 \Rightarrow \lambda = \lambda_0$.

Remark 4.19 Since the components Σ_∞ and Σ_0 are connected and for $\lambda = 0$ the equation (4.14) has only the trivial solution, it follows that both Σ_∞ and Σ_0 are contained in $]0, \infty) \times X$ and hence correspond to positive solutions of (4.14). ∎

The next theorem studies the relationships between Σ_∞ and Σ_0.

Theorem 4.20 *Suppose that the same assumptions made in Theorems 4.14 and 4.18 hold.*

(a) *If $\exists \alpha > 0$ such that $f(u) \geq \alpha u$, $\forall u \geq 0$, then setting $\Lambda = \lambda_1/\alpha$ one has that $\Sigma_0 \subset]0, \Lambda]$. As a consequence, $\Sigma_0 = \Sigma_\infty$.*

(b) *If $\exists s_0 > 0$ such that $f(s_0) \leq 0$, then $S_\lambda(u) \neq 0$ for all $u \in X$ with $\|u\|_\infty = s_0$. As a consequence, $\Sigma_0 \cap \Sigma_\infty = \emptyset$.*

Proof. (a) If $(\lambda, u) \in \Sigma$, with $\lambda > 0$, then $-\Delta u = \lambda f(u) \geq \lambda \alpha u$. Multiplying by φ_1 and integrating, we find

$$\lambda_1 \int_\Omega u\varphi_1 dx \geq \lambda \alpha \int_\Omega u\varphi_1 dx.$$

Since $u > 0$ it follows that $\lambda \leq \lambda_1/\alpha = \Lambda$. Moreover, when $\lambda = 0$ equation (4.14) has only the trivial solution and this implies that there exists $c(\lambda) > 0$ such that $\lambda \geq c(\lambda) > 0$ for all $(\lambda, u) \in \Sigma_0$. Let us remark that $c(\lambda) \to 0$ as $\lambda \to 0+$, while the fact that λ_∞ is the only bifurcation from infinity for S_λ yields the existence of $c_0 > 0$ such that $c(\lambda) \geq c_0$ for all λ bounded away from zero.

(b) Let $(\lambda, u) \in \Sigma$, with $\lambda > 0$ and $\|u\|_\infty = s_0$. Then $0 \leq u \leq s_0$ in Ω. Let $m > 0$ be such that $f(u) + mu$ is monotone increasing for $u \in [0, s_0]$. From

$$-\Delta u + \lambda mu = \lambda(f(u) + mu)$$

and the fact that $\lambda m s_0 \geq \lambda(f(s_0) + m s_0)$, we get

$$(-\Delta + \lambda m)(s_0 - u) \geq \lambda \left[f(s_0) + m s_0 - f(u) - mu \right] \geq 0, \quad x \in \Omega,$$

as well as $s_0 - u > 0$ on $\partial\Omega$. By the maximum principle, we deduce that $s_0 - u > 0$ in Ω. This implies $\|u\|_\infty < s_0$, a contradiction, proving that $S_\lambda(u) \neq 0$ for all $u \in X$ with $\|u\|_\infty = s_0$. Now, let us point out that both Σ_0 and Σ_∞ are connected, Σ_0 contains a sequence (λ_i, u_i) with $\|u_i\|_\infty \to 0$ and Σ_∞ contains a sequence (λ_i, u_i) with $\|u_i\|_\infty \to \infty$. If $\Sigma_0 \cap \Sigma_\infty \neq \emptyset$ then they contain a point (λ, u) with $\|u\|_\infty = s_0$, a contradiction. ∎

Finally, it is possible to give conditions that allow us to describe in a precise way the behaviour of the branch bifurcating from infinity. They are the counterparts of the conditions that provide a subcritical or a supercritical bifurcation from the trivial solution, see Remark 2.9(b) in Chapter 2.

Lemma 4.21 *Suppose that either*

$$\gamma' := \liminf_{u \to +\infty} g(u) > 0, \tag{4.18}$$

or

$$\gamma'' := \limsup_{u \to +\infty} g(u) < 0. \tag{4.19}$$

Then Σ_∞ bifurcates to the left, respectively to the right, of (λ_∞, ∞).

Proof. We will show that if (4.18) holds, then we can sharpen the statement (a) of Lemma 4.15 by taking $J = [\lambda_\infty, b]$ with $b > \lambda_\infty$ (the new feature here is that

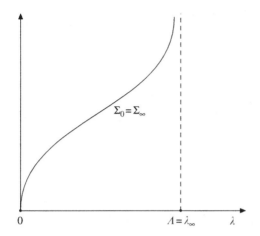

Figure 4.6 Bifurcation diagram in case of Theorem 4.21 (a), with $f(0) > 0$ and $\lambda_\infty = \Lambda$.

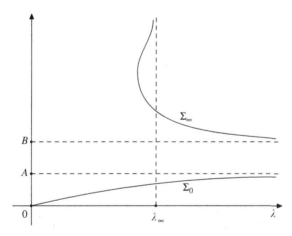

Figure 4.7 Bifurcation diagram in case of Theorem 4.21(b), with $f(0) > 0$. The interval $[A, B]$ is such that $f(u) \leq 0$ if and only if $u \in [A, B]$.

now we allow λ_∞ to be the left end side of the compact interval J). Otherwise, there exists $\lambda_j \downarrow \lambda_\infty$ and $\|u_j\| \to \infty$ such that $S(\lambda_j, u_j) = 0$. Repeating the arguments carried out in Lemma 4.15, we would find a sequence $v_j = u_j \|u_j\|^{-1}$ which converges strongly to some $\bar{v}$ such that

$$-\Delta \bar{v} = \lambda_\infty m \bar{v} = \lambda_1 \bar{v}, \qquad \bar{v} \geq 0, \quad \|\bar{v}\| = 1.$$

This now implies $\bar{v} = a\varphi_1$ for some $a > 0$. Hence $\bar{v} > 0$ in Ω and $u_j(x) = \|u_j\| v_j(x) \to +\infty$, for all $x \in \Omega$. Writing again $u_j = s_j \varphi_1 + w_j$, one has

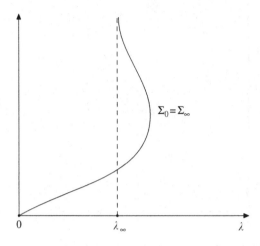

Figure 4.8 Bifurcation diagram when $f(0) > 0$ and $\gamma'' < 0$.

that $s_j \lambda_1 \varphi_1 - \Delta w_j = \lambda_j (m u_j + g(u_j))$ and therefore (we take φ_1 such that $\int_\Omega \varphi_1^2 \, dx = 1$)

$$s_j \lambda_1 = \lambda_j m + \int_\Omega g(u_j(x)) \varphi_1(x) \, dx.$$

Since $\lambda_j > \lambda_\infty$ then it follows that

$$\int_\Omega g(u_j(x)) \varphi_1(x) \, dx < 0.$$

Since $u_j(x) \to +\infty$ on Ω, using the Fatou lemma we infer

$$\gamma' \int_\Omega \varphi_1 \, dx \leq \liminf \int_\Omega g(u_j(x)) \varphi_1(x) \, dx \leq 0.$$

This is in contradiction to the assumption (4.18). In the case that (4.19) holds, we take $J = [0, \lambda_\infty]$ and repeat the preceding arguments. ∎

Remark 4.22 The preceding results have been extended in great generality by D. Arcoya and J. Gamez [32], where the relationships between the bifurcation from infinity and the anti-maximum principle are also investigated. ∎

4.5 Exercises

(i) Let $X = \mathbb{R}^2$, and consider the system

$$\begin{cases} x_1 = \lambda x_1 - x_2^3, \\ x_2 = 2\lambda x_2 + x_1^3. \end{cases}$$

Show that in this case the alternative (b) of the Rabinowitz global bifurcation theorem holds true.

(ii) Consider the problem

$$\begin{cases} -u'' = \lambda u + h(x)u^3 & x \in (0, \pi) \\ u(0) = u(\pi) = 0, \end{cases}$$

where h is Lipschitz such that $h(x) > 0$, respectively $h(x) < 0$, in $[0, \pi]$, and let $\mathcal{C}_k$ be the unbounded connected components bifurcating from $\lambda_k = k^2$ (see Theorem 4.12).

(a) Prove that if $(\lambda, u) \in \mathcal{C}_k$, $\lambda > 0$, then $\lambda < k^2$, respectively $\lambda > k^2$.
(b) Prove that the projection on $\mathbb{R}$ of $\mathcal{C}_k \subset \mathbb{R} \times E$ is $(-\infty, k^2)$, respectively $(k^2, +\infty)$.
(c) Draw the bifurcation diagrams and evaluate the number of nontrivial solutions of the equations $-u'' = u + h(x)u^3$ in $(0, \pi)$, $u(0) = u(\pi) = 0$ in both the cases $h > 0$ and $h < 0$. [Hint: find the intersections of $\mathcal{C}_k$ with $\{1\} \times E \subset \mathbb{R} \times E$.]

(iii) Consider the problem

$$\begin{cases} -\Delta u = \lambda u - h(x)u^3 & x \in \Omega \\ u(x) = 0 & x \in \partial\Omega, \end{cases}$$

where h is Lipschitz, $h \geq 0$ in Ω, and suppose that there exists an open set $\Omega_0 \subset \Omega$ such that $h(x) = 0$ if and only if $x \in \Omega_0$. Prove that there is a connected component of positive solutions Σ_0 such that $(\lambda_1, 0) \in \overline{\Sigma}_0$ with the property that

(a) $\overline{\Sigma}_0 \subset [\lambda_1, \lambda_1(\Omega_0)[\times L^2(\Omega)$, where $\lambda_1(\Omega_0)$ denotes the first eigenvalue of $-\Delta$ on Ω_0 with zero Dirichlet boundary conditions;
(b) $\overline{\Sigma}_0$ bifurcates from $(\lambda_1(\Omega_0), \infty)$. [Hint: show that if $(\lambda_j, u_j) \in \overline{\Sigma}_0$ and $\|u_j\| \to \infty$, then $\lambda_j \uparrow \lambda_1(\Omega_0)$].

(iv) Extend Lemma 4.21 to the case in which $g = g(x, u)$, $\gamma'(x) = \liminf_{u \to +\infty} g(x, u)$, $\gamma''(x) = \limsup_{u \to +\infty} g(x, u)$, assuming that $\int_\Omega \gamma'(x)\varphi_1(x)\,dx > 0$, or that $\int_\Omega \gamma'(x)\varphi_1(x)\,dx < 0$.

(v) Let $u^+ = u \wedge 0$. Consider the problem

$$\begin{cases} -\Delta u = (u - \lambda)^+ & \text{in } \Omega \\ u = 0 & \text{on } \partial\Omega, \end{cases}$$

where $\lambda \in \mathbb{R}$. The set of solutions (λ, u) with $u \not\equiv 0$ is denoted by Σ. Prove that there exists an unbounded component $\Sigma^* \subset \Sigma$ such that $(0, 0) \in \overline{\Sigma}^*$.

PART II

Variational methods, I

5

Critical points: extrema

In this chapter we will discuss the existence of maxima and minima for a functional on a Hilbert or Banach space.

5.1 Functionals and critical points

Let E be a Banach space. A *functional* on E is a continuous real valued map $J : E \to \mathbb{R}$.

More in general, one could consider functionals defined on open subsets of E. But, for the sake of simplicity, in the sequel we will always deal with functionals defined on all of E, unless explicitly remarked.

Let J be (Fréchet) differentiable at $u \in E$ with derivative $\mathrm{d}J(u) \in L(E, \mathbb{R})$. Recall that (see Section 1.1):

- if J is differentiable on E, namely at every point $u \in E$ and the map $E \mapsto L(E, \mathbb{R})$, $u \mapsto \mathrm{d}J(u)$, is continuous, we say that $J \in C^1(E, \mathbb{R})$;
- if J is k times differentiable on E with kth derivative $\mathrm{d}^k J(u) \in L_k(E, \mathbb{R})$ (the space of k-linear maps from E to $\mathbb{R}$) and the application $E \mapsto L_k(E, \mathbb{R})$, $u \mapsto \mathrm{d}^k J(u)$, is continuous, we say that $J \in C^k(E, \mathbb{R})$.

Definition 5.1 *A critical, or stationary, point of $J : E \to \mathbb{R}$ is a $z \in E$ such that J is differentiable at z and $\mathrm{d}J(z) = 0$. A critical level of J is a number $c \in \mathbb{R}$ such that there exists a critical point $z \in E$ with $J(z) = c$. The set of critical points of J will be denoted by Z, while Z_c will indicate the set of critical points at level c: $Z_c = \{z \in Z : J(z) = c\}$.*

According to the definition, a critical point z satisfies

$$\mathrm{d}J(z)[v] = 0, \quad \forall \, v \in E.$$

We will see that, in the applications, critical points turn out to be weak solutions of differential equations. Let us illustrate this claim immediately with a rather elementary example.

77

Let $E = H_0^{1,2}(0, 1) = \{u \in H^{1,2}(0, 1) : u(0) = u(1) = 0\}$ and define the functional $J : E \mapsto \mathbb{R}$ by setting

$$J(u) = \int_0^1 \left[\frac{1}{2} \dot{u}^2 + \frac{1}{2} u^2 - \frac{1}{4} u^4 \right] dt, \qquad \dot{u} := \frac{du}{dt}.$$

One has that $J \in C^\infty(E, \mathbb{R})$ and there holds

$$dJ(u)[v] = \int_0^1 [\dot{u}\dot{v} + uv - u^3 v] dt.$$

Then a critical point of J on E is an element $z \in E$ such that

$$\int_0^1 [\dot{z}\dot{v} + zv - z^3 v] dt = 0, \qquad \forall\, v \in H_0^{1,2}(0, 1)$$

and this means that z is a weak (and by regularity, see Theorem 1.16, classical) solution of the two point problem

$$\ddot{u} - u + u^3 = 0, \qquad u(0) = u(1) = 0. \tag{5.1}$$

Let us remark that the boundary conditions $z(0) = z(1) = 0$ are automatically satisfied because $z \in H_0^{1,2}(0, 1)$. If we take $E = H^{1,2}(0, 1)$, respectively $E = \{u \in H^{1,2}(\mathbb{R}) : u(t + 1) = u(t), \forall t \in \mathbb{R}\}$, then a critical point z of J on E satisfies $\ddot{z} - z + z^3 = 0$ together with the Neumann boundary conditions $\dot{z}(0) = \dot{z}(1) = 0$, respectively z is a 1-periodic solution of $\ddot{u} - u + u^3 = 0$. Of course, the choice of Sobolev spaces like $H^{1,2}$ is also related to the fact that (5.1) is a second order equation and the term $\frac{1}{2} \int \dot{u}^2\, dt$ makes sense.

The preceding example is a model of problems we will discuss later on. Roughly, we will look for solutions of *boundary value problems* consisting of a differential equation together with some boundary conditions. These equations will have a *variational structure*, namely they will be the Euler–Lagrange equation of a functional J on a suitable space of functions E, chosen depending on the boundary conditions, and the critical points of J on E will give rise to solutions of these boundary value problems.

5.2 Gradients

Let $J \in C^1(E, \mathbb{R})$. If E is a Hilbert space, by the *Riesz theorem* there exists a unique $J'(u) \in E$ such that

$$(J'(u) \mid v) = dJ(u)[v], \qquad \forall\, v \in E.$$

$J'(u)$ is called the *gradient of J* at u (in some cases we will also use the notation $\nabla J(u)$ to denote the gradient of $J'(u)$). With this notation, a critical point of J is a solution of the equation $J'(u) = 0$.

In a quite similar way, dealing with the second derivative $d^2 J(u)$ (which is a symmetric bilinear map : $E \times E \to \mathbb{R}$), we can define the operator $J''(u)$: $E \to E$ by setting

$$(J''(u)v \mid w) = d^2 J(u)[v, w], \quad \forall v, w \in E.$$

Example 5.2

(i) If $J(u) = \frac{1}{2}(Au \mid u)$, where $A \in L(E)$ is symmetric, then one has $dJ(u)[v] = (Au \mid v)$. Moreover, from

$$(J'(u) \mid v) = (Au \mid v), \quad \forall v \in E,$$

we deduce that $J' = A$. In particular, if $J(u) = \frac{1}{2}\|u\|^2$ then $J'(u) = u$. Similarly, from $d^2 J(u)[v, w] = (Av \mid w)$ we infer that $J''(u) : v \mapsto Av$.

(ii) Let Ω be a bounded domain in $\mathbb{R}^n$ with smooth boundary $\partial\Omega$ and set $E = L^2(\Omega)$. Given $h \in E$, consider the linear functional $J(u) = \int_\Omega hu \, dx$. Of course, $J \in C^\infty(E, \mathbb{R})$ and

$$dJ(u)[v] = \int_\Omega hv \, dx.$$

In this case the gradient $J'(u)$ verifies

$$(J'(u) \mid v) = \int_\Omega hv \, dx.$$

Hence $J'(u) = h$.

(iii) Consider, as before, the same linear functional $J(u) = \int_\Omega hu \, dx$, but let us now take the Sobolev space $E = H_0^1(\Omega)$ endowed with the scalar product and norm, respectively,

$$(u \mid v) \quad = \int_\Omega \nabla u \cdot \nabla v \, dx,$$
$$\|u\|^2 \quad = \int_\Omega |\nabla u|^2 \, dx.$$

In the present case $w = J'(u)$ verifies

$$(w \mid v) = \int_\Omega hv \, dx, \quad \forall v \in E,$$

namely

$$\int_\Omega \nabla w \cdot \nabla v \, dx = \int_\Omega hv \, dx, \quad \forall v \in E.$$

This means that, in contrast with the case discussed in point (ii) above, w is not equal to h but is the weak solution of the Dirichlet boundary value problem (BVP in short)

$$\begin{cases} -\Delta w(x) = h(x) & x \in \Omega \\ w(x) = 0 & x \in \partial\Omega. \end{cases}$$

In other words, letting K denote the Green operator of $-\Delta u$ in $H_0^1(\Omega)$, one has that $J'(u) = K(h)$.

(iv) We use the notation introduced before as well as in Sections 1.2 and 1.3. Suppose that $f \in \mathbb{F}_p$ with $1 < p \le (n+2)/(n-2)$, and let Φ denote the functional defined on $E = H_0^1(\Omega)$ by setting $\Phi(u) = \int_\Omega F(x, u) \, dx$, where $\partial_u F = f$. According to Theorem 1.8, $\Phi \in C^1(E, \mathbb{R})$ and

$$d\Phi(u)[v] = \int_\Omega f(x, u)v \, dx.$$

Here the gradient Φ' becomes $K \circ f$ and satisfies

$$(\Phi'(u) \mid v) = \int_\Omega f(x, u)v \, dx.$$

Moreover, the functional

$$J(u) = \tfrac{1}{2}\|u\|^2 - \Phi(u)$$

is of class $C^1(E, \mathbb{R})$ and there holds

$$dJ(u)[v] = (u \mid v) - \int_\Omega f(x, u)v \, dx.$$

Therefore the critical points z of J satisfy

$$\int_\Omega [\nabla z \cdot \nabla v - f(x, z)v] dx = 0, \quad \forall v \in E,$$

and hence are nothing but the weak solutions of the semilinear Dirichlet BVP

$$\begin{cases} -\Delta u(x) = f(x, u(x)) & x \in \Omega \\ u(x) = 0 & x \in \partial\Omega. \end{cases}$$

As for the gradient, one finds $J'(u) = u - Kf(u)$. If f is of class C^1 with respect to u, then $\Phi \in C^2(E, \mathbb{R})$ and one finds $J''(u) : v \mapsto v - Kf'(u)v$. In particular, the kernel of $J''(u)$ consists of the solutions of the linear Dirichlet problem $-\Delta v = f_u(x, u)v$, $v \in H_0^1(\Omega)$. ∎

5.3 Existence of extrema

We say that $z \in E$ is a local *minimum*, respectively *maximum* of the functional $J \in C(E, \mathbb{R})$ if there exists a neighbourhood $\mathcal{N}$ of z such that

$$J(z) \le J(u), \qquad \text{respectively } J(z) \ge J(u), \quad \forall u \in \mathcal{N} \setminus \{z\}. \tag{5.2}$$

If the above inequalities are strict we say that z is a *strict local minimum (maximum)*. If (5.2) holds for every $u \in E$, not merely on $\mathcal{N} \setminus \{z\}$, z is a *global minimum (maximum)*. It is immediate to check that if $z \in E$ is a local minimum (maximum) and if J is differentiable at z, then z is a stationary point of J, namely $dJ(z) = 0$ or else $J'(z) = 0$.

Next, we state some results dealing with the existence of minima or maxima. We begin with a classical result dealing with functionals which are coercive and weakly lower semi-continuous (w.l.s.c. in short). Let us recall that $J \in C(E, \mathbb{R})$ is *coercive* if

$$\lim_{\|u\| \to +\infty} J(u) = +\infty.$$

J is w.l.s.c. if for every sequence $u_n \in E$ such that $u_n \rightharpoonup u$ one has that

$$J(u) \le \liminf J(u_n).$$

Lemma 5.3 *Let E be a reflexive Banach space and let $J : E \to \mathbb{R}$ be coercive and w.l.s.c.*
Then J is bounded from below on E, namely there exists $a \in \mathbb{R}$ such that $J(u) \ge a$ for all $u \in E$.

Proof. Arguing by contradiction, let $u_n \in E$ be such that $J(u_n) \to -\infty$. Since J is coercive it follows there is $R > 0$ such that $\|u_n\| \le R$. Hence there exists $u \in E$ such that (without relabelling) $u_n \rightharpoonup u$. Since J is w.l.s.c. we infer that

$$J(u) \le \liminf J(u_n) = -\infty,$$

a contradiciton, proving the lemma. ■

Remark 5.4 The same arguments show that a w.l.s.c. functional is bounded from below on any ball $B_r = \{u \in E : \|u\| \le r\}$. ■

Theorem 5.5 *Let E be a reflexive Banach space and let $J : E \to \mathbb{R}$ be coercive and w.l.s.c.*
Then J has a global minimum, namely there is $z \in E$ such that $J(z) = \min\{J(u) : u \in E\}$. If J is differentiable at z, then $dJ(z) = 0$.

Proof. From the preceding lemma it follows that $m \equiv \inf\{J(u) : u \in E\}$ is finite. Let u_n be a *minimizing sequence*, namely such that $J(u_n) \to m$. Again, the coercivity of J implies that $\|u_n\| \le R'$, and $u_n \rightharpoonup z$ for some $z \in E$. Since J is w.l.s.c. it follows that

$$J(z) \le \liminf J(u_n) = m.$$

Of course, $J(z)$ cannot be strictly smaller than m and thus J achieves its infimum at z: $J(z) = m$. ∎

Remark 5.6 Since z is a maximum for J if and only if it is a minimum for $-J$, a similar result holds for the existence of maxima, provided $-J$ is coercive and w.l.s.c. ∎

Remark 5.7 Let E be a Hilbert space. Using Remark 5.4 and repeating the preceding arguments one shows that *any w.l.s.c. functional J achieves its minimum on the ball B_r*. It is worth pointing out that if $J \in C^1(E, \mathbb{R})$ and if the minimum m is achieved at a point z on the boundary $S_r = \partial B_r$ of the ball, then z is not necessarily a stationary point of J but there exists $\lambda \leq 0$ such that $J'(z) = \lambda z$. In fact, z is a constrained critical point of J on ∂B_r. Since $j(t) := J(tz)$, $t \in [0, 1]$, has a minimum at $t = 1$, then

$$j'(1) = (J'(z) \mid z) \leq 0,$$

and from $\lambda = (J'(z) \mid z)$ it follows that $\lambda \leq 0$. The same argument holds for the maximum of $-J$. In particular, if J is weakly continuous both the maximum and the minimum are achieved. ∎

The easiest example of w.l.s.c. functional is the map $u \mapsto \|u\|^2$. Conditions that imply the w.l.s. continuity of functionals are of great importance and have been broadly studied, beginning with the pioneering works of L. Tonelli. But this question is beyond the scope of this book and will not be discussed here. The interested reader is referred for example to [84, 102].

5.4 Some applications

Here we show some applications of the preceding results. First of all, let us consider the case that E is a Hilbert space and

$$J(u) = \tfrac{1}{2}\|u\|^2 - \Phi(u). \tag{5.3}$$

Theorem 5.8 *Let J be of the form (5.3) and suppose $\Phi \in C^1(E, \mathbb{R})$ is weakly continuous (namely $u_n \rightharpoonup u \Rightarrow \Phi(u_n) \to \Phi(u)$) and satisfies*

$$|\Phi(u)| \leq a_1 + a_2\|u\|^\alpha,$$

with $a_1, a_2 > 0$ and $\alpha < 2$.
Then J achieves its global minimum at some $z \in E$ and there holds $J'(z) = 0$, namely $\Phi'(z) = z$.

Proof. One has

$$J(u) \geq \tfrac{1}{2}\|u\|^2 - a_1 - a_2\|u\|^\alpha$$

and hence J is coercive, because $\alpha < 2$. Since $u \mapsto \|u\|^2$ is w.l.s.c. and Φ is weakly continuous, then J is w.l.s.c. and Theorem 5.5 yields the existence of a global minimum z of J satisfying $J'(z) = 0$, namely $z - \Phi'(z) = 0$. ∎

Theorem 5.8 can be used, for example, to handle Dirichlet boundary value problems, see Section 1.4 to which we refer for notation,

$$\begin{cases} -\Delta u(x) = f(x, u(x)) & x \in \Omega \\ u(x) = 0 & x \in \partial\Omega, \end{cases} \tag{D}$$

with $f : \Omega \times \mathbb{R} \to \mathbb{R}$ sublinear at infinity. Precisely, we assume that f is locally Hölder continuous and that there exists $a_1 \in L^2(\Omega)$, $a_2 > 0$ and $0 < q < 1$ such that

$$|f(x, u)| \leq a_1(x) + a_2|u|^q, \qquad \forall\, (x, t) \in \Omega \times \mathbb{R}. \tag{5.4}$$

We set $E = H_0^1(\Omega)$, with norm $\|u\|^2 = \int_\Omega |\nabla u|^2 \, dx$. Taking into account that E is compactly embedded in $L^2(\Omega)$ and repeating the arguments carried out in Theorem 1.8, one readily finds that

$$\Phi(u) := \int_\Omega F(x, u) \, dx, \qquad \text{where} \quad F(x, u) = \int_0^u f(x, s) \, ds$$

is $C^1(E)$ and is weakly continuous. We will show the following.

Theorem 5.9 *Let f be locally Hölder continuous and suppose that (5.4) holds. Then (D) has a solution.*

Proof. Consider the functional $J \in C^1(E, \mathbb{R})$,

$$J(u) = \tfrac{1}{2}\|u\|^2 - \Phi(u) = \tfrac{1}{2}\int_\Omega |\nabla u|^2 \, dx - \int_\Omega F(x, u) \, dx,$$

whose critical points are the solutions of (D). Using (5.4), we find constants $a_i > 0$ such that

$$|\Phi(u)| \leq a_3\|u\|_{L^2} + a_4\|u\|_{L^{q+1}}^{q+1}$$

$$\leq a_5\|u\| + a_6\|u\|^{q+1}.$$

Since $q < 1$ one infers that J is coercive on E. Furthermore, Φ is weakly continuous and thus Theorem 5.8 applies yielding a minimum z such that $J'(z) = z - \Phi'(z) = 0$, which gives rise to a solution of (D). ∎

Remark 5.10 Small modifications of the preceding arguments allow us to prove the existence of a solution of (**D**) provided that f is locally Hölder continuous and such that $f(x, s)/s \to 0$ as $|s| \to \infty$, uniformly with respect to $x \in \Omega$. ∎

Theorem 5.9 applies to the following example.

Example 5.11 Consider the BVP

$$\begin{cases} -\Delta u(x) = \lambda u - f(u) & x \in \Omega \\ u(x) = 0 & x \in \partial\Omega, \end{cases} \tag{5.5}$$

where λ is a *given* real parameter (notice the difference with the eigenvalue problems, where λ is an unknown) and $f : [0, +\infty) \mapsto \mathbb{R}$ is locally Hölder continuous and such that

$$\lim_{u \to 0+} \frac{f(u)}{u} = 0, \qquad \lim_{u \to +\infty} \frac{f(u)}{u} = +\infty.$$

We want to show that (5.5) has a positive solution whenever $\lambda > \lambda_1$, where λ_1 is the first eigenvalue of the Laplace operator with zero Dirichlet boundary conditions.

From the assumption on f it follows that there exists $\xi = \xi_\lambda > 0$ such that $\lambda\xi = f(\xi)$ and $\lambda u - f(u) > 0$ for all $0 < u < \xi$. Let $g_\lambda : \mathbb{R} \mapsto \mathbb{R}$ denote the function

$$g_\lambda(u) = \begin{cases} 0 & \text{if } u < 0 \\ \lambda u - f(u) & \text{if } 0 \leq u \leq \xi \\ 0 & \text{if } u > \xi. \end{cases}$$

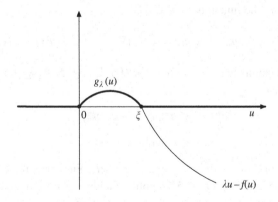

Figure 5.1 Graph of g_λ (in bold) versus the graph of $\lambda u - f(u)$.

Consider the auxiliary boundary value problem

$$\begin{cases} -\Delta u(x) = g_\lambda(u) & x \in \Omega \\ u(x) = 0 & x \in \partial\Omega. \end{cases} \tag{D}$$

By the maximum principle, any non-trivial solution u of (D) is positive. Moreover, repeating the arguments of Theorem 4.19 (b), one finds that $0 < u(x) < \xi_\lambda$, for all $x \in \Omega$, and hence is a positive solution of (5.5). Since g_λ is locally Hölder continuous and bounded, Theorem 5.9 applies to

$$J_\lambda(u) = \tfrac{1}{2} \|u\|^2 - \lambda \int_\Omega G_\lambda(u) \, dx, \qquad G_\lambda(u) = \int_0^u g_\lambda(s) \, ds.$$

If $\lambda > \lambda_1$ we claim that $\min J_\lambda < 0$. Actually, let $\varphi_1 \in E$ be such that $\varphi_1 > 0$ in Ω and

$$-\Delta\varphi_1(x) = \lambda_1\varphi_1(x), \qquad x \in \Omega, \qquad \int_\Omega \varphi_1^2 \, dx = 1.$$

For $t > 0$ small, one has that $g_\lambda(t\varphi_1) = \lambda t\varphi_1 - f(t\varphi_1)$. Since $f(u) = o(u)$, we easily deduce that

$$J_\lambda(t\varphi_1) = \tfrac{1}{2} \lambda_1 t^2 - \tfrac{1}{2} \lambda t^2 + o(t^2).$$

Then, if $\lambda < \lambda_1$, it follows that $J_\lambda(t\varphi_1) < 0$ for $t > 0$ small. This proves that $\min_E J_\lambda < 0$ and that (5.5) has a positive solution. ∎

Remark 5.12 If f is defined on all $\mathbb{R}$ and $\lim_{u\to 0} f(u)/u = 0$, $\lim_{|u|\to+\infty} f(u)/u = +\infty$, we can also find, in addition to the preceding positive solution, a negative solution. It suffices to consider the negative part of $\lambda u - f(u)$ and argue as before. ∎

As a further application let us consider the nonlinear eigenvalue problem

$$\begin{cases} -\lambda \, \Delta u(x) = f(x, u) & x \in \Omega \\ u(x) = 0 & x \in \partial\Omega, \end{cases} \tag{EP_λ}$$

where $f \in \mathbb{F}_p$ with $1 < p < (n+2)/(n-2)$, see Section 1.3, f is locally Hölder continuous and satisfies the growth condition

$$|f(x, u)| \leq a_1(x) + a_2|u|^p, \qquad \forall (x, u) \in \Omega \times \mathbb{R}, \tag{5.6}$$

where $1 < p < (n+2)/(n-2)$.

Proposition 5.13 *Let $f \in \mathbb{F}_p$, $1 < p < (n+2)/(n-2)$, satisfy*

$$f(x,0) = 0, \qquad uf(x,u) > 0, \qquad \forall\, u \neq 0. \tag{5.7}$$

*Then the nonlinear eigenvalue problem (**EP**$_\lambda$) has a nontrivial solution $z > 0$, with $\|z\|^2 = \int_\Omega |\nabla z|^2 \, dx = 1$ and $\lambda = \int_\Omega f(x,z)z \, dx > 0$.*

Proof. As usual, we set $F(x,u) = \int_0^u f(x,s) \, ds$ and

$$\Phi(u) = \int_\Omega F(x, u(x)) \, dx.$$

Let us apply Remark 5.7, with $E = H_0^1(\Omega)$ and $J(u) = \Phi(u)$. Since Φ is weakly continuous, see Theorem 1.8 in Section 1.3, the supremum $m = \sup\{J(u) : u \in \overline{B}_1\}$ is achieved at some $z \in \overline{B}_1$, see also Remark 5.7. Moreover m is positive because $uf(x,u) > 0$ for any $u \neq 0$. If $\|z\| < 1$ then $\Phi'(z) = 0$, namely $\int_\Omega f(x,z)v \, dx = 0$ for all $v \in E$ and, in view of (5.7), this implies $z = 0$. Hence $m = \Phi(z) = \Phi(0) = 0$, a contradiction. Thus the maximum is achieved on the unit sphere ∂B_1, so z is a constrained maximum of Φ on ∂B_1 and $J'(z) = \lambda z$. It follows that $z \in \partial B_1$ gives rise to a solution of (**EP**$_\lambda$) with $\lambda = \lambda\|z\|^2 = \int_\Omega f(x,z)z \, dx > 0$, because $z \in \partial B_1$. Finally, using the maximum principle, it is easy to see that $z > 0$ in Ω. ∎

5.5 Linear eigenvalues

Here we discuss the specific case of the linear eigenvalue problem

$$Au = \mu u, \qquad u \in E, \qquad \mu \in \mathbb{R}, \tag{5.8}$$

where E is a Hilbert space and $A \in L(E)$ is symmetric. Let us define the quadratic functional

$$J(u) = \tfrac{1}{2}(Au \mid u),$$

whose gradient is given by $J' = A$, see Section 5.1.

Lemma 5.14 *Suppose that A is compact and positive definite, namely $(Au \mid u) > 0$ for all $u \neq 0$.*
Then J achieves its maximum on the unit ball $\overline{B}_1 = \{u \in E : \|u\| \leq 1\}$ at a point z such that $\|z\| = 1$ and satisfying (5.8) with

$$\mu = \max\{(Au \mid u) : u \in B\} = (Az \mid z) > 0.$$

Proof. It suffices to repeat, with obvious changes, the arguments carried out in the proof of Proposition 5.13, taking into account that $J(u) = \frac{1}{2}(Au \mid u)$ is weakly continuous because A is compact. ∎

Remark 5.15 From the fact that $\mu = \max\{(Au \mid u) : u \in B\}$ it immediately follows that μ is the largest eigenvalue of A. μ is called the *principal eigenvalue* of A. ∎

Taking $E = H_0^1(\Omega)$, with scalar product $(u \mid v) = \int_\Omega (\nabla u \cdot \nabla v)\, dx$ and norm $\|u\|^2 = (u \mid u), A = K$, the Green operator of $-\Delta$ with zero Dirichlet boundary conditions, see Section 5.1, and $J(u) = \frac{1}{2}(Ku \mid u) = \frac{1}{2}\int_\Omega |u|^2\, dx$, the preceding lemma yields the existence of $\mu_1 > 0$ and $z = \varphi_1 \in E$, with $\|\varphi_1\| = 1$, such that

$$K(\varphi_1) = \mu_1 \varphi_1.$$

Setting $\lambda_1 = 1/\mu_1$, we find that

$$\begin{cases} -\Delta\varphi_1(x) = \lambda_1 \varphi_1(x) & x \in \Omega \\ \varphi_1(x) = 0 & x \in \partial\Omega, \end{cases} \tag{5.9}$$

namely that λ_1 is an eigenvalue of $-\Delta$ on $H_0^1(\Omega)$, with eigenfunction φ_1. Actually, from Remark 5.15 it follows that λ_1 is the smallest eigenvalue of $-\Delta$ on $H_0^1(\Omega)$. Let us recall, see Theorem 1.13 in Section 1.4, that λ_1 is a *simple* eigenvalue, namely the geometric and the algebraic multiplicity of λ_1 is one. Furthermore, the eigenfunctions of $-\Delta$ on $H_0^1(\Omega)$ do not change sign in Ω and hence we can choose φ_1 to be positive in Ω.

Let us conclude this subsection with some further remarks.

Remark 5.16
(i) A direct application of the regularity results discussed in Theorem 1.10 yields that the eigenfunctions φ of (5.9) are smooth (C^∞).
(ii) By homogeneity one has that

$$\frac{1}{\lambda_1} = \max_{H_0^1(\Omega)\setminus\{0\}} \frac{\int_\Omega u^2\, dx}{\|u\|^2}$$

and hence one infers

$$\int_\Omega u^2\, dx \leq \frac{1}{\lambda_1} \int_\Omega |\nabla u|^2\, dx, \qquad \forall\, u \in H_0^1(\Omega),$$

which is nothing but the *Poincaré inequality*.
(iii) In the preceding application to (5.9) we can consider the equation $-\Delta u(x) = \lambda a(x)u(x)$ with $a \in L^\infty(\Omega)$ and $a(x) > 0$ in Ω. The case that a changes sign has also been studied in [127].

(iv) In all the preceding examples, the differential operator $-\Delta$ can be substituted by any second order uniformly elliptic operator in a divergence form with smooth coefficients, see Section 1.4.   ■

5.6 Exercises

(i) Let $A \in L(E)$ be symmetric and positive definite. For $\lambda \in \mathbb{R}$, consider a smooth functional $J_\lambda : E \mapsto \mathbb{R}$ such that $J(u) = \|u\|^2 + C(u) - \lambda(Au \mid u)$, where $C : E \mapsto E$ is compact, positive and satisfies $C(t\,u) = t^4 C(u)$, for all $u \in E$, for all $t \in \mathbb{R}$. Prove that J_λ has a minimum $u_\lambda \neq 0$ provided $\lambda > \mu^{-1}$, where μ denotes the principal eigenvalue of A.

(ii) Prove the result claimed in Remark 5.10.

(iii) Consider the problem (see Example 5.11)

$$\begin{cases} -\Delta u(x) &=& \lambda u - u^3, & x \in \Omega, \\ u(x) &=& 0, & x \in \partial\Omega, \end{cases}$$

and prove that it has exactly one positive and one negative solutions for all $\lambda_1 < \lambda < \lambda_2$. [*Hint*: show that any nontrivial solution which does not change sign in Ω has Leray- Schauder index equal to 1.] Extend the result to $\lambda = \lambda_2$.

6

Constrained critical points

In this chapter we introduce the notion of critical points of a functional constrained on a manifold.

6.1 Differentiable manifolds, an outline

This preliminary section is devoted to give an outline on differentiable manifolds. The reader will find a more complete treatment of this topic in, for example, [113, 160].

Let X be a Hilbert space and $\mathcal{I}$ a set of indices. A topological space M is a C^k *Hilbert manifold* modelled on X, if there exist an open covering $\{U_i\}_{i \in \mathcal{I}}$ of M and a family $\psi_i : U_i \to X$ of mappings such that the following conditions hold:

- $V_i = \psi_i(U_i)$ is open in X and ψ_i is a homeomorphism from U_i onto V_i;
- $\psi_j \circ \psi_i^{-1} : \psi_i(U_i \cap U_j) \to \psi_j(U_i \cap U_j)$ is of class C^k.

Each pair (U_i, ψ_i) is called a *chart*. If $p \in U_i$, (U_i, ψ_i) is a chart at p. The maps $\psi_j \circ \psi_i^{-1}$ are the *changes of charts*. The pair (V_i, ψ_i^{-1}) is called a *local parametrization* of M. If $X = \mathbb{R}^n$ we say that M is n-dimensional.

If, in the preceding definition, X is a Banach space, we will say that M is a Banach manifold modelled on X. Moreover, in more general situations, each map ψ_i could map U_i in a possibly different Hilbert space X_i. However, on any connected component of M, each X_i will be isomorphic to a given Hilbert space X and we will still say that M is modelled on X.

For the applications we will discuss in the rest of this book, it suffices to consider the very specific case in which M is a subset of a Hilbert space E and is modelled on a Hilbert subspace $X \subset E$. We will limit ourselves to giving an outline of this case, referring for more details to [113], p. 23 and following.

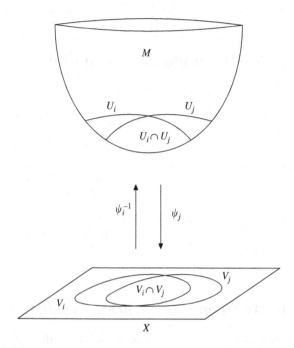

Figure 6.1

We will suppose that for any point $p \in M$ there are

- an open set $\widetilde{U} \subset E$, with $p \in \widetilde{U}$,
- an open set $\widetilde{V} \subset E$,
- a C^k diffeomorphism $\widetilde{\varphi} : \widetilde{V} \mapsto \widetilde{U}$,

such that, setting $V = \widetilde{V} \cap X$, $U := \widetilde{U} \cap M$ and denoting by φ the restriction of $\widetilde{\varphi}$ to V, one has that $x := \varphi^{-1}(p) \in V$ and $\varphi(V) = U$. Clearly, the pair (V, φ) is a *local parametrization* of M.

Definition 6.1 *The tangent space T_pM to M at p is defined as the image of X through the linear map $d\widetilde{\varphi}(x) \in L(X, E)$:*

$$T_pM = d\widetilde{\varphi}(x)[X]. \tag{6.1}$$

In order to justify the preceding definition, let us show that (6.1) does not depend on the choice of the local parametrization. Actually, let (V_i, φ_i), $i = 1, 2$, be two local parametrizations and consider the commutative diagram

$$
\begin{array}{ccc}
M \subset E & \simeq & M \subset E \\
\uparrow \widetilde{\varphi}_1 & & \uparrow \widetilde{\varphi}_2 \\
\widetilde{V}_1 \subset E & \xrightarrow{\widetilde{\varphi}_2^{-1} \circ \widetilde{\varphi}_1} & \widetilde{V}_2 \subset E.
\end{array}
$$

Recall that $\widetilde{\varphi}_2^{-1} \circ \widetilde{\varphi}_1$ is a diffeomorphism. Taking the derivatives, one immediately finds that $d\widetilde{\varphi}_1[X] = d\widetilde{\varphi}_2[X]$.

From the definition it immediately follows that $T_p M$ is a Hilbert space homeomorphic to X. We anticipate that $T_p M$ coincides with the space of 'tangent vectors' at p to the smooth curves on M (see below for a precise statement). Moreover, when $M = G^{-1}(c)$ with $G \in C^1(E, \mathbb{R})$, and $G'(u) \neq 0$ for all $u \in M$ (see Section 6.3 later on), then

$$
T_p M = \{v \in E : (G'(p) \mid v) = 0\}.
$$

Let $M_i \subset E_i$, $i = 1, 2$, be two C^1 Hilbert manifolds modelled on $X_i \subset E_i$. We want to define the differential of a map $f : M_1 \mapsto M_2$. Once more, we will consider a special situation that suffices for our purposes. Precisely, we will assume that there exist

- an open set $U_1 \subset E_1$ containing M_1,
- a differentiable map $\widetilde{f} : U_1 \mapsto E_2$,

such that $f = \widetilde{f}$ on M_1.

Let $p \in M_1$. It is easy to see that if $\widehat{f}$ is any other differentiable map which coincides with f on M_1, then $d\widetilde{f}(p)[v] = d\widehat{f}(p)[v]$ for all $v \in T_p M_1$. This allows us to give the following definition.

Definition 6.2 *The differential of $f : M_1 \mapsto M_2$ at p is the restriction to $T_p M_1$ of the linear map $d\widetilde{f}(p) \in L(E_1, E_2)$.*

We use the notation $d_{M_1} f(p)$ to denote the differential of f at p. When there is no possible misunderstanding, we will omit the subscript M_1 writing simply $df(p)$ instead of $d_{M_1} f(p)$. We say that $f : M_1 \mapsto M_2$ is *differentiable* (on M_1) if it is differentiable at any point of M_1 and we say that $f \in C^1(M_1, M_2)$ if f is differentiable on M_1 and the map $df : M_1 \to L(T_p M_1, E_2)$ is continuous. Similarly, if M_i are C^k Hilbert manifolds, $k \geq 1$, we can define higher differentials and C^k maps.

For example, let us consider a smooth local parametrization (V, φ) of M at p. Here, φ is the restriction to $V \subset X$ of a smooth map $\widetilde{\varphi} : \widetilde{V} \mapsto E$, where $\widetilde{V} \subset E$, and we are in the preceding situation, with $E_1 = E_2 = E$, $M_1 = X$, $M_2 = M$ and $f = \varphi$. Let $x \in V$ be such that $p = \varphi(x)$. Then $d\varphi(x)$ is the restriction of $d\widetilde{\varphi}(x)$ to X, and hence $d\varphi(x) \in T_p M$.

More in general, let $q = f(p)$ and let φ_i and $V_i \subset X_i$, $i = 1, 2$, be local parametrizations with $\varphi_1(x_1) = p$ and $\varphi_2(x_2) = q$. Consider the following commutative diagram

$$
\begin{array}{ccc}
M_1 \subset E_1 & \xrightarrow{\ \ f\ \ } & M_2 \subset E_2 \\
\uparrow{\varphi_1} & & \uparrow{\varphi_2} \\
V_1 \subset X_1 & \xrightarrow{\ \ g\ \ } & V_2 \subset X_2
\end{array}
$$

where $g = \varphi_2^{-1} \circ f \circ \varphi_1$. Taking the differentials we find the commutative diagram

$$
\begin{array}{ccc}
T_p M_1 & \xrightarrow{\ \ d_{M_1} f(p)\ \ } & E_2 \\
\uparrow{d\varphi_1(x_1)} & & \uparrow{d\varphi_2(x_2)} \\
X_1 & \xrightarrow{\ \ dg(x_i)\ \ } & X_2
\end{array}
$$

and this makes it clear that $d_{M_1} f(p)$ is a linear map from $T_p M_1$ into $T_q M_2$, with $q = f(p)$: $d_{M_1} f(p) \in L(T_p M_1, T_{f(p)} M_2)$.

As an important example, consider a smooth functional $J : E \to \mathbb{R}$ (or else defined on an open subset of E containing M). In this case $d_M J(p)$ is the linear map from $T_p M$ to $\mathbb{R}$ defined as the restriction to $T_p M$ of $dJ(p) \in L(E, \mathbb{R})$. Similar to Section 5.2, $d_M J(p)$ defines the *constrained gradient* of J at p by setting

$$(\nabla_M J(p) \mid v) = d_M J(p)[v], \quad \forall v \in T_p M.$$

Let us emphasize that the relationship between the *free* gradient J' and the constrained gradient $\nabla_M J$ is given by

$$(\nabla_M J(p) \mid v) = (J'(p) \mid v), \quad \forall v \in T_p M.$$

Hence $\nabla_M J(p)$ is nothing but the projection of $J'(p)$ on $T_p M$.

A final remark concerning the tangent space is in order. Consider a smooth curve $\gamma : [a, b] \mapsto M$, with $a < 0 < b$, and let $\gamma(0) = p$. According to the preceding discussion, $d\gamma(t)$ is a linear map from $\mathbb{R}$ to the tangent space $T_{\gamma(t)} M$. As usual, we will identify $L(\mathbb{R}, T_{\gamma(t)} M)$ with $T_{\gamma(t)} M$ and write $\gamma'(t)$ for $d\gamma(t)[1]$. The vector $\gamma'(t)$ is called the tangent vector to the curve γ at $\gamma(t)$. In particular, for any smooth curve γ on M, the tangent vector $\gamma'(0)$ at p belongs to $T_p M$. Conversely, let us show that for every $v \in T_p M$ there is a smooth curve γ such that $\gamma'(0) = v$. For this, let (V, φ) be a local parametrization of M at p with, say, $\varphi^{-1}(p) = 0$, and let $w = d\widetilde{\varphi}^{-1}(p)[v]$ (we are using the same notation employed before). Consider the straight line $t \mapsto tw$ and the curve

$\gamma(t) = \widetilde{\varphi}(tw)$. This γ is a smooth curve, $\gamma(0) = p$ and

$$\gamma'(0) = d\widetilde{\varphi}(0)[w] = d\widetilde{\varphi}[d\widetilde{\varphi}^{-1}(p)[v]] = v.$$

In conclusion, we can say that T_pM is nothing but the space of the tangent vectors to the smooth curves on M.

Remark 6.3 It is worth pointing out explicitly that E induces on M a natural *Riemannian structure*. More precisely, let $v_i \in T_pM$, $i = 1, 2$ and set

$$\langle v_1 \mid v_2 \rangle_p = (v_1 \mid v_2)_E,$$

where $(\cdot \mid \cdot)_E$ denotes the scalar product in E. Obviously, $\langle \cdot \mid \cdot \rangle_p$ is a symmetric, positive definite, bilinear form on T_pM which defines a topology equivalent to the one induced on T_pM by the norm $\| \cdot \|_E$. Furthermore, given any smooth curve $\gamma : [a, b] \to M$, we define the length of γ by

$$\ell(\gamma) = \int_a^b \langle \gamma'(t) \mid \gamma'(t) \rangle_{\gamma(t)}^{1/2} \, dt.$$

If M is arcwise connected, for every $p, q \in M$ there is a smooth curve γ joining p and q and we can define a distance on M by setting

$$d(p, q) = \inf \ell(\gamma),$$

where the infimum is taken on the set of all the smooth curves γ joining p and q. We will always assume that M is arcwise connected and complete under the metric d defined above. ∎

6.2 Constrained critical points

Let $J : E \to \mathbb{R}$ be a differentiable functional and let $M \subset E$ be a smooth Hilbert manifold. A *constrained critical point* of J on M is a point $z \in M$ such that $d_M J(z) = 0$, namely

$$dJ(z)[v] = 0, \quad \forall v \in T_zM.$$

Using the constrained gradient, we can say that a constrained critical point z of J on M satisfies

$$(\nabla_M J(z) \mid v) = 0, \quad \forall v \in T_zM.$$

Since $(\nabla_M J(z) \mid v) = (J'(z) \mid v)$, for all $v \in T_zM$, one finds that z is a constrained critical point of J on M whenever $(J'(z) \mid v) = 0$, for all $v \in T_zM$, namely whenever $J'(z)$ is orthogonal to T_zM.

We will use the notation Z to denote the set of critical points of J constrained on M; we also put $Z_c = \{z \in Z : J(z) = c\}$. Once more, using the same notation

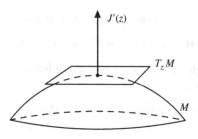

Figure 6.2

for critical points of J on M and stationary points of J (with no constraints, see the definition in Section 5.1) should not cause any misunderstanding.

Remark 6.4 Let $\gamma : [-a, a] \to M$, $a > 0$, be any smooth curve such that $\gamma(0) = z$ and consider the real valued function $\phi(t) = J(\gamma(t))$. One has that

$$\phi'(0) = J'(z)[\gamma'(0)],$$

where $\gamma'(0)$ belongs to the tangent space $T_z M$. Hence, if z is a critical point of J constrained on M, $t = 0$ is a critical point of $\phi(t) = J(\gamma(t))$, for any curve γ on M. Conversely, let φ be a local parametrization of M such that $\varphi(0) = z$ and let $v \in T_z M$. Then there exists $w \in X$ such that $v = d\varphi(0)w$ and, setting $\gamma(t) = \varphi(tw)$, $\phi(t) = J(\gamma(t))$, one has

$$\phi'(0) = J'(z)[\gamma'(0)] = J'(z)[v].$$

Thus, if $t = 0$ is a critical point for the real valued function $\phi(t) = J(\gamma(t))$ for every curve γ on M then $J'(z)[v] = 0$ for all $v \in T_z M$. This implies that z is a critical point of J on M.   ∎

Examples of constrained critical points are local constrained minima, respectively maxima. Let $M \subset E$ be a smooth manifold modelled on the Hilbert space $X \subset E$ and let $J \in C(E, \mathbb{R})$. We say that $z \in M$ is a local *constrained minimum*, resp. *maximum*, of J on M if there exists a neighbourhood $\mathcal{V}$ of z such that

$$J(z) \leq J(u), \qquad \text{respectively } J(z) \geq J(u), \qquad \forall\, u \in \mathcal{V} \cap M. \tag{6.2}$$

Let φ be a local parametrization of M with $\varphi(0) = z$. The above definition is equivalent to saying that there exists a neighbourhood $\mathcal{N} \subset X$ of $0 \in X$ such that

$$J(z) = J(\varphi(0)) \leq J(\varphi(\xi)), \qquad \text{respectively } J(z) = J(\varphi(0)) \geq J(\varphi(\xi)),$$
$$\forall\, \xi \in \mathcal{N}.$$

In other words, $z \in M$ is a local constrained minimum, respectively maximum, if and only if $0 \in X$ is a local minimum, respectively maximum, of $J \circ \varphi : \mathcal{N} \mapsto \mathbb{R}$. It follows that $0 \in X$ is a stationary point of $J \circ \varphi$, namely $\mathrm{d}(J \circ \varphi)(0)[\xi] = 0$ for every $\xi \in X$, or else:

$$\mathrm{d}J(z)[\mathrm{d}\varphi(0)[\xi]] = 0, \qquad \forall\, \xi \in X.$$

Taking into account the definition of tangent space $T_z M$, it follows that $\mathrm{d}J(z)[v] = 0$, for all $v \in T_z M$, namely that z is a constrained critical point of J on M.

6.3 Manifolds of codimension one

If the manifold $M \subset E$ is modelled on a subspace of codimension one in E, we say that M is a manifold of codimension one in E. For our purposes, the case $M = G^{-1}(0)$ where $G \in C^1(E, \mathbb{R})$, E is a Hilbert space and $G'(u) \neq 0$ for all $u \in M$, is particularly interesting. For example, if $G(u) = \|u\|^2 - r^2$ then M turns out to be the sphere $S_r = \{\|u\| = r\}$ in E.

Let $p \in M$ and consider the linear subspace

$$X_p = \{v \in E : (G'(p) \mid v) = 0\}.$$

Setting $w = \|G'(p)\|^{-2} G'(p)$, one has that $E = X_p \oplus \langle w \rangle$. Let us define the map $\psi : E \to E$ by

$$\psi(u) = u - p - (G'(p) \mid u - p)w + G(u)w.$$

It is easy to check that

$$(G'(p) \mid \psi(u)) = G(u)$$

and hence

$$\psi(u) \in X_p \quad \Longleftrightarrow \quad u \in M.$$

Moreover, $\psi(p) = 0$, $\psi \in C^1$ and $\mathrm{d}\psi(p) = \mathrm{Id}$. It follows that ψ is locally invertible at p and that it induces a diffeomorphism between a neighbourhood $\widetilde{U}$ of p and a neighbourhood $\widetilde{V}$ of 0. Moreover, its inverse $\widetilde{\varphi} = \psi^{-1}$ maps the neighbourhood $V = \widetilde{V} \cap X_p$ of 0 onto the neighbourhood $U = \widetilde{U} \cap M$ of p. Thus, letting φ denote the restriction of $\widetilde{\varphi}$ to V, it follows that M is a C^1 manifold with local parametrization at p given by (V, φ), with $V \subset X_p$. Let us remark that all the X_p are isomorphic to a fixed subspace $X \subset E$ of codimension one (i.e. $E = X \oplus \langle w^* \rangle$, for some $w^* \in E \setminus \{0\}$). Let us also point out that since $\mathrm{d}\widetilde{\varphi}(0) = (\mathrm{d}\psi(p))^{-1} = \mathrm{Id}$, then the tangent space $T_p M$ is nothing but

$$T_p M = \mathrm{d}\widetilde{\varphi}(0)[X_p] = X_p.$$

Let J be a C^1 functional on E. Since $T_p M = \{v \in E : (G'(p) \mid v) = 0\}$, a constrained critical point is an element $z \in M$ such that

$$(J'(z) \mid v) = 0, \quad \forall v \in E \text{ such that } (G'(z) \mid v) = 0,$$

and thus there exists $\lambda \in \mathbb{R}$ such that

$$J'(z) = \lambda G'(z).$$

This is nothing but the extension to the infinite dimensional case of the classical Lagrange multiplier rule.

The constrained gradient (is the projection of $J'(u)$ on X_p and hence) has the form

$$\nabla_M J(u) = J'(u) - \frac{(J'(u) \mid G'(u))}{\|G'(u)\|^2} \, G'(u).$$

So, if $z \in M$ is a constrained critical point of J, from $\nabla_M J(z) = 0$ it follows that $J'(z) = \lambda G'(z)$ with

$$\lambda = \frac{(J'(z) \mid G'(z))}{\|G'(z)\|^2}.$$

The following examples highlight that, in applications, constrained critical points correspond to solutions of eigenvalue problems.

Example 6.5

(i) If $J \in C^1(E, \mathbb{R})$ and M is the unit sphere in the Hilbert space E, then a critical point of J on M is a solution of the eigenvalue problem

$$J'(u) = \lambda u, \quad \|u\| = 1.$$

If z is such a critical point, there holds $\lambda = (J'(z) \mid z)$. In particular, if $J(u) = \frac{1}{2}(Au \mid u)$, where A is a symmetric linear operator, constrained critical points on the Hilbert sphere become solutions of the linear eigenvalue problem $Au = \lambda u$.

(ii) Setting $E = H_0^1(\Omega)$ with scalar product $(u \mid v) = \int_\Omega \nabla u \cdot \nabla v \, dx$, norm $\|u\|^2 = \int_\Omega |\nabla u|^2 \, dx$ and $M = S = \{u \in E : \|u\| = 1\}$, let $J(u) = \frac{1}{2}(K(u) \mid u)$, where, as usual, K denotes the Green operator of $-\Delta$ on E. The solutions z of $J'(u) = \lambda u$ satisfy $K(z) = \lambda z$, namely $-\lambda \Delta z = z$, with $\|z\| = 1$ and $\lambda = \int_\Omega z^2 > 0$. Then $1/\lambda$ is an eigenvalue with corresponding eigenfunction z of the Laplacian with zero Dirichlet boundary conditions.

(iii) Setting again $E = H_0^1(\Omega)$ and $M = S = \{u \in E : \|u\| = 1\}$, let $\Phi(u)$ be the functional defined in Example 5.2(iv), see also Section 1.3. A critical point of Φ constrained on S is a point $z \in S$ such that $\Phi'(z) = \lambda z$, namely $(\Phi'(z) \mid v) = \lambda(z \mid v)$ or else

$$\lambda \int_\Omega \nabla z \cdot \nabla v \, dx = \int_\Omega f(x, z) v \, dx, \quad \forall v \in E.$$

Thus z is a solution of the nonlinear eigenvalue problem

$$\begin{cases} -\lambda \Delta u(x) = f(x, u(x)) & x \in \Omega \\ u(x) = 0 & x \in \partial\Omega. \end{cases}$$

Let us remark that if z is a solution of the preceding eigenvalue problem and $f(x, \cdot)$ is p-homogeneous $(p > 1)$, then the function $v(x) = \lambda^{1/(p-1)} z(x)$ solves the BVP

$$\begin{cases} -\Delta v(x) = f(x, v(x)) & x \in \Omega \\ v(x) = 0 & x \in \partial\Omega. \end{cases} \qquad \blacksquare$$

6.4 Natural constraints

Let E be a Hilbert space and $J \in C^1(E, \mathbb{R})$. A manifold M is called a *natural constraint* for J if there exists a functional $\widetilde{J} \in C^1(E, \mathbb{R})$ with the property that every constrained critical point of $\widetilde{J}$ on M is indeed a stationary point of J, namely

$$\nabla_M \widetilde{J}(u) = 0, \quad u \in M \iff J'(u) = 0.$$

The last remark in the previous section shows that the unit sphere in $E = H_0^1(\Omega)$ is an example of natural constraint for $J : E \mapsto \mathbb{R}, J(u) = \frac{1}{2}\|u\|^2 - (1/(p+1)) \int_\Omega |u|^{p+1} \, dx$.

Remark 6.6 When $\inf_E J = -\infty$ and $\sup_E J = +\infty$, it might be convenient to look for critical points of J on a natural constraint M, because it could happen that J is bounded on M, see Example 6.8 below. In such a case, one can try to find stationary points of J as *minima* (or *maxima*) of J constrained on M. $\blacksquare$

Another example of a natural constraint is the so-called Nehari manifold[1]

$$M = \{u \in E \setminus \{0\}; (J'(u) \mid u) = 0\}.$$

Precisely, let us show the following.

Proposition 6.7 *Let $J \in C^2(E, \mathbb{R})$ and let $M = \{u \in E \setminus \{0\} : (J'(u)|u) = 0\}$ be non-empty. Moreover, let us assume [formulas (6.3) and (6.4)]. Then*

$$\exists r > 0 : B_r \cap M = \emptyset, \tag{6.3}$$

---

[1] The name comes from Z. Nehari who first introduced this manifold.

and that

$$(J''(u)u \mid u) \neq 0, \quad \forall u \in M. \tag{6.4}$$

$M = \{u \in E \setminus \{0\} : (J'(u) \mid u) = 0\}$ *is non empty. Then M is a natural constraint for J.*

Proof. We will take $\tilde{J} = J$ and prove that the critical points of J constrained on M are indeed stationary points of J. Set $G(u) = (J'(u) \mid u)$, in such a way that $M = G^{-1}(0) \setminus \{0\}$. Clearly, $G \in C^1(E, \mathbb{R})$ and for $u \in M$ one has

$$(G'(u) \mid u) = (J''(u)u \mid u) + (J'(u) \mid u) = (J''(u)u \mid u) \neq 0. \tag{6.5}$$

Thus $G'(u) \neq 0$ for all $u \neq 0$ and this, jointly with (6.3), implies that M is a C^1 manifold of codimension one. If z is a critical point of J on M there holds

$$J'(z) = \lambda G'(z).$$

Taking the scalar product with z we find

$$(J'(z) \mid z) = \lambda(G'(z) \mid z).$$

One has that $(J'(z) \mid z) = G(z) = 0$, while $(G'(z) \mid z) \neq 0$ by (6.5). Then it follows that $\lambda = 0$ and hence that $J'(z) = 0$. Conversely, it is obvious that every stationary point $u \neq 0$ of J belongs to M. ∎

Example 6.8 Let $E = H_0^1(\Omega)$ with norm $\|u\|^2 = \int_\Omega |\nabla u|^2 \, dx$ and

$$J(u) = \frac{1}{2}\|u\|^2 - \frac{1}{p+1}\int_\Omega |u|^{p+1} \, dx$$

with $1 < p + 1 < 2^*$. Here $G(u) = \|u\|^2 - \int_\Omega |u|^{p+1} \, dx$ and

$$M = \left\{ u \in E \setminus \{0\} : \|u\|^2 = \int_\Omega |u|^{p+1} \, dx \right\}.$$

It is easy to check that (6.3) holds. Moreover, $(G'(u) \mid u) = 2\|u\|^2 - (p + 1) \int_\Omega |u|^{p+1} \, dx$ and hence for $u \in M$ one has $(G'(u) \mid u) = (1 - p)\|u\|^2 < 0$. Finally, $J_{|M}$ becomes $J(u) = (\frac{1}{2} - (1/p + 1))\|u\|^2$. Let us point out that, if $2 < p + 1 < 2^*$, then $J_{|M} > 0$, while $\inf_E J = -\infty$ (it suffices to fix $v \in E \setminus \{0\}$ and to remark that $\lim_{t \to +\infty} J(tv) = -\infty$). ∎

6.5 Exercises

(i) Let J_λ denote the Euler functional corresponding the problem (see Exercise (iii))

$$\begin{cases} -\Delta u(x) = \lambda u - |u|^{p-1}u, & x \in \Omega, \\ u(x) = 0, & x \in \partial\Omega, \end{cases}$$

where $1 < p < (n+2)/(n-2)$, and consider its Nehari manifold
$M_\lambda = \{u \in H_0^1(\Omega) \setminus \{0\} : (J_\lambda'(u) \mid u) = 0\}$.

(a) Show that $M_\lambda = \emptyset$ if and only if $\lambda \leq \lambda_1$.

(b) Show that M_λ is a natural constraint.

(c) Extend the discussion to the case in which $|u|^{p-1}u$ is replaced by a general function $f(u) \sim |u|^{p-1}u$. Under which assumptions on f, is M_λ a natural constraint?

(ii) Let $J(u) = \frac{1}{2}\int_\Omega |\nabla u|^2 \, dx + (1/\alpha) \int_\Omega |u|^\alpha \, dx - (1/\beta) \int_\Omega |u|^\beta \, dx$, where $u \in H_0^1(\Omega)$ and $2 \leq \alpha < \beta < 2^*$. Find a natural constraint for J.

(iii) Consider the BVP

$$\begin{cases} -\Delta u(x) &= \lambda |u|^{q-1}u + |u|^{p-1}u, \quad x \in \Omega, \\ u(x) &= 0, \qquad\qquad\qquad\quad x \in \partial\Omega, \end{cases}$$

where $1 < q < 2 < p < \frac{n+2}{n-2}$ and set

$$M_\lambda = \left\{ u \in H_0^1(\Omega) \setminus \{0\} : \|u\|^2 = \lambda \int_\Omega |u|^{q+1} + \int_\Omega |u|^{p+1} \right\}.$$

Show that there exists λ_0 such that for $0 < \lambda < \lambda_0$ there holds

(a) M_λ is a natural constraint and there exists $\rho_\lambda > 0$ such that
$$u \in M_\lambda \implies \|u\| \geq \rho_\lambda.$$

(b) M_λ is the union of two disjoint manifolds.

7

Deformations and the Palais–Smale condition

The existence of constrained critical points is closely related to the deformation of sublevels. On the other hand, to carry out deformations one needs some compactness condition, like the Palais–Smale condition. These two basic tools will be discussed in the present chapter.

7.1 Deformations of sublevels

In this section we will deal with some preliminary results that will be used extensively in the rest of the book.

Let $J : M \subset E \mapsto \mathbb{R}$ and let $a \in \mathbb{R}$. The set

$$M^a = \{u \in M : J(u) \leq a\},$$

is a sublevel of J on M. The main goal is to carry out deformations of sublevels. By deformation, we mean the following.

Definition 7.1 A deformation *of* $A \subset M$ *in* M *is a map* $\eta \in C(A, M)$ *which is homotopic to the identity: there exists* $H \in C([0, 1] \times A, M)$ *such that*

$$H(0, u) = u, \quad H(1, u) = \eta(u), \quad \forall u \in M.$$

Roughly, if A can be deformed in A', then A and A' have the same topological properties. We will see that if the interval $[a, b]$ does not contain any critical point of J on M and M is compact, then M^b can be deformed into M^a. On the other hand, the presence of critical levels in $[a, b]$ might prevent the possibility of deforming M^b into M^a. Before proving these statements, let us consider a couple of elementary examples.

Let M be a compact hyper-surface in $\mathbb{R}^n$ and suppose that b is not a critical level for J on M. Then the level set $\{p \in M : J(p) = b\}$ is a smooth submanifold N_b of M and at any point of N_b the vector $-\nabla_M J(p)$ is different from zero. By compactness, $\min_{p \in N_b} |\nabla_M J(p)| > 0$ and hence, by means of these gradient vectors, M^b is 'deformed' into the level $M^{b-\varepsilon}$, for some $\varepsilon > 0$. We can repeat

100

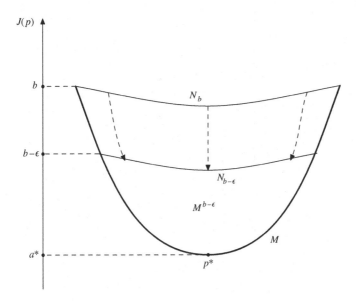

Figure 7.1

this procedure a finite number of times until we find a critical level and so, if the interval $[a, b]$ contains no critical levels, the sublevel M^b can be 'deformed' into M^a. As a consequence, the topological properties of the sublevels M^a do not change when a does not cross a critical level. On the other hand, if $p^* \in M$ is the minimum of J on M, with $a^* = J(p^*)$, we have that $M^a \neq \emptyset$ if $a > a^*$, while $M^a = \emptyset$ if $a < a^*$ so that they cannot be deformed one into the other.

As a second example, let M be a two dimensional torus in $\mathbb{R}^3$ and let $J(x, y, z) = z$. The critical points of J on M are the four points p_i where the gradient of J, namely the vector $(0, 0, 1)$, is orthogonal to M.
For example, all the sublevels M^b are diffeomorphic to a cylinder $S^1 \times [0, 1]$ provided that $b \in (c_2, c_3)$. On the other hand, if $a \in (c_1, c_2)$ then M^a is diffeomorphic to the unit ball $B_1 \subset \mathbb{R}^2$. Hence M^b cannot be deformed into M^a. Actually, M^b contains closed curves that cannot be contracted in a point on the torus, while M^a does not.

7.2 The steepest descent flow

To extend the preceding procedure to the general case, we will use flows of differential equations, in particular the *steepest descent flow*. First some preliminaries are in order.

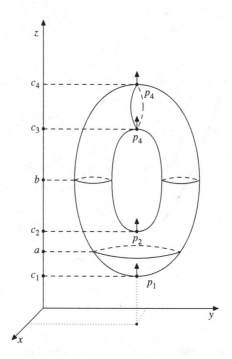

Figure 7.2

Given a map $W \in C^{0,1}(E, E)$, let $\alpha(t) = \alpha(t, u)$ denote the solution of the Cauchy problem

$$\begin{cases} \alpha'(t) = W(\alpha(t)), \\ \alpha(0) = u \in E. \end{cases} \tag{7.1}$$

Since $W \in C^{0,1}$, a standard application of the contraction mapping theorem shows that (7.1) has a unique solution $\alpha(t, u)$, defined for t in a neighbourhood of $t = 0$ and depending continuously on u on the compact subsets of $\mathbb{R}$. We will denote by (t_u^-, t_u^+) the maximal interval of existence of α; namely, $t_u^{\pm}$ are such that there are no solutions of (7.1) defined on an interval which contains strictly (t_u^-, t_u^+).

For what follows it is important that the solutions of (7.1) are globally defined for positive t, namely that $t^+ = +\infty$. Let us begin with a well known result.

Lemma 7.2 *If* $t_u^+ < +\infty$ *(respectively* $t_u^- > -\infty$*) then* $\alpha(t, u)$ *has no limit points as* $t \uparrow t_u^+$ *(respectively* $t \downarrow t_u^-$*).*

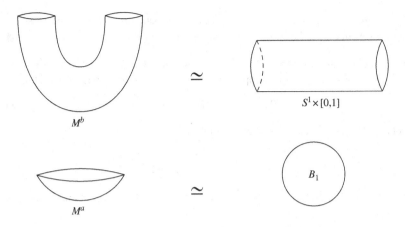

Figure 7.3

Proof. If not, let $v \in E$ be such that $v = \lim_{t \uparrow t^+} \alpha(t, u)$ and let β denote the solution of the Cauchy problem

$$\begin{cases} \beta'(t) = W(\beta(t)), \\ \beta(t_u^+) = v \in E, \end{cases}$$

defined in a neighbourhood $(t^+ - \epsilon, t^+ + \epsilon)$ (the dependence on u is understood). The function

$$\widetilde{\alpha}(t, u) = \begin{cases} \alpha(t, u) & \text{for } t \in (t^-, t^+) \\ \beta(t, v) & \text{for } t \in [t^+, t^+ + \epsilon) \end{cases}$$

is a solution of (7.1) defined in the interval $(t^-, t^+ + \epsilon)$, a contradiction. The same argument holds for t^-. ∎

Furthermore, one has the following result.

Lemma 7.3 *Let $A \subseteq E$ be closed and suppose there exists $C > 0$ such that $\|W(u)\| \leq C$ for all $u \in A$. Let $u \in A$ be such that $\alpha(t, u) \in A$ for all $t \in [0, t_u^+)$. Then $t_u^+ = +\infty$.*

Proof. Suppose that $t^+ < +\infty$. For all $t_i, t_j \in [0, t^+)$ there holds

$$\alpha(t_i, u) - \alpha(t_j, u) = \int_{t_j}^{t_i} \frac{\mathrm{d}}{\mathrm{d}t} \alpha(s, u) \, \mathrm{d}s = \int_{t_j}^{t_i} W(\alpha(s, u)) \, \mathrm{d}s.$$

Since $\|W\|$ is bounded on A and $\alpha(s, u) \in A$ for all $s \in [0, t^+)$, it follows that

$$\|\alpha(t_i, u) - \alpha(t_j, u)\| \leq \int_{t_j}^{t_i} \|W(\alpha(s, u))\| \, \mathrm{d}s \leq C |t_i - t_j|.$$

Therefore, as $t_i \uparrow t^+$, $\alpha(t_i, u)$ is a Cauchy sequence that converges to some point in A, in contradiction with the preceding lemma.   ∎

We will now investigate more closely the specific case in which $W(u) = -\nabla_M J(u)$. Precisely, let us suppose that there exists $G \in C^{1,1}(E, \mathbb{R})$ such that

$$M = G^{-1}(0), \quad \text{with} \quad G'(u) \neq 0, \forall \quad u \in M. \tag{M}$$

Let $J \in C^{1,1}(E, \mathbb{R})$ and consider the function

$$W(u) = -\left[J'(u) - \frac{(J'(u) \mid G'(u))}{\|G'(u)\|^2} G'(u) \right],$$

which is well defined in a neighbourhood of M, is of class $C^{0,1}$ and coincides with $-\nabla_M J(u)$ for all $u \in M$.

The corresponding solution $\alpha(t, u)$ of (7.1) will be called the *steepest descent flow* of J. Let us point out explicitly that $\alpha(t) = \alpha(t, u) \in M$ for all $t \in (t_u^-, t_u^+)$, whenever $u \in M$. Actually, one has that

$$\frac{\mathrm{d}}{\mathrm{d}t} G(\alpha(t)) = (G'(\alpha(t)) \mid \alpha'(t)) = (G'(\alpha(t)) \mid W(\alpha(t)))$$

$$= -(G'(\alpha(t)) \mid J'(\alpha(t))) + \frac{(J'(\alpha(t)) \mid G'(\alpha(t)))}{\|G'(\alpha(t))\|^2}$$

$$\times (G'(\alpha(t)) \mid G'(\alpha(t)))$$

$$= 0.$$

It follows that $G(\alpha(t))$ is constant:

$$G(\alpha(t)) = G(\alpha(0)) = G(u), \quad \forall\, t \in (t_u^-, t_u^+).$$

Then, $G(\alpha(t)) = 0$ if and only if $G(u) = 0$, namely $\alpha(t) \in M$ whenever $u \in M$.

Lemma 7.4 *Suppose that (M) holds and that $J \in C^{1,1}(E, \mathbb{R})$. Then the steepest descent flow of J verifies:*

(i) *the function $t \mapsto J(\alpha(t, u))$, $t \in [0, t_u^+)$, is nonincreasing;*

(ii) *for $\tau, t \in [0, t_u^+)$ there holds*

$$J(\alpha(t, u)) - J(\alpha(\tau, u)) = -\int_\tau^t \|\nabla_M J(\alpha(s, u))\|^2 \, \mathrm{d}s; \tag{7.2}$$

(iii) *if J is bounded from below on M, then $t_u^+ = +\infty$, for all $u \in M$.*

Proof. One has (the dependence on u is understood):

$$\frac{\mathrm{d}}{\mathrm{d}t} J(\alpha(t)) = (J'(\alpha(t)) \mid \alpha'(t)) = -(J'(\alpha(t)) \mid \nabla_M J(\alpha(t))).$$

Since $\nabla_M J(\alpha)$ is the projection of $J'(\alpha)$ on $T_\alpha M$, then

$$(J'(\alpha(t)) \mid \nabla_M J(\alpha(t))) = \|\nabla_M J(\alpha(t))\|^2$$

and (i) and (ii) follow immediately.

As for (iii), we argue by contradiction. Let $u \in M$ be such that $t_u^+ < +\infty$. Using (7.2) with $\tau = 0$ we infer

$$J(\alpha(t)) - J(u) = -\int_0^t \|\nabla_M J(\alpha(s))\|^2 \, ds.$$

Since J is bounded from below on M and $\alpha(t) \in M$, this implies that there exists $a > 0$ such that

$$\int_0^t \|\nabla_M J(\alpha(s))\|^2 \, ds \le a < +\infty. \tag{7.3}$$

Let $t_i \uparrow t^+$. As in the proof of Lemma 7.3 we have

$$\|\alpha(t_i) - \alpha(t_j)\| \le \int_{t_j}^{t_i} \|\nabla_M J(\alpha(s))\| \, ds.$$

Using the Hölder inequality and (7.3) we deduce

$$\|\alpha(t_i, u) - \alpha(t_j, u)\| \le |t_i - t_j|^{1/2} \cdot \left\{ \int_{t_j}^{t_i} \|\nabla_M J(\alpha(s, u))\|^2 \, ds \right\}^{1/2}$$

$$\le \sqrt{a} \, |t_i - t_j|^{1/2}.$$

Thus $\alpha(t_i, u)$ is a Cauchy sequence, contrary to Lemma 7.2. ∎

Remark 7.5 We anticipate that the condition $J \in C^{1,1}$ can be weakened by requiring that $J \in C^1(E, \mathbb{R})$, only. For this, one uses the *pseudo gradient vector fields*. This topic will be discussed in Section 8.1, see in particular the proof of Lemma 8.4 and Section 8.1.1. ∎

7.3 Deformations and compactness

In this section it is understood that $M = G^{-1}(0)$, where $(\boldsymbol{M})$ holds, and that $J \in C^{1,1}(E, \mathbb{R})$, unless a different assumption is explicitly made.

The following lemma shows how the steepest descent flow allows us to deform sublevels.

Lemma 7.6 *Suppose there are $\delta > 0$ and $c \in \mathbb{R}$ such that*

$$\|\nabla_M J(u)\| \ge \delta, \qquad \forall u \in M \text{ for which } |J(u) - c| \le \delta. \tag{7.4}$$

Then there exists a deformation η in M such that

$$\eta(M^{c+\delta}) \subset M^{c-\delta}. \tag{7.5}$$

Proof. Let us first suppose that J is bounded from below on M, Lemma 7.4(iii) implies that the steepest descent flow α is globally defined on $[0, +\infty)$ for all $u \in M$. Let $T = 2/\delta$ and consider the deformation $\eta(u) := \alpha(T, u)$. We claim that $J(\eta(u)) \leq c - \delta$ for all $u \in M^{c+\delta}$. Otherwise, $J(\alpha(T, U)) > c - \delta$ for some $u \in M^{c+\delta}$. Since $J(\alpha(s))$ is not increasing, see Lemma 7.4(i), it follows that

$$c - \delta < J(\alpha(s, u)) \leq c + \delta, \quad \forall s \in [0, T].$$

Therefore (7.4) implies that

$$\|\nabla_M J(\alpha(s, u))\| \geq \delta, \quad \forall t \in [0, T]. \tag{7.6}$$

Using (7.2) with $t = T$ and $\tau = 0$ we infer

$$J(a(T, u)) - J(u) = -\int_0^T \|\nabla_M J(\alpha(s, u))\|^2 \, ds.$$

From (7.6) and the choice of $T = 2/\delta$ it follows that

$$J(a(T, u)) \leq J(u) - \delta^2 T = J(u) - 2\delta. \tag{7.7}$$

Since $J(u) \leq c + \delta$ we find that $J(a(T, u)) \leq c + \delta - 2\delta = c - \delta$, a contradiction.

If J is not bounded from below, we substitute J with a truncated functional

$$\widehat{J}(u) = h(J(u)),$$

where $h \in C^\infty(\mathbb{R}, \mathbb{R})$ is strictly increasing, bounded from below and such that $h(s) = s$ if $s \geq c - d$. Since $\widehat{J}$ is bounded from below and $\widehat{J} \equiv J(u)$ on $\{u \in M : J(u) \geq c - \delta\}$, we can repeat the preceding arguments and the result follows. ∎

When M is compact and c is not a critical level for J on M then (7.4) holds and hence the preceding lemma allows us to deform $M^{c+\delta}$ into $M^{c-\delta}$. On the other hand, if M is not compact this might not be possible. For example, in Figure 7.4, $M = \mathbb{R}$, $M^{c+\delta} = (-\infty, \beta] \cup [\gamma, +\infty)$ is disconnected while $M^{c-\delta} = (-\infty, \alpha]$ is connected and hence the former cannot be deformed into the latter.

When M is not compact we need to make some further assumption that allows us to perform deformations, avoiding situations like that shown in Figure 7.4.

The following compactness condition was first introduced by *Palais* and *Smale*.

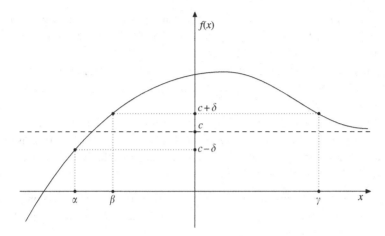

Figure 7.4 $M = \mathbb{R}, M^{c-\delta} = (-\infty, \alpha], M^{c+\delta} = (-\infty, \beta] \cup [\gamma, +\infty)$.

7.4 The Palais–Smale condition

A sequence $u_n \in M$ is called a *Palais–Smale* sequence on M (*PS* in short) if $J(u_n)$ is bounded and $\nabla_M J(u_n) \to 0$. If $J(u_n) \to c$ the *PS*-sequence will be called a PS_c-sequence.

Definition 7.7 *We say that J satisfies the (PS), respectively $(PS)_c$, condition on M, if every PS-sequence, respectively PS_c-sequence, has a converging subsequence.*

Remark 7.8

(a) If J satisfies $(PS)_c$ then every PS_c-sequence converges (up to a subsequence) to some u^* and, by continuity, one has that $J(u^*) = c$ and $\nabla_M J(u^*) = 0$. In other words, u^* is a critical point of J on M and c is a critical level of J. In particular, the set of critical points at level c, $\{z \in M : J(z) = c, \nabla_M J(z) = 0\}$ is compact.

(b) If $J \in C^1(\mathbb{R}^n, \mathbb{R})$ is bounded from below and coercive, then (PS) holds. Actually, from $J(u_j) \to c$ it follows that $|u_j| \leq$ constant and the Bolzano–Weierstrass theorem implies that u_j converges, up to a subsequence. On the other hand, if $J \in C^1(E, \mathbb{R})$ and E is infinite dimensional, J might be bounded from below and coercive without satisfying the (PS) condition. For example, letting $g : \mathbb{R}^+ \mapsto \mathbb{R}$ denote a smooth function such that $g(s) = 0$ if $s \in [0, 2]$ and $g(s) = s$ if $s \geq 3$, the functional $J(u) = g(\|u\|)$ is bounded from below and coercive but $(PS)_c$

does not hold at level $c = 0$. To see this, notice that any sequence $u_j \in E$ with $\|u_j\| = 1$ is such that $J(u_j) = 0$ and $J'(u_j) = 0$. If $(PS)_0$ holds then the unit sphere in E would be compact.

(c) It is possible to show that if $J \in C^1(E, \mathbb{R})$ is bounded from below, satisfies the (PS) condition and the set of its critical points is bounded, then J is coercive, see [116].

(d) The fact that in order to verify the (PS) condition one has to consider sequences u_j which satisfy *both* $J(u_j) \to c$ as well as $J'(u_j) \to 0$ turns out to be an important advantage when one deals with elliptic BVP. See also Remark 8.12 in the next chapter. ∎

The next lemma shows that if c is not a critical level, the $(PS)_c$ condition provides a uniform bound from below for the gradient of J, avoiding a situation like that in Figure 7.4.

Lemma 7.9 *Suppose that $c \in \mathbb{R}$ is not a critical level for $J \in C^1(E, \mathbb{R})$ and that $(PS)_c$ holds. Then there exist $\delta > 0$ such that $\|\nabla_M J(u)\| \geq \delta$ for all $u \in M$ with $|J(u) - c| \leq \delta$.*

Proof. Arguing by contradiction, there exists a sequence $u_n \in M$ such that $J(u_n) \to c$ and $\nabla_M J(u_n) \to 0$. Since J satisfies $(PS)_c$ then (up to a subsequence) u_n converges to a critical point u^* of J with $J(u^*) = c$, namely c is a critical level of J, contrary to the assumption. ∎

We are now ready to prove the deformation lemmas we were looking for. The first one is the counterpart of Lemma 7.6.

Lemma 7.10 *Suppose that $c \in \mathbb{R}$ is not a critical level for J on M and that $(PS)_c$ holds. Then there exists $\delta > 0$ and a deformation η in M such that $\eta(M^{c+\delta}) \subset M^{c-\delta}$.*

Proof. It suffices to apply Lemma 7.6 taking into account that if c is not a critical level for J and $(PS)_c$ holds, then (7.4) follows from Lemma 7.9. ∎

Similarly, we also get the following result.

Lemma 7.11 *Let $a, b \in \mathbb{R}$, $b > a$, be such that J has no critical levels in the interval $[a, b]$ and satisfies $(PS)_c$ for all $c \in [a, b]$. Then there is a deformation η in M such that $\eta(M^b) \subset M^a$.*

Proof. We can either apply the previous lemma a finite number of times or else we can argue directly as in Lemma 7.6. It suffices to notice that the fact that

$(PS)_c$ holds for all $c \in [a, b]$ and that J has no critical level in $[a, b]$, imply there exists $\delta > 0$ such that $\|\nabla_M J(u)\| \geq \delta$ for all $u \in M$ with $a \leq J(u) \leq b$. Then using the steepest descent flow $\alpha(T, u)$ with $T = (b - a)/\delta^2$, we deduce as in (7.7) that $J(a(T, u)) \leq J(u) - \delta^2 T \leq b - \delta^2 T = a$. ∎

7.5 Existence of constrained minima

We are now in a position to state a result concerning the existence of constrained extrema. Actually, the preceding discussion allows us to prove the following abstract result.

Theorem 7.12 *Let $J \in C^{1,1}(E, \mathbb{R})$, and let (**M**) hold. Moreover, suppose that J is bounded from below on M and satisfies $(PS)_m$ where*

$$m := \inf_{u \in M} J(u) > -\infty.$$

Then the infimum m is achieved. Precisely, there exists $z \in M$ such that $J(z) = m$ and $\nabla_M J(z) = 0$.

Proof. Let $u_i \in M$ be a minimizing sequence: $J(u_i) \to m$. We can assume that $\|\nabla_M J(u_i)\| \to 0$, otherwise an application of Lemma 7.6 would yield a positive δ and a deformation η in M such that $\eta(M^{m+\delta}) \subset M^{m-\delta}$, a contradiction because $M^{m-\delta}$ is empty while $M^{c+\delta}$ is not. Therefore u_i is a $(PS)_m$ sequence and (up to a subsequence) u_i converges to some $z \in M$. Obviously one has that $J(z) = m$ and $\nabla_M J(z) = 0$. ∎

Remark 7.13 Completing Remark 7.5, we point out that the assumption that J is of class $C^{1,1}$ can be weakend and one can handle functionals $J \in C^1(E, \mathbb{R})$, see also Remark 10.11. Moreover, the preceding proof shows that if $J \in C^{1,1}(M, \mathbb{R})$ is bounded from below on M then there exists a minimizing sequence $u_i \in M$ such that $\|\nabla_M J(u_i)\| \to 0$. These sequences, even if J is not smooth, have been studied by I. Ekeland [92]. ∎

7.6 An application to a superlinear Dirichlet problem

Following [9], we will apply the preceding theorem to prove the existence of a positive solution of a class of superlinear Dirichlet boundary value problem:

$$\begin{cases} -\Delta u(x) = f(u(x)) & x \in \Omega \\ u(x) = 0 & x \in \partial\Omega. \end{cases} \qquad (D)$$

Here Ω is a bounded domain in $\mathbb{R}^n$ and $f \in C^2(\mathbb{R}, \mathbb{R})$ satisfies the following assumptions: there exist $a_1, a_2 > 0$ and $p \in (1, (n+2)/(n-2))$, such that

$$|f(u)| \le a_1 + a_2|u|^p, \tag{7.8}$$

$$|uf'(u)| \le a_1 + a_2|u|^p, \quad |u^2 f''(u)| \le a_1 + a_2|u|^p. \tag{7.9}$$

Moreover, assume that $f(u) = uh(u)$, where

(h1) h is convex,
(h2) $uh'(u) > 0, \quad \forall u \ne 0$,
(h3) $h(0) = 0$,
(h4) $\lim_{u \to +\infty} h(u) = +\infty$.

For example, $h(u) = |u|^{p-1}$, $1 < p < (n+2)/(n-2)$ satisfies $(h1)$–$(h4)$. Remark that in such a case a nontrivial solution of (D) can be found, after rescaling, looking for the min $\int_\Omega |u|^{p+1} \, dx$ on the sphere $\{u \in H_0^1(\Omega) : \|u\| = 1\}$, see Example 6.5(iii). The following theorem shows that this result can be extended to handle (possibly nonhomogeneous) nonlinearities like $uh(u)$ satisfying the preceding assumptions.

Theorem 7.14 *Suppose that $f \in C^2(\mathbb{R}, \mathbb{R})$ satisfies* (7.8), (7.9) *and* **(h1)–(h4)**. *Then (D) has a positive solution.*

The proof will make use of the method of the natural constraint. First, some notation is in order. We take $E = H_0^1(\Omega)$ with scalar product $(u \mid v) = \int_\Omega \nabla u \cdot \nabla v \, dx$ and norm $\|u\|^2 = (u \mid u)$. Furthermore, we set

$$F(u) = \int_0^u f(s) \, ds = \int_0^1 f(su)u \, ds,$$

$$\Phi(u) = \int_\Omega F(u(x)) \, dx = \int_0^1 ds \int_\Omega u(x)f(su(x)) \, dx,$$

$$\Psi(u) = (\Phi'(u) \mid u) = \int_\Omega u(x)f(u(x)) \, dx.$$

Since f satisfies (7.8) and (7.9), Theorem 1.8 implies that $\Phi \in C^3(E, \mathbb{R})$ and $\Psi \in C^2(E, \mathbb{R})$. Moreover, Φ and Ψ are weakly continuous, while Φ' and Ψ' are compact.

The solutions of (D) are the critical points of

$$J(u) = \tfrac{1}{2}\|u\|^2 - \Phi(u).$$

Let us set

$$G(u) = (J'(u) \mid u) = \|u\|^2 - \Psi(u)$$

and
$$M = \{u \in E \setminus \{0\} : \|u\|^2 = \Psi(u)\}.$$

Our first goal is to show that M is a natural constraint for J, namely that there exists a (smooth) functional $\widetilde{J}$ such that

$$\nabla_M \widetilde{J}(u) = 0, \qquad u \in M \iff J'(u) = 0.$$

Once this is proved, we will check that $\widetilde{J}$ achieves a minimum on M, giving rise to a solution of (D).

First of all, the following lemma holds.

Lemma 7.15 $G \in C^2(E, \mathbb{R})$. *Moreover,*

(i) $M \neq \emptyset$,
(ii) $\exists \rho > 0$ *such that* $\|u\| \geq \rho$, *for all* $u \in M$,
(iii) $(G'(u) \mid u) < 0$, *for all* $u \in M$.

Proof. The regularity of G immediately follows from fact that $\Psi \in C^2(E, \mathbb{R})$. To prove (i) we take any $u \in E$ with $u > 0$, $\|u\| = 1$ and evaluate

$$G(tu) = t^2 - t^2 \int_\Omega u^2 h(tu) \, dx.$$

Using ($h3$) and ($h4$) we find

$$\lim_{t \to 0} \frac{G(tu)}{t^2} = 1, \qquad \lim_{t \to +\infty} \frac{G(tu)}{t^2} = -\infty.$$

Hence there exists $\bar{t} = \bar{t}(u) > 0$ such that $\bar{t}u \in M$.

From ($h3$) it follows that

$$(G''(0)v \mid v) = 2\|v\|^2 - \int_\Omega f'(0)v^2 \, dx = 2\|v\|^2,$$

and this implies that (ii) holds.

For $u \in M$ one finds

$$(G'(u) \mid u) = 2\|u\|^2 - (\Psi'(u) \mid u) = 2\Psi(u) - (\Psi'(u) \mid u). \qquad (7.10)$$

One also has

$$\begin{aligned}
2\Psi(u) - (\Psi'(u) \mid u) &= 2 \int_\Omega uf(u) \, dx - \left[\int_\Omega uf(u) \, dx + \int_\Omega u^2 f'(u) \, dx \right] \\
&= \int_\Omega u^2 h(u) \, dx - \int_\Omega u^2 (h(u) + uh'(u)) \, dx \\
&= -\int_\Omega u^3 h'(u) \, dx.
\end{aligned}$$

Since $0 \notin M$, using ($h2$) we infer that (iii) holds true. $\blacksquare$

Let us remark explicitly that from (iii) it follows that M is a C^2 submanifold of codimension one in E.

Next, let us now introduce the functional $\widetilde{J} \in C^2(E, \mathbb{R})$, defined by setting

$$\widetilde{J}(u) = \tfrac{1}{2}\Psi(u) - \Phi(u).$$

Let us point out that, since $\Psi(u) = \|u\|^2$ on M, it follows that $\widetilde{J}(u) = J(u)$ for all $u \in M$. But it is more convenient to use $\widetilde{J}$ instead of J because $\widetilde{J}$ shares the same properties as Ψ and Φ, namely $\widetilde{J}$ is weakly continuous and $\widetilde{J}'$ is compact.

Lemma 7.16 *If $z \in M$ is a constrained critical point of $\widetilde{J}$ on M, then z is a nontrivial critical point of J.*

Proof. If $z \in M$ is a constrained critical point of $\widetilde{J}$ on M, then $z \neq 0$ and there exists $\lambda \in \mathbb{R}$ such that

$$\widetilde{J}'(z) = \lambda G'(z).$$

Taking the scalar product with z we get

$$(\widetilde{J}'(z) \mid z) = \lambda(G'(z) \mid z). \qquad (7.11)$$

On the other hand, we have $(\widetilde{J}'(z) \mid z) = \tfrac{1}{2}(\Psi'(z) \mid z) - (\Phi'(z) \mid z) = \tfrac{1}{2}(\Psi'(z) \mid z) - \Psi(z)$ and hence, using (7.10) we deduce

$$(\widetilde{J}'(z) \mid z) = -\tfrac{1}{2}(G'(z) \mid z).$$

From this and (7.11) we infer that $-\tfrac{1}{2}(G'(z) \mid z) = \lambda(G'(z) \mid z)$ and since $(G'(z) \mid z) < 0$, see Lemma 7.15(iii), then $\lambda = -\tfrac{1}{2}$. Finally, we have that $\widetilde{J}'(z) = \tfrac{1}{2}\Psi'(z) - \Phi'(z)$, and $G'(z) = 2z - \Psi'(z)$. Thus, the equation $\widetilde{J}'(z) = -\tfrac{1}{2}G'(z)$ becomes

$$\tfrac{1}{2}\Psi'(z) - \Phi'(z) = -z + \tfrac{1}{2}\Psi'(z).$$

Therefore $z = \Phi'(z)$, namely $J'(z) = 0$. ■

It remains to show that $\widetilde{J}$ achieves the minimum on M. In view of this, let us first prove that $\widetilde{J}$ is bounded from below on M.

Lemma 7.17 *For all $u \in M$ one has that $\widetilde{J}(u) \geq \tfrac{1}{6}\|u\|^2$.*

Proof. With easy calculations we get

$$\widetilde{J}(u) = \frac{1}{2}\int_\Omega uf(u)\,dx - \int_0^1 ds \int_\Omega u(x)f(su(x))\,dx$$

$$= \int_0^1 ds \int_\Omega [suf(u) - uf(su)]\,dx = \int_0^1 ds \int_\Omega su^2[h(u) - h(su)]\,dx.$$

Since h is convex, then $h(u) - h(su) \geq (1 - s)h(u)$, $0 \leq s \leq 1$, and hence

$$\widetilde{J}(u) \geq \int_0^1 ds \int_\Omega s(1 - s)u^2h(u)\,dx = \frac{1}{6}\int_\Omega u^2h(u)\,dx = \frac{1}{6}\|u\|^2. \quad \blacksquare$$

In order to prove that $(PS)_c$ holds for every $c > 0$, we collect some properties of $(PS)_c$ sequences for $\widetilde{J}$ on M.

Lemma 7.18 *Let $u_i \in M$ be a $(PS)_c$ sequence for $\widetilde{J}$ on M. Then*

(i) $\|u_i\|$ is bounded and there exists $\overline{u} \neq 0$ such that, up to a subsequence, $u_i \rightharpoonup \overline{u}$;
(ii) there exists $k > 0$ (depending on c), such that $\|\widetilde{J}'(u_i)\| \geq k$.

Proof. Since $\widetilde{J}(u_i) \to c > 0$ and $u_i \in M$, Lemma 7.17 implies that $\|u_i\| \leq$ constant and, up to a subsequence, $u_i \rightharpoonup \overline{u}$. Moreover, one has that $\|u_i\|^2 = \Psi(u_i)$. Using Lemma 7.15(ii), we get $\Psi(u_i) \geq \rho^2 > 0$. Passing to the limit we find $\Psi(\overline{u}) \geq \rho^2 > 0$ and thus $\overline{u} \neq 0$, proving (i).

Arguing by contradiction, suppose that $\widetilde{J}'(u_i) \to 0$. Since $\widetilde{J}'$ is compact, we infer that $\widetilde{J}'(\overline{u}) = \lim \widetilde{J}'(u_i) = 0$. This implies

$$0 = (\widetilde{J}'(\overline{u}) \mid \overline{u}) = \frac{1}{2}(\Psi'(\overline{u}) \mid \overline{u}) - \Psi(\overline{u}) = \frac{1}{2}\int_\Omega \overline{u}^3 h'(\overline{u})\,dx > 0,$$

a contradiction, and (ii) follows. $\quad\blacksquare$

Lemma 7.19 *The functional $\widetilde{J}$ satisfies $(PS)_c$ on M, for every $c > 0$.*

Proof. We have to show that any sequence $u_i \in M$, for which

(a) $\widetilde{J}(u_i) \to c > 0$,
(b) $\nabla_M \widetilde{J}(u_i) \to 0$,

has a converging subsequence. By Lemma 7.18(i) we infer that, up to a subsequence $u_i \rightharpoonup \overline{u}$. One has that

$$\nabla_M \widetilde{J}(u_i) = \widetilde{J}'(u_i) - \alpha_i G'(u_i), \qquad \text{where} \qquad \alpha_i = \frac{(\widetilde{J}'(u_i) \mid G'(u_i))}{\|G'(u_i)\|^2}.$$

Since $\|G'(u_i)\| \leq$ constant and $\|\widetilde{J}'(u_i)\| \geq k > 0$, see Lemma 7.18(ii), we deduce that there exists $\alpha^* > 0$ such that $|\alpha_i| \geq \alpha^*$ for $i \gg 1$. Then we can write, for i large,

$$G'(u_i) = \frac{1}{\alpha_i}[\widetilde{J}'(u_i) - \nabla_M \widetilde{J}(u_i)],$$

namely

$$2u_i = \Psi'(u_i) + \frac{1}{\alpha_i}[\widetilde{J}'(u_i) - \nabla_M \widetilde{J}(u_i)].$$

Since Ψ', Φ' are compact, $\nabla_M \widetilde{J}(u_i) \to 0$ (see (b) above) and $|\alpha_i| \geq \alpha^*$, we deduce that, up to a subsequence, u_i converges strongly. ∎

Proof of Theorem 7.14. Since $\widetilde{J}$ is bounded from below on M and satisfies the $(PS)_c$ condition for all $c > 0$, we can apply Theorem 7.12 yielding a $z \in M$ such that $\nabla_M \widetilde{J}(z) = 0$. By Lemma 7.16 such a z is a nontrivial solution of $(\mathbf{D})$. To find a positive solution, it suffices to substitute f with its positive part f^+. It is easy to check that all the preceding arguments can be carried out with small modifications, yielding a $z \in E$, $z \neq 0$, such that $-\Delta z = f^+(z)$. By the maximum principle, $z > 0$. ∎

Remark 7.20 Since $\widetilde{J}_{|M} = J_{|M}$, and $\widetilde{J}(z) = \min_M \widetilde{J}$, then $J(z) \leq J(u)$ for any $u \in M$. In particular, since all the possible nontrivial critical points of J belong to M, we infer that $J(z) \leq J(z')$, for all nontrivial critical points z' of J. Moreover, the solution z found above is not a local minimum of J. Actually,

$$J''(z)[z, z] = \|z\|^2 - \int_\Omega f'(z)z^2 \, \mathrm{d}x$$

$$= \int_\Omega [f(z)z - f'(z)z^2] \, \mathrm{d}x = -\int_\Omega z^3 h'(z) \, \mathrm{d}x < 0. \qquad ∎$$

For other results that use the Nehari manifold, see for example **??**.

7.7 Exercises

(i) Prove the result claimed in Remark 5.10.

(ii) (See Exercise 4.5(iv) in Chapter 4.) Consider the problem

$$\begin{cases} -\Delta u(x) = \lambda u - u^3 & x \in \Omega \\ u(x) = 0 & x \in \partial\Omega, \end{cases}$$

and prove that it has exactly three solutions for all $\lambda_1 < \lambda \leq \lambda_2$. [Hint: show that any nontrivial solution which does not change sign in Ω has Leray–Schauder index equal to 1.]

(iii) We use the notation introduced in Section 7.2. Suppose that J is bounded below on M and that (PS) holds. Let $\alpha(t, u)$ denote the steepest descent flow of J on M. Prove that $\lim_{t \to +\infty} \alpha(t, u)$ exists and is a critical point of J on M. Show an example in which this is not true when the (PS) fails.

(iv) Same notation as in the preceding exercise. In addition, suppose that J is even and M is symmetric, namely $u \in M \iff -u \in M$. Prove that $u \mapsto \alpha(t, u)$ is odd.

(v) More in general, suppose there is a group G acting on E through isometries such that $J(gu) = J(u)$ for all $g \in G$ and $u \in E$. Moreover, let

M be such that $gu \in M$ for all $g \in G$ and $u \in E$. Prove that the steepest descent flow α satisfies $\alpha(t, gu) = g\alpha(t, u)$ for all $g \in G$ and $u \in E$.

(vi) Completing Exercise 6-(i), prove that for $\lambda > \lambda_1$, J_λ has a minimum on the natural constraint M_λ. Compare this result with the one discussed in Example 5.10.

(vii) Let $E = W^{1,2}(\mathbb{R})$ and define $J : E \mapsto \mathbb{R}$ by setting $J(u) = \frac{1}{2} \int_\mathbb{R} (|u'|^2 + u^2)dx - \frac{1}{4} \int_\mathbb{R} u^4 dx$. Show that (PS) does not hold. [Hint: take, for example, the level $c = 4/3$.]

8

Saddle points and min-max methods

In this chapter we will discuss the existence of stationary points of a functional J on a Hilbert space E different from minima or maxima, which are found by means of appropriate min-max procedures. The results are particularly important for functionals that are not bounded from below, nor from above.

Notation 8.1 In this and in the subsequent Chapters 9 and 10, the Sobolev space $E = H_0^1(\Omega)$ will be endowed with the scalar product $(u \mid v) = \int_\Omega (\nabla u \cdot \nabla v) \, dx$ and norm $\|u\|^2 = \int_\Omega |\nabla u|^2 \, dx$.

8.1 The mountain pass theorem

We have seen in the preceding chapter that the Dirichlet BVP

$$\begin{cases} -\Delta u = f(x, u), & x \in \Omega \\ u = 0, & x \in \partial\Omega, \end{cases} \tag{D}$$

has a positive solution z provided $f \sim |u|^{p-1}u$, $1 < p < (n+2)/(n-2)$. We have already seen that this is the case when $f = |u|^{p-1}u$ (see Example 6.5(iii)) or when $f(u) = uh(u)$ and h satisfies suitable assumptions, including a convexity condition (see Theorem 7.14). A natural question is to establish whether a solution exists making *only* assumptions on the behaviour of $f(u)$ at $u = 0$ and at infinity. For this purpose, it is convenient to consider the corresponding functional

$$J(u) = \tfrac{1}{2} \|u\|^2 - \int_\Omega F(u) \, dx, \qquad F(u) = \int_0^u f(x, s) \, ds \qquad u \in E = H_0^1(\Omega).$$

The geometry of this functional is easily understood. If $f'(0) = 0$ then $(J''(0)v \mid v) = \|v\|^2$ and hence J has a proper local minimum at $u = 0$. Moreover, taking a fixed $\bar{u} \in E \setminus \{0\}$ and assuming that $F(u) \sim |u|^{p+1}$ ($1 < p < (n+2)/(n-2)$) as $|u| \to +\infty$, one readily finds that

$$\lim_{t \to +\infty} J(tu) = \lim_{t \to +\infty} \left[\frac{t^2}{2} \|u\|^2 - \int_\Omega F(t\bar{u}) \, dx \right] = -\infty.$$

116

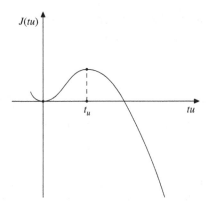

Figure 8.1 Behaviour of $J(tu)$, $t > 0$.

In particular, J is not bounded from below on E. Let us also remark that $\sup_E J = +\infty$. Actually, one can easily find a sequence u_i such that $\|u_i\| \to +\infty$ while $|\int_\Omega F(u_i)\,dx| \le$ constant, so that $J(u_i) \to +\infty$.

Let us make some more comments on the critical level $J(z)$. Consider the model case in which $f(u) = |u|^{p-1}u$, with $1 < p < (n+2)/(n-2)$. For each $u \in E \setminus \{0\}$ we have that

$$J(tu) = \frac{1}{2}t^2\|u\|^2 - \frac{1}{p+1}|t|^{p+1}\int_\Omega |u|^{p+1}\,dx.$$

Therefore, the real valued map $t \mapsto J(tu)$ achieves its maximum at a unique $t = t_u > 0$. Let us remark that this t_u is such that $t_u u \in M$, where $M = \{u \in E \setminus \{0\} : (J'(u) \mid u) = 0\}$ denotes the natural constraint introduced in Section 7.6. Since $J(z) = \min_M J$, see Remark 7.20, then

$$J(z) = \min_{u \in E \setminus \{0\}} \max_{t \in \mathbb{R}} J(tu).$$

The preceding discussion suggests a *min-max* procedure that will allow us to find critical points of a wide class of functionals J, which can be found at a level of *mountain pass*, namely the lowest level among all the paths leaving the well around a local minimum. In the sequel, to fix the notation, we will suppose that this local minimum arises at $u = 0$ and that $J(0) = 0$.

More precisely, we consider a class of functionals with the following geometric features:

(MP-1) $J \in C^1(E, \mathbb{R})$ $J(0) = 0$ and $\exists r, \rho > 0$ such that $J(u) \ge \rho$ for all $u \in S_r = \{u \in E : \|u\| = r\}$;

(MP-2) $\exists e \in E$ with $\|e\| > r$ such that $J(e) \le 0$.

It is worth pointing out that J might be unbounded from below (this actually is the case in most of the applications): what we require is only that it is *bounded below on S_r*.

Let Γ denote the class of all paths joining $u = 0$ and $u = e$,

$$\Gamma = \{\gamma \in C([0,1], E) : \gamma(0) = 0, \ \gamma(1) = e\}. \tag{8.1}$$

Of course, $\Gamma \neq \emptyset$ because $\gamma(t) = te$ belongs to Γ. We set

$$c = \inf_{\gamma \in \Gamma} \max_{t \in [0,1]} J(\gamma(t)). \tag{8.2}$$

Roughly, the level c is just the minimal level one has to reach to get out from the well around $u = 0$. Precisely, since any path $\gamma \in \Gamma$ crosses S_r then (MP-1) implies

$$c \geq \min_{u \in S_r} J(u) \geq \rho > 0. \tag{8.3}$$

However, elementary examples in finite dimension (see Remark 8.6 later on) show that the level c might not be critical for J because, roughly speaking, the points where the maxima in (8.2) are achieved are not necessarily bounded. To overcome this problem, we will use the *Palais–Smale* compactness condition *(PS)* introduced in Section 7.4.

Theorem 8.2 (Mountain pass) [21] *Suppose that $J \in C^1(E, \mathbb{R})$ satisfies (MP-1) and (MP-2). Let c be defined as in (8.2) and suppose that $(PS)_c$ holds. Then c is a positive critical level for J, namely there exists $z \in E$ such that $J(z) = c$ and $J'(z) = 0$. In particular, $z \neq 0$, e.*

The level c given by (8.3) is called the *mountain pass* (MP in short) critical level. Points on $Z_c = \{z \in E : J'(z) = 0, \ J(z) = c\}$ will be called *mountain pass* (MP in short) critical points.

Remark 8.3 For future reference, let us point out that the assumptions (MP-1) and (MP-2) can be substituted by

(MP-1) $J(0) = 0$ and $\exists r, \rho > 0$ such that $J(u) > 0$ for all $u \in B_r \setminus \{0\}$, and $J(u) \geq \rho$ for all $u \in S_r$;

(MP-2) $\exists e \in E$ such that $J(e) \leq 0$. ∎

For the proof we need a deformation lemma which is slightly different from Lemma 7.10. We denote by $J^a = \{u \in E : J(u) \leq a\}$, the sublevel of J on E.

Lemma 8.4 *Let $J \in C^1(E, \mathbb{R})$. Suppose $c \in \mathbb{R}$ is not a critical level of J, namely that $Z_c = \emptyset$, and that $(PS)_c$ holds.*

Then there exists $\delta > 0$, with $c - 2\delta > 0$, and a deformation η in E such that

(i) $\eta(J^{c+\delta}) \subset J^{c-\delta}$;
(ii) $\eta(u) = u$ for all u such that $J(u) \leq c - 2\delta$.

Proof. We will divide the proof into two parts. In the former we will assume that $J \in C^{1,1}(E, \mathbb{R})$. In the latter we will drop this restriction by introducing a *pseudogradient vector field.*

Step 1. The case $J \in C^{1,1}(E, \mathbb{R})$. Since $Z_c = \emptyset$ and $(PS)_c$ holds then Lemma 7.9 yields the existence of $\delta > 0$ such that

$$\|J'(u)\| \geq \delta, \quad \forall u \in E : |J(u) - c| \leq \delta.$$

Without loss of generality we can assume that $c - 2\delta > 0$ (recall that $c > 0$, see (8.3)). Let $b \in C^{0,1}(\mathbb{R}^+, \mathbb{R}^+)$ satisfy

$$b(\xi) = 1, \qquad \text{for } \xi \in [0, 1],$$

$$b(\xi) = t^{-1}, \qquad \text{for all } \xi \geq 1.$$

Moreover, we set

$$A = \{u \in E : c - \delta \leq J(u) \leq c + \delta\},$$
$$B = \{u \in E : J(u) \leq c - 2\delta\} \cup \{u : J(u) \geq c + 2\delta\},$$

and define $g \in C^{0,1}(E, \mathbb{R})$ by setting

$$g(u) = \frac{\text{dist}(u, B)}{\text{dist}(u, B) + \text{dist}(u, A)}.$$

Let us remark that $0 \leq g(u) \leq 1$ and

$$g(u) = \begin{cases} 0 & \forall u \in B \\ 1 & \forall u \in A. \end{cases}$$

Define $W(u) = -h(u)J'(u)$, where $h(u) = g(u)b(\|J'(u)\|) \geq 0$, and consider the steepest descent flow α, see Section 7.1. From the definition of g and b it follows that $\|W(u)\| \leq 1$ and hence we can use Lemma 7.3 to infer that $\alpha(t, u)$ is globally defined for all $t \geq 0$. Take $T = 2/\delta$ and set $\eta(u) = \alpha(T, u)$. Property (i) immediately follows as in Lemma 7.10. For the reader's convenience, let us repeat the outline of the argument. Let $u \in J^{c+\delta}$. If $J(\eta(u)) > c - \delta$, then $J(\alpha(t, u)) \in (c - \delta, c + \delta]$ and hence $h(\alpha(t, u)) = b(\|J'(\alpha(t, u))\|)$. Then

$$J(\eta(u)) - J(u) = -\int_0^T b(\|J'(\alpha(t, u))\|)\|J'(\alpha(t, u))\|^2 \, dt.$$

Since $\|J'(\alpha(t,u)\| \geq \delta$ and $b(\xi)\xi^2 \geq \delta^2$, we infer

$$J(\eta(u)) \leq J(u) - T\delta^2 \leq c - \delta.$$

Finally property (ii) follows directly because $g(u) \equiv 0$, and hence $W(u) \equiv 0$ on $\{u \in E : J(u) \leq c - 2\delta\}$. This completes the proof of step 1.

Step 2. The case $J \in C^1(E,\mathbb{R})$. In this case we cannot use directly the steepest descent flow of J'. Following R. Palais [141], we introduce the notion of pseudogradient vector field (PGVF in short). Let $E_0 = \{u \in E : J'(u) \neq 0\}$. A PGVF for J on E_0 is a map $X \in C^{1,1}(E_0, E)$ such that

$$\|X(u)\| \leq 2\|J'(u)\|, \qquad \forall\, u \in E_0, \tag{8.4}$$

$$(J'(u) \mid X(u)) \geq \|J'(u)\|^2, \qquad \forall\, u \in E_0. \tag{8.5}$$

It is possible to show that if $J \in C^1(E, \mathbb{R})$ then there exists a PGVF for J on E_0. The proof of this result is technical and is postponed to the end of this section. Let us show how we can use the PGVF to find a deformation η satisfying (i) and (ii). Let X be the PGVF on E_0 and set $\widetilde{W}(u) = -g(u)b(\|J'(u)\|)X(u)$. Since E_0 contains the set $E \setminus B$ and $g(u) \equiv 0$ for $u \in B$, we can assume that $\widetilde{W}$ is defined on E. Of course $\widetilde{W} \in C^{0,1}(E, E)$. Consider the flow α defined as the solution of the Cauchy problem $\alpha' = \widetilde{W}(\alpha)$, $\alpha(0) = u$. Since $\widetilde{W} \equiv 0$ on B, and (from (8.4))

$$\|\widetilde{W}(u)\| \leq b(\|J'(u)\|)\|X(u)\| \leq 2b(\|J'(u)\|)\|J'(u)\| \leq 2, \qquad \forall\, u \in E \setminus B,$$

we infer that $\alpha(t,u)$ is globally defined for all $t \geq 0$, as before. Moreover, one has that

$$\frac{\mathrm{d}}{\mathrm{d}t} J(\alpha(t,u)) = (J'(\alpha(t,u)) \mid \alpha'(t,u))$$

$$= -g(\alpha(t,u))b(\|J'(\alpha(t,u))\|)(J'(\alpha(t,u)) \mid X(\alpha(t,u))) \leq 0,$$

by (8.5). This shows that $t \mapsto J(\alpha(t,u))$ is nonincreasing. Finally, take once more $T = 2/\delta$ and set $\eta(u) = \alpha(T,u)$. As before, if $J(\eta(u)) > c - \delta$ one has that

$$J(\eta(u)) - J(u) = -\int_0^T b(\|J'(\alpha(t,u))\|)(J'(\alpha(t,u)) \mid X(\alpha(t,u)))\,\mathrm{d}t.$$

Using (8.5), we get

$$J(\eta(u)) - J(u) \leq -\int_0^T b(\|J'(\alpha(t,u))\|)\|J'(\alpha(t,u))\|^2\,\mathrm{d}t,$$

and the conclusion follows as in step 1. This completes the proof of the lemma. ∎

Remark 8.5 For future reference let us point out that the preceding proof provides a deformation η satisfying (i) for some $\delta > 0$ and

(ii') $\eta(u) = u$ for all u such that $J(u) \leq \beta$, for any $\beta < c$.

Of course, we will take δ such that $c - 2\delta > \beta$. ∎

Proof of Theorem 8.1. By contradiction, let c be a noncritical level and let η be the deformation found in the preceding lemma. According to the definition of c there exists $\overline{\gamma} \in \Gamma$ such that

$$\max_{t \in [0,1]} J(\overline{\gamma}(t)) \leq c + \delta.$$

We claim that

$(*)$ $\eta \circ \gamma \in \Gamma, \quad \forall \gamma \in \Gamma.$

Actually, one has that $\eta \circ \gamma \in C([0, 1], E)$. Moreover, since both $0, e \in J^0$ and $\eta(u) = u$ on J^0 (see (ii) above) then

- $\eta \circ \gamma(0) = \eta(0) = 0$,
- $\eta \circ \gamma(1) = \eta(e) = e$.

This proves $(*)$. On the other hand, from (i) it follows that

$$\max_{t \in [0,1]} J(\eta \circ \overline{\gamma}(t)) \leq c - \delta.$$

This is in contradiction with the fact that c is the infimum on Γ of $\max_{t \in [0,1]} J(\gamma(t))$. The proof of the theorem is complete. ∎

Remark 8.6 The following elementary example (due to Brezis and Nirenberg) shows that the geometric assumptions (MP-1) and (MP-2) alone (i.e. without any compactness condition) do not suffice for the existence of a MP critical point. Consider $E = \mathbb{R}^2$ and $J(x, y) = x^2 + (1 - x)^3 y^2$. The graph of the level curves of J is reported in Figure 8.2.
A straight calculation shows that (MP-1) holds with $r = \frac{1}{2}$ and $\rho = \frac{1}{32}$; moreover, taking $e = (2, 2)$ one has that $J(e) = 0$ and hence (MP-2) holds. But the only critical point of J is $(0, 0)$. See also Exercise 8.5(iv). ∎

In order to apply Theorem 8.2 one has to check that J satisfies

- a geometric property, namely the assumptions (MP-1) and (MP-2),
- the (PS) condition at level c.

Fortunately, both are quite general conditions and so the MP theorem applies to a large variety of equations. Actually, many variational problems fit in the abstract frame of the MP theorem.

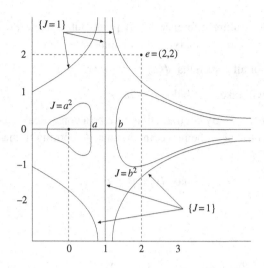

Figure 8.2 Level curves of the function $J(x, y) = x^2 + (1 - x)^3 y^2$.

Remark 8.7

(a) It is clear that in order to prove the mountain pass theorem (as well as Theorem 7.12) it suffices to show that if $u_j \in E$ is such that $J(u_j) \to c$ and $J'(u_j) \to 0$, then $Z_c \neq \emptyset$.

(b) A different proof of the MP theorem and some generalizations, using the Ekeland ε-variational principle, can be found in [93, Chapter IV, Section 1].   ∎

8.1.1 Existence of pseudogradients

Here we prove the existence of a PGVF for J on E_0. Given $u \in E_0$, from $J'(u) \neq 0$ and the definition of $\|J'(u)\| = \sup_{\|v\|=1} (J'(u) \mid v)$, it follows that there exists $w = w(u) \in E$ such that

$$\|w\| = 1, \qquad (J'(u) \mid w) > \tfrac{2}{3}\|J'(u)\|. \tag{8.6}$$

Set $\widetilde{X}(u) = \tfrac{3}{2}\|J'(u)\| w(u)$. Then, using again the fact that $u \in E_0$, one finds

$$\|\widetilde{X}(u)\| = \tfrac{3}{2}\|J'(u)\| < 2\|J'(u)\|.$$

Moreover, using (8.6),

$$(J'(u) \mid \widetilde{X}(u)) = \tfrac{3}{2}\|J'(u)\| (J'(u) \mid w) > \|J'(u)\|.$$

Since J' is continuous, there exists a ball $B_r(u)$ with radius $r = r(u) > 0$

such that

$$\|\widetilde{X}(u)\| < 2\|J'(u)\|, \qquad \forall\, z \in B_r(u), \tag{8.7}$$

$$(J'(u) \mid \widetilde{X}(u)) > \|J'(u)\|, \qquad \forall\, z \in B_r(u). \tag{8.8}$$

Of course, $\cup_{u \in E_0} B_r(u) \supseteq E_0$ and we can take a locally finite covering $U_i := B_{r_i}(u_i)$. For $u \in E_0$ we set $d_i(u) = \mathrm{dist}(u, E \setminus U_i)$, $\widetilde{X}_i = \widetilde{X}(u_i)$ and

$$X(u) = \sum_i \frac{d_i(u)}{\sum_j d_j(u)} \widetilde{X}_i.$$

One has that X is of class $C^{1,1}$. Moreover, using (8.7) and (8.8), a direct calculation shows that X verifies (8.4) and (8.5) and hence it is a PGVF for J on E_0.

8.2 Applications

The MP theorem is well suited to semilinear elliptic boundary value problems. We will show below a couple of such examples.

Our first application of the MP theorem is to find *positive solutions* of the following superlinear BVP

$$\begin{cases} -\Delta u = f(x, u) & x \in \Omega \\ u = 0 & x \in \partial\Omega, \end{cases} \tag{D}$$

where $f : \Omega \times \mathbb{R} \to \mathbb{R}$ behaves like $\lambda u + u^p$ as $u \searrow 0$ and as $u \to +\infty$. Since we are interested in positive solutions of (D), we can assume without loss of generality that $f(x, u) \equiv 0$ for all $u < 0$. We will suppose that $f \in \mathbb{F}_p$ with $1 < p < (n+2)/(n-2)$, and that:

$$\lim_{u \to 0^+} f(x, u)/u = \lambda > 0, \qquad \text{for a.e. } x \in \Omega; \tag{8.9}$$

$$\exists\, r > 0,\ \theta \in (0, \tfrac{1}{2}) \ : \ 0 < F(x, u) \le \theta u f(x, u) \quad \forall x \in \Omega,\ u \ge r. \tag{8.10}$$

Here, as usual, $F(x, u) = \int_0^u f(x, s)\, \mathrm{d}s$.

Remark 8.8 If $f(u) = u^p$, or else if $f(u) = u^{p_1} + \cdots + u^{p_k}$ with $1 < p_i < (n+2)/(n-2)$ for all $i = 1, \ldots, k$, condition (8.10) is obviously satisfied with $\theta = 1/(p+1)$, respectively $\theta = 1/(\min p_i + 1)$. In general, integrating (8.10) it follows that

$$\exists\, c > 0 \text{ such that } F(u) \ge c \cdot u^{1/\theta} \quad \forall\, u \ge r. \tag{8.11}$$

In this sense we say that f is *superlinear*. ∎

As usual, we take $E = H_0^1(\Omega)$ and set $\Phi(u) = \int_\Omega F(x, u)\, dx$. With this notation, solutions of (D) are the critical points of the functional

$$J(u) = \frac{1}{2}\|u\|^2 - \Phi(u).$$

We first prove that J has the MP geometry.

Lemma 8.9 *If* (8.9) *holds with* $\lambda < \lambda_1$, *then* J *satisfies* (MP-1).

Proof. Fix $\varepsilon = \frac{1}{2}(\lambda_1 - \lambda)$. Using (8.9) and the fact that $f \in \mathbb{F}_p$, there exists $A \in L^2(\Omega)$ such that $|F(x, u)| \leq \frac{1}{2}(\lambda+\varepsilon)\, u^2 + A|u|^{p+1}$. From this it follows that

$$\Phi(u) \leq \tfrac{1}{2}(\lambda + \varepsilon) \int_\Omega |u|^2\, dx + \int_\Omega A|u|^{p+1}\, dx.$$

Take r and $s = r/(r-1)$ such that $r \geq 2$ and $s(p+1) \leq 2^*$. Then the Hölder inequality implies

$$\int_\Omega A|u|^{p+1}\, dx \leq \left(\int_\Omega A^r\, dx\right)^{1/r} \left(\int_\Omega |u|^{s(p+1)}\, dx\right)^{1/s}$$

$$\leq c_1 \left(\int_\Omega |u|^{s(p+1)}\, dx\right)^{1/s}.$$

Using this, the Poincaré inequality and the Sobolev embedding theorem we find

$$|\Phi(u)| \leq \frac{\lambda + \varepsilon}{2\lambda_1}\|u\|^2 + c_2\|u\|^{p+1},$$

which implies

$$J(u) \geq \left(\frac{1}{2} - \frac{\lambda + \varepsilon}{2\lambda_1}\right)\|u\|^2 - c_2\|u\|^{p+1} = \frac{\lambda_1 - \lambda}{4\lambda_1}\|u\|^2 - c_2\|u\|^{p+1}.$$

This immediately shows that (MP-1) holds provided $\lambda < \lambda_1$. ∎

Lemma 8.10 *Under the above assumptions,* J *satisfies* (MP-2).

Proof. Fix any $e \in E$ with $e(x) > 0$ in Ω. From (8.11) it follows that

$$J(te) = \tfrac{1}{2} t^2 \|e\|^2 - \int_\Omega F(x, te)\, dx$$

$$\leq \tfrac{1}{2} t^2 \|e\|^2 - a_3\, t^{1/\theta} \int_\Omega |e|^{1/\theta} - a_4.$$

Since $1/\theta > 2$ we infer that $\lim_{t\to+\infty} J(te) = -\infty$. ∎

It remains to show that $(PS)_c$ holds (actually, we will show that, more in general, (PS) holds). For this, we first prove that

(∗) *every (PS) sequence is bounded.*

Let u_n be a (PS) sequence. From (8.10) we deduce

$$\Phi(u_n) = \int_{u(x) \leq r} F(x, u) \, dx + \int_{u(x) \geq r} F(x, u) \, dx$$

$$\leq a_5 + \theta \int_{u(x) \geq r} f(x, u) u \, dx$$

$$\leq a_6 + \theta \int_{\Omega} f(x, u) u \, dx. \tag{8.12}$$

Since $J(u_n) \leq a_7$ it follows that

$$\|u_n\|^2 \leq 2a_7 + 2\Phi(u_n)$$

and hence (8.12) implies

$$\|u_n\|^2 \leq a_8 + 2\theta \int_{\Omega} f(x, u) u \, dx. \tag{8.13}$$

From

$$\left| \|u\|^2 - \int_{\Omega} f(x, u) u \, dx \right| = |(J'(u_n) \mid u_n)| \leq \|J'(u_n)\| \|u_n\|$$

it follows that

$$\int_{\Omega} f(x, u) u \, dx \leq \|u\|^2 + \|J'(u_n)\| \|u_n\|. \tag{8.14}$$

Combining (8.13) and (8.14) we have

$$\|u_n\|^2 \leq a_8 + 2\theta \|u\|^2 + 2\theta \|J'(u_n)\| \|u_n\|.$$

Since $2\theta < 1$ this implies $\|u_n\| \leq$ constant, proving (∗). It is now standard to prove that (PS) holds: up to subsequences, $u_n \rightharpoonup u$ as well as $\Phi'(u_n) \to \Phi'(u)$ and from $J'(u_n) = u_n - \Phi'(u_n)$ it follows that $u_n = J'(u_n) + \Phi'(u_n)$ converges strongly to $\Phi'(u)$.

From the MP Theorem it follows that (D) has a non-trivial solution u. Since $f(u) \equiv 0$ for $u < 0$ it follows that $u \geq 0$ in Ω. Finally, the assumptions made on f allow us to find $\delta > 0$ such that $f(u) + \delta u > 0$ for all $u > 0$. Then the maximum principle applied to $-\Delta u + \delta u = f(u) + \delta u$ implies that $u > 0$ in Ω.

In conclusion we have the following theorem.

Theorem 8.11 *Let $f \in \mathbb{F}_p$ with $1 < p < (n + 2)/(n - 2)$, satisfy (8.9) with $\lambda < \lambda_1$ and (8.10). Then (D) has a positive solution.*

Remark 8.12 The preceding proof highlights why the *(PS)* condition is quite appropriate for applications to semilinear BVP in the case in which there is no a priori bound. The reason is that in the *(PS)* condition we do not limit ourselves to consider sequences such that $J'(u_n) \to 0$, but we also require that $J(u_n)$ is bounded. It is just the combination of the two properties that allows us to show that the sequence u_n is bounded. For example, $J(u) = \frac{1}{2}\|u_n\|^2 - \int_\Omega |u|^4 \, dx$, $u \in H_0^1(\Omega)$, has an unbounded sequence of critical points u_m, see Theorem 10.23 and Remark 10.24, such that $J(u_m) \to +\infty$. On the other hand, the *(PS)* holds, as we have seen before. ∎

Remark 8.13 As we will see in Remark A1.11 one can give another proof of Theorem 8.11 (under slightly different assumptions on f) by using bifurcation arguments jointly with some a priori estimates for positive solutions. ∎

Let us recall that in [2] the existence of positive solutions of BVP such as

$$\begin{cases} -\Delta u = \lambda u + b(x)f(u) & \text{in } \Omega \subset \mathbb{R}^n \\ u = 0 & \text{on } \partial\Omega, \end{cases} \tag{8.15}$$

when b has a nontrivial positive and a negative part, has been studied. Roughly, suppose that

$$\begin{cases} f(u) \sim |u|^{q-1}u & \text{as } u \to 0, \quad 1 < q < \dfrac{n+2}{n-2} \\ f(u) \sim |u|^{p-1}u & \text{as } |u| \to \infty, \quad 1 < p < \dfrac{n+2}{n-2} \\ |f(u)u - pF(u)| \le c_1|u|^2 + c_2 & \text{where } F'(u) = f(u), \end{cases}$$

and

$$\int_\Omega b(x)\varphi_1^{q-1}(x) \, dx < 0, \tag{8.16}$$

where φ_1 denotes the positive, normalized eigenfunction associated to λ_1. Then there exists $\Lambda > \lambda_1$ such that (see Figure 8.3)

(a) for all $\lambda \in (\lambda_1, \Lambda)$, (8.15) has at least two positive solutions;
(b) for $\lambda \le \lambda_1$ and $\lambda = \Lambda$, (8.15) has at least a positive solution;
(c) for $\lambda \ge \Lambda$, (8.15) has no positive solution.

The reader should notice that condition (8.16) implies that the branch $\mathcal{C}$ bifurcating from λ_1 is supercritical (see Remark 2.9 (b)).

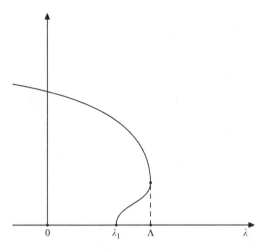

Figure 8.3 Bifurcation diagram in the case of problem (8.15).

We conclude this section by discussing an application to the following model problem:

$$\begin{cases} -\Delta u = \mu \left[u^s - u^p \right] & x \in \Omega \\ u > 0 & x \in \Omega \\ u = 0 & x \in \partial\Omega, \end{cases} \tag{8.17}$$

with $1 < s < p$. As in Example 5.11, we consider

$$\bar{f}(x, u) = \begin{cases} 0 & \text{if } u \le 0 \\ u^s - u^p & \text{if } 0 < u < 1 \\ 0 & \text{if } u \ge 1. \end{cases} \tag{8.18}$$

By the maximum principle, it follows that solutions of

$$\begin{cases} -\Delta u = \mu \bar{f}(x, u) & x \in \Omega \\ u = 0 & x \in \partial\Omega, \end{cases} \tag{8.19}$$

satisfy $0 \le u(x) \le 1$ and hence solve (8.17). Solutions of (8.17) are critical points of

$$J_\mu(u) = \tfrac{1}{2} \|u\|^2 - \mu \int_\Omega \bar{F}(x, u) \, dx, \qquad u \in E = H_0^1(\Omega), \tag{8.20}$$

where $\bar{F}(x, u) = \int_0^u \bar{f}(x, s) \, ds$. Since $\bar{f}(u)/u \to 0$ as $u \to 0+$, it follows that J_μ satisfies (MP-1). It is also easy to check that J_μ is bounded from below and coercive and the latter property readily implies that (PS) holds. Moreover, fixing any smooth $e \in E$, $e(x) > 0$, from $te(x) < 1$ for $t \ll 1$,

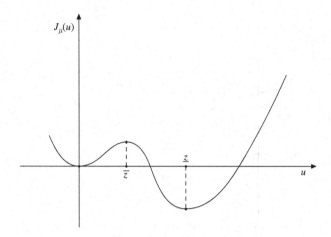

Figure 8.4 Behaviour of J_μ for $\mu > \underline{\mu}$.

we get

$$J_\mu(t\,e) = \frac{1}{2}t^2\|e\|^2 - \mu\left[\frac{1}{s+1}t^{s+1}\|e\|_{L^{s+1}}^{s+1} - \frac{1}{p+1}t^{p+1}\|e\|_{L^{p+1}}^{p+1}\right].$$

Since $(1 <)\,s < p$ we infer that there exists $t^* > 0$ sufficiently small and $\underline{\mu} > 0$ such that $J_\mu(t^*\,e) < 0$ for all $\mu > \underline{\mu}$. As a first consequence, we infer that $\min J_\mu < 0$ for $\mu > \underline{\mu}$ and hence a first solution $\underline{z} > 0$ with $J_\mu(\underline{z}) < 0$ can be found by taking the minimum of J_μ on E. Moreover, for $\mu > \underline{\mu}$ we can also apply the mountain pass theorem yielding a second positive solution $\overline{z}$ such that $J_\mu(\overline{z}) > 0$.

The preceding arguments can be carried out in greater generality to prove the following.

Theorem 8.14 *Suppose that $f : \Omega \times \mathbb{R}^+ \mapsto \mathbb{R}$ is locally Hölder continuous, $f(x,0) = 0$, and that f satisfies:*

(a_1) $\lim_{u \to 0+} f(x,u)/u = 0$, *uniformly with respect to $x \in \Omega$,*
(a_2) $f(x,u) > 0$ *for $u > 0$ in a deleted neighbourhood of $u = 0$,*
(a_3) *there exists $\overline{u} > 0$ such that $f(x,\overline{u}) < 0$ for all $x \in \overline{\Omega}$.*

Then there exists $\underline{\mu} > 0$ such that

$$\begin{cases} -\Delta u = \mu f(x,u) & x \in \Omega \\ u = 0 & x \in \partial\Omega, \end{cases} \tag{8.21}$$

has a pair of positive solutions $\underline{z}, \overline{z}$.

The details are left to the reader.

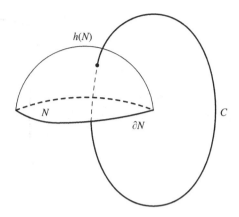

Figure 8.5

8.3 Linking theorems

The MP theorem turns out to be a particular case of a more general result. Roughly, suppose that $J \in C^1(E, \mathbb{R})$ and let $\mathcal{C}$ be a class of subsets of $A \subset E$ such that

- $\mathcal{C} \neq \emptyset$,
- $c = \inf_{A \in \mathcal{C}} \sup_{u \in A} J(u)$ is finite,
- if η denotes a deformation obtained through the steepest descent flow of J (or a PGVF of J), then $\eta(A) \in \mathcal{C}$ for all $A \in \mathcal{C}$.

Then c is a critical level of J, provided $(PS)_c$ holds.

To construct such a class, let us give the following definition. Let

- N be a manifold with boundary ∂N,
- C be a subset of E,
- $\mathcal{H} = \{h \in C(N, E) : h(u) = u, \ \forall u \in \partial N\}$.

Definition 8.15 *We say that ∂N and C link if*

$$C \cap h(N) \neq \emptyset, \ \forall h \in \mathcal{H}.$$

In other words, ∂N and C link if C meets every continuous surface spanned by ∂N, see Figure 8.5. To illustrate the preceding definition some examples are in order.

Example 8.16 The case covered by the MP theorem fits in this setting. Actually, let C be the sphere $S_r = \{u \in E : \|u\| = r\}$ and let $\partial N = \{0, e\}$, with $e \in E$.

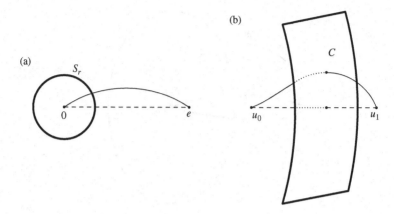

Figure 8.6

Here the manifold N can be taken as the segment $[0, e] = \{u = te \; : \; 0 \leq t \leq 1\}$ and $\mathcal{H} = \{h \in C([0, e], E) \; : \; h(0) = 0, h(e) = e\}$ is nothing but the class of paths Γ introduced in the preceding section. It is clear that ∂N and C link whenever $\|e\| > r$. See Figure 8.6(a). ∎

Example 8.17 More in general, let C be a manifold of codimension one in E and suppose that u_0, u_1 are points of $E \setminus C$ belonging to two distinct connected components of $E \setminus C$. Taking $\partial N = \{u_0, u_1\}$, $N = [u_0, u_1]$ and letting $\mathcal{H}$ be the class of paths joining u_0 and u_1, one has that ∂N and C link. See Figure 8.6(b). ∎

Example 8.18 Let $E = V \oplus W$, where V, W are orthogonal closed subspaces and $\dim(V) = k < +\infty$. Given $e \in W$ and $R > 0$ we consider the $k + 1$ dimensional manifold with boundary

$$N = \{u = v + se \; : \; v \in V, \|v\| \leq R, \; s \in [0, 1]\}.$$

Let $C = \{w \in W \; : \; \|w\| = r\}$ be the sphere of radius r on W, see Figure 8.7. We claim that ∂N and C link provided $\|e\| > r$.

To prove this fact we will use the topological degree. We have to show that, for every map $h \in C(N, E)$ such that $h(u) = u$ for every $u \in \partial N$, there exists $u^* \in N$ such that $\|h(u^*)\| = r$. Let us identify N with the set $\{u = (v, s) \in V \times [0, 1] \; : \; \|v\| \leq R\}$ and let $P : N \mapsto V$ denote the projection onto V. Consider the map $\widetilde{h} \in C(N, V \times \mathbb{R})$ defined by setting

$$\widetilde{h}(v, s) = (Ph(u), \|h(u) - Ph(u)\| - r).$$

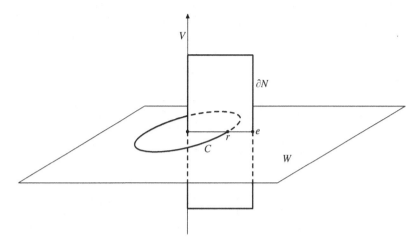

Figure 8.7

Since for $(v, s) \in \partial N$ we have that $h(u) = u$ and $Ph(u) = v$, then

$$\widetilde{h}(v, s) = (v, s\|e\| - r), \qquad \forall \, (v, s) \in \partial N.$$

Since $\|e\| > r$, it follows that $\widetilde{h}(v, s) \neq (0, 0)$, $\forall \, (v, s) \in \partial N$. In particular, it makes sense to consider the topological degree $\deg(\widetilde{h}, N_0, \mathbf{0})$, where N_0 is the interior of N and $\mathbf{0} := (0, 0)$ is the origin in $V \times \mathbb{R}$. Moreover, by the properties of the degree, see Theorem 3.2, $\deg(\widetilde{h}, N_0, \mathbf{0})$ is equal to the degree of any continuous $g \in C(N, V \times \mathbb{R})$ which coincides with $\widetilde{h}$ on ∂N. Taking, for example, $g(v, s) = (v, s\|e\| - r)$ and using again the fact that $\|e\| > r$, we infer that $\deg(\widetilde{h}, N_0, \mathbf{0}) = \deg(g, N, \mathbf{0}) = 1$. By the solution property of the degree, there exists $u^* \in N$ such that $\widetilde{h}(u^*) = (0, 0)$, namely

$$\begin{cases} Ph(u^*) = 0, \\ \|h(u^*) - Ph(u^*)\| = r. \end{cases}$$

Thus one finds that $\|h(u^*)\| = r$, as required. ∎

Example 8.19 As before suppose that $E = V \oplus W$, where V, W are closed subspaces and $\dim(V) = k < +\infty$. Let $C = W$ and $\partial N = \{v \in V : \|v\| = r\}$, see Figure 8.8. Here N is the ball of radius r in V and the sets $h(N)$, $h \in \mathcal{H}$, are the k dimensional surfaces spanned by ∂N, the sphere of radius r in V. Using arguments similar to those of Example 8.18, one can show that ∂N and C link. The details are left to the reader. ∎

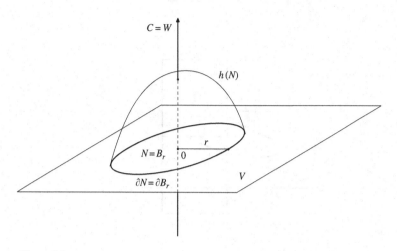

Figure 8.8

8.3.1 Linking and critical points

Let $J \in C^1(E, \mathbb{R})$ and let N, C be two subsets of E such that ∂N and C link. We will assume that

(L-1) J is bounded from below on C

$$\rho := \inf_{u \in C} J(u) > -\infty,$$

(L-2) $\rho > \beta := \sup_{u \in \partial N} J(u)$.

Remark 8.20 In the case discussed in Example 8.16 related to the MP theorem, (L-1) and (L-2) are nothing but the assumptions (MP-1) and (MP-2) of the preceding section (where $\sup_{\partial N} J(u) = 0$). ■

Let us set

$$c = \inf_{h \in \mathcal{H}} \sup_{u \in N} J(h(u)). \tag{8.22}$$

The level c is called the *linking* level of J (corresponding to N and C).

Lemma 8.21 *If ∂N and C link and (L-1) holds then $c \geq \rho$.*

Proof. By the definition of linking one has that $C \cap h(N) \neq \emptyset$ for all $h \in \mathcal{H}$. Then it follows that

$$\sup_{u \in N} J(h(u)) \geq \inf_{u \in C \cap h(N)} J(u) \geq \inf_{u \in C} J(u) = \rho.$$

Since this is true for any $h \in \mathcal{H}$, the lemma follows. ■

Theorem 8.22 *Let ∂N and C link and let $J \in C^1(E, \mathbb{R})$ satisfy* (L-1) *and* (L-2). *Furthermore, letting c be defined as in* (8.22), *suppose that* $(PS)_c$ *holds. Then $c \geq \rho$ is a critical level of J.*

Proof. First of all, let us remark that from Lemma 8.21 and (L-2) it follows that $c \geq \rho > \beta$. Suppose by contradiction that $Z_c = \emptyset$. We can use Lemma 8.4 and Remark 8.5 to get a deformation η satisfying (i) and (ii$'$) therein, namely

(i) $\eta(J^{c+\delta}) \subset J^{c-\delta}$,
(ii$'$) $\eta(u) = u$ for all u such that $J(u) \leq \beta$.

In particular from the latter property it follows that $\eta \circ h \in \mathcal{H}$ for all $h \in \mathcal{H}$. Actually, if $u \in \partial N$ one has that $J(u) \leq \beta$ and hence (ii$'$) implies $\eta \circ h(u) = u$. Let $h \in \mathcal{H}$ be such that

$$\sup_{u \in N} J(h(u)) \leq c + \delta.$$

From (i) we infer that

$$\sup_{u \in N} J(\eta \circ h(u)) \leq c - \delta.$$

Since $\eta \circ h \in \mathcal{H}$, this is in contradiction with the definition of the linking level c and the proof is complete. ∎

The preceding theorem is a general result that contains as particular cases several others, including the MP theorem (for the latter, see Example 8.16 and Remark 8.20).

Below, we state some specific existence results following Examples 8.17–8.19. The level c is the linking level.

Theorem 8.23 *Let C be a manifold of codimension one in E and suppose that u_0, u_1 are points of $E \setminus C$ belonging to two distinct connected components of $E \setminus C$. Let $J \in C^1(E, \mathbb{R})$ satisfy*

(L-3) $\inf_C J(u) > \max\{J(u_0), J(u_1)\}$

and $(PS)_c$.
Then J has a critical point z at level c. In particular $z \neq u_0, u_1$.

The following result is due to P. Rabinowitz [150, 151].

Theorem 8.24 *Let $E = V \oplus W$, where V, W are closed subspaces and $\dim(V) = k < +\infty$. Suppose $J \in C^1(E, \mathbb{R})$ satisfies*

(L-4) *there exist $r, \rho > 0$ such that*

$$J(w) \geq \rho, \qquad \forall w \in W, \qquad \|w\| = r,$$

(L-5) *there exist $R > 0$ and $e \in W$, with $\|e\| > r$ such that, letting $N = \{u = v + te : v \in V, \|v\| \leq R, t \in [0, 1]\}$, one has that*

$$J(u) < 0, \quad \forall u \in \partial N.$$

Moreover, suppose that $(PS)_c$ holds.
Then J has a critical point z at level $c > 0$. In particular $z \neq 0$.

Remark 8.25 The two preceding results improve the MP theorem. If in the former we take $C = S_r$, $u_0 = 0$ and $u_1 = e$ we just obtain the MP theorem. The latter corresponds to the case that $u = 0$ is not a local minimum but a saddle point with finite Morse index. If $V = \{0\}$ we find the MP theorem. ∎

Remark 8.26 A result like Theorem 8.24, in which V can be infinite dimensional, has been proved by V. Benci and P. Rabinowitz [52]. For further related results, see for example [48, 51]. ∎

Theorem 8.27 *Let $E = V \oplus W$, where V, W are closed subspaces and $0 < \dim(V) = k < +\infty$. Suppose $J \in C^1(E, \mathbb{R})$ satisfies*

(L-6) *there exist $\rho > 0$ such that*

$$J(w) \geq \rho, \qquad \forall w \in W,$$

(L-7) *there exist $r > 0, \beta < \rho$ such that*

$$J(v) \leq \beta, \quad \forall u \in V, \qquad \|v\| = r.$$

Moreover, suppose that $(PS)_c$ holds.
Then J has a critical point z at level c.

As a straight application of Theorem 8.24 we will improve Theorem 8.11. The following result is essentially due to Ahmad, Lazer and Paul [1].

Theorem 8.28 *Let $f \in \mathbb{F}_p$ with $1 < p < (n + 2)/(n - 2)$, and suppose f satisfies (8.9) for any $\lambda \in \mathbb{R}$, and (8.10). Then (D) has a nontrivial solution.*

Remark 8.29 The new feature of the preceding result is that we do not require $\lambda < \lambda_1$. On the other hand, we will no longer find a positive solution. ∎

Proof of Theorem 8.28. For simplicity, we will carry out the proof in the case of the model problem

$$
\begin{cases}
-\Delta u = \lambda u + |u|^{p-1}u & x \in \Omega \\
u = 0 & x \in \partial\Omega.
\end{cases}
\tag{8.23}
$$

The general case requires only minor changes.

Using the same notation as in Section 8.2, let $E = H_0^1(\Omega)$ and

$$
J(u) = \frac{1}{2}\|u\|^2 - \frac{1}{2}\lambda \int_\Omega u^2\, dx - \frac{1}{p+1}\int_\Omega |u|^{p+1}\, dx.
$$

If $\lambda < \lambda_1$ we have seen in Section 8.2 that J can be handled by the MP theorem. Now we want to show that if $\lambda_k \le \lambda < \lambda_{k+1}$ we can apply Theorem 8.24 with $V = \text{span}\{\varphi_1, \ldots, \varphi_k\}$ and $W = V^\perp$, the L^2 complement of V. Indeed, if $w = \sum_{i=k+1}^{\infty} a_i \varphi_i \in W$ and $\|w\| \to 0$, one finds:

$$
J(w) = \frac{1}{2}\sum_{i=k+1}^{\infty} a_i^2\left(1 - \frac{\lambda}{\lambda_i}\right) + o(\|w\|^2) \ge \frac{1}{2}\left(1 - \frac{\lambda}{\lambda_{k+1}}\right)\|w\|^2 + o(\|w\|^2)
$$

and this shows that (L-4) holds. To prove (L-5) we first take any finite dimensional subspace $\widetilde{V}$ of E. For $\widetilde{v} \in \widetilde{V}$, $\|\widetilde{v}\| = 1$ one has

$$
J(R\widetilde{v}) = \frac{1}{2}R^2 - \frac{1}{2}\lambda^2 R^2 |\widetilde{v}|_{L^2}^2 - \frac{1}{p+1}R^{p+1}|\widetilde{v}|_{L^{p+1}}^{p+1}.
$$

Since $p > 1$ and $\widetilde{V}$ is finite dimensional it follows that there exists $R > 0$ such that $J(R\widetilde{v}) < 0$, for all $\widetilde{v} \in \widetilde{V}$ with $\|\widetilde{v}\| = 1$. In particular, we can find $R > r$ and $e \in W$, with $\|e\| = R$, such that $J(v + te) < 0$ provided $\|v + te\| \ge R$. Then on the three sides of ∂N given by $\{u = v + te : \|v\| = R\} \cup \{u = v + Re\}$ one has that $J(u) < 0$. On the other hand for $v = \sum_1^k a_i \varphi_i \in V$ we also have $|v|_{L^2}^2 = \sum_1^k \lambda_i^{-1}a_i^2 \ge \lambda_k^{-1}\|v\|^2$ and thus

$$
J(v) \le \frac{1}{2}\|v\|^2 - \frac{1}{2}\lambda |v|_{L^2}^2 \le \frac{1}{2}\left(1 - \frac{\lambda}{\lambda_k}\right)\|v\|^2 \le 0.
$$

This shows that (L-5) holds. The proof of $(PS)_c$ is the same as in the application discussed in Section 8.2. In conclusion we can apply Theorem 8.24 obtaining a solution of (8.23) for all $\lambda \in \mathbb{R}$. ∎

8.4 The Pohozaev identity

This last section is devoted to proving a celebrated identity due to S. Pohozaev [143], see also a remarkable extension due to P. Pucci and J. Serrin [146], dealing with solutions of nonlinear Dirichlet boundary value problems like

$$\begin{cases} -\Delta u = f(u) & x \in \Omega \\ u = 0 & x \in \partial\Omega. \end{cases} \tag{D}$$

Let $F(u) = \int_0^u f(s)\, ds$.

Theorem 8.30 *Let Ω be a bounded domain in $\mathbb{R}^n$ and let v denote the unit outer normal at $\partial\Omega$. If u is any classical solution ($u \in H^2(\Omega) \cap H_0^1(\Omega)$ would suffice see [54]) of* (**D**) *then the following identity holds* ˙

$$n \int_\Omega F(u)\, dx - \frac{n-2}{2} \int_\Omega uf(u)\, dx = \frac{1}{2} \int_{\partial\Omega} u_v^2 (x \cdot v)\, d\sigma, \tag{P}$$

where $u_v = \partial u / \partial v$.

Proof. Setting $\Theta(x) = (x \cdot \nabla u(x))\nabla u(x)$, one has

$$\text{div } \Theta = \Delta u(x \cdot \nabla u) + \sum_k \frac{\partial u}{\partial x_k} \frac{\partial}{\partial x_k} \left(\sum_i x_i \frac{\partial u}{\partial x_i} \right)$$

$$= \Delta u(x \cdot \nabla u) + \sum_k \left(\frac{\partial u}{\partial x_k} \right)^2 + \sum_{i,k} \frac{\partial u}{\partial x_k} x_i \frac{\partial^2 u}{\partial x_i \partial x_k}$$

$$= \Delta u(x \cdot \nabla u) + |\nabla u|^2 + \frac{1}{2} \sum_i x_i \frac{\partial}{\partial x_i} |\nabla u|^2.$$

Then an application of the divergence theorem yields

$$\int_\Omega \Delta u(x \cdot \nabla u)\, dx + \int_\Omega |\nabla u|^2\, dx + \frac{1}{2} \int_\Omega \sum_i x_i \frac{\partial}{\partial x_i} |\nabla u|^2\, dx$$

$$= \int_{\partial\Omega} (x \cdot \nabla u)(\nabla u \cdot v)\, d\sigma.$$

As for the boundary term, since $u = 0$ on $\partial\Omega$ one has that $\nabla u(x) = u_v v$ and thus the preceding equation becomes

$$\int_\Omega \Delta u(x \cdot \nabla u)\, dx + \int_\Omega |\nabla u|^2\, dx + \frac{1}{2} \int_\Omega \sum_i x_i \frac{\partial}{\partial x_i} |\nabla u|^2\, dx = \int_{\partial\Omega} (x \cdot v) u_v^2\, d\sigma.$$

$$\tag{8.24}$$

Next, let $\Theta_1(x) = \frac{1}{2} |\nabla u|^2 x$. Since

$$\text{div } \Theta_1 = \frac{n}{2} |\nabla u|^2 + \frac{1}{2} \sum_i x_i \frac{\partial}{\partial x_i} |\nabla u|^2,$$

another application of the divergence theorem yields

$$\frac{n}{2}\int_\Omega |\nabla u|^2\,dx + \frac{1}{2}\int_\Omega \sum_i x_i \frac{\partial}{\partial x_i}|\nabla u|^2\,dx = \frac{1}{2}\int_{\partial\Omega}(x\cdot v)u_v^2\,d\sigma. \quad (8.25)$$

Substituting (8.25) into (8.24) we find

$$\int_\Omega \Delta u(x\cdot \nabla u)\,dx + (1-\tfrac{n}{2})\int_\Omega |\nabla u|^2\,dx = \tfrac{1}{2}\int_{\partial\Omega}(x\cdot v)u_v^2\,d\sigma. \quad (8.26)$$

As for the first integral in (8.26), using the fact that u solves (D), we find

$$-\int_\Omega \Delta u\,(x\cdot \nabla u)\,dx = \int_\Omega f(u)(x\cdot \nabla u)\,dx = \int_\Omega f(u)\sum_i x_i \frac{\partial u}{\partial x_i}\,dx$$
$$= \int_\Omega \sum_i x_i \frac{\partial F(u)}{\partial x_i}\,dx.$$

Integrating by parts, we get

$$\int_\Omega \sum_i x_i \frac{\partial F(u)}{\partial x_i}\,dx = -n\int_\Omega F(u)\,dx$$

and thus

$$\int_\Omega \Delta u\,(x\cdot \nabla u)\,dx = n\int_\Omega F(u)\,dx. \quad (8.27)$$

Finally, from (D) one infers that

$$\int_\Omega |\nabla u|^2\,dx = \int_\Omega uf(u)\,dx.$$

Substituting this and (8.27) into (8.26) we find (P). ∎

As a consequence, we will find that, in general, the growth restriction $p < (n+2)/(n-2)$ cannot be eliminated, if we want to find nontrivial solutions of

$$\begin{cases} -\Delta u &= |u|^{p-1}u \quad x\in\Omega \\ u &= 0 \qquad\quad x\in\partial\Omega. \end{cases} \quad (8.28)$$

Actually the following corollary holds.

Corollary 8.31 *If Ω is star shaped with respect to the origin $0 \in \mathbb{R}^n$, i.e. if $x \cdot \nu > 0$ on $\partial\Omega$, then any smooth solution of* (8.28) *satisfies*

$$n \int_\Omega F(u) \, dx - \frac{n-2}{2} \int_\Omega u f(u) \, dx > 0.$$

In particular, if $f(u) = |u|^{p-1}u$, then we find

$$\left(\frac{n}{p+1} - \frac{n-2}{2} \right) \int_\Omega |u|^{p+1} \, dx > 0,$$

and hence $u \neq 0$ implies that $p < (n+2)/(n-2)$.

The previous considerations highlight that the exponent $(n+2)/(n-2)$ is *critical* not only from the point of view of Sobolev's embedding, but also from that of the existence of nontrivial solutions to (D).

Remark 8.32 In contrast with Corollary 8.31, we will show that the equation $-\Delta u = \lambda u + u^{(n+2)/(n-2)}$, $u \in H_0^1(\Omega)$, might have a positive solution for any bounded domain $\Omega \subset \mathbb{R}^n$, for suitable values of $\lambda > 0$, see Section 11.2 later on. ∎

8.5 Exercises

(i) Let $\Phi \in C^2(E, \mathbb{R})$ be such that $\Phi(tu) = t^4 \Phi(u)$, $\forall u \in E, \forall t \in \mathbb{R}$. Show that $J(u) = \frac{1}{2}\|u\|^2 - \Phi(u)$ satisfies (MP-1) and (MP-2). Denoting by c the MP level (8.2), prove that $c = \inf_M J$, where $M = \{u \in E \setminus \{0\} : \|u\|^2 = (\Phi'(u) \mid u)\}$ (the Nehari manifold).

(ii) Let $k > 0$ and consider the BVP $(*)$ $-\Delta u = \lambda(u - k)^+$, $u \in H_0^1(\Omega)$, where $\lambda \in \mathbb{R}$ and $v^+ = \max\{v(x), 0\}$. Prove that if $\lambda > \lambda_1$ then $(*)$ has a positive solution whose maximum is greater than k, by showing:

   (a) if $\lambda > \lambda_1$, then the Euler functional J_λ corresponding to $(*)$ has the MP geometry;

   (b) if u_j is any (PS) sequence, then, $\|u_j\|$ is bounded [hint: letting $z_j = u_j \|u_j\|^{-1}$ one has that $z_j \to z^*$ strongly in $H_0^1(\Omega)$; hence, using also the maximum principle it follows that $z^* > 0$ and satisfies $-\Delta z^* = \lambda(z^*)$, a contradiction];

   (c) (PS) holds.

(iii) Let $1 < p < (n+2)/(n-2)$ and let β denote the conjugate exponent of $p + 1$. Consider in $X = L^\beta(\Omega)$ the functional $J_\lambda(v) = \int_\Omega |v|^\beta \, dx - \frac{1}{2} \int_\Omega v A_\lambda v \, dx$, where λ is not an eigenvalue of $-\Delta$

with zero Dirichlet boundary conditions on Ω, and A_λ is defined by setting $A_\lambda v = u$ if and only if $u \in H_0^{2,\beta}(\Omega)$ satisfies $-\Delta u - \lambda u = v$.

(a) Show that the MP theorem applies to J_λ.

(b) Let $v \in X$ be a critical point of J_λ. Show that $u = A_\lambda v$ is a solution of
$$-\Delta u = \lambda u + |u|^{\alpha-2}u, \ u \in H_0^1(\Omega).$$

(iv) Consider the function $J(x, y) = x^2 + (1 - x)^3 y^2$, $(x, y) \in \mathbb{R}^2$ introduced in Remark 8.6. Show that the MP level is $c = 1$ and that (PS) fails at that level.

PART III

Variational methods, II

9

Lusternik–Schnirelman theory

In this chapter we discuss an elegant theory, introduced by Lusternik and Schnirelman, that allows us to find critical points of a functional J on a manifold M, in connection with the topological properties of M. In particular, this theory enables us to obtain multiplicity results.

General remark 9.1 In the sequel we will always understand that

$$M = G^{-1}(0), \quad \text{where} \quad G \in C^{1,1}(E, \mathbb{R}), \quad \text{and} \quad G'(u) \neq 0, \ \forall \, u \in M. \quad (M)$$

9.1 The Lusternik–Schnirelman category

The main ingredient of the Lusternik–Schnirelman (L-S, for short) theory is a topological tool, the L-S category, that we are going to define.

Let M be a topological space. A subset A of M is *contractible* in M if the inclusion $i : A \to M$ is homotopic to a constant $p \in M$, namely if there exists $H \in C([0, 1] \times A, M)$ such that $H(0, u) = u$ and $H(1, u) = p$.

Definition 9.2 *The (L-S) category of A with respect to M (or simply the category of A with respect to M), denoted by* $\mathrm{cat}(A, M)$, *is the least integer k such that* $A \subset A_1 \cup \ldots \cup A_k$, *with* A_i ($i = 1, \ldots, k$) *closed and contractible in M. We set* $\mathrm{cat}(\emptyset, M) = 0$ *and* $\mathrm{cat}(A, M) = +\infty$ *if there are no integers with the above property. We will use the notation* $\mathrm{cat}(M)$ *for* $\mathrm{cat}(M, M)$.

Remark 9.3 From the definition it follows that $\mathrm{cat}(A, M) = \mathrm{cat}(\overline{A}, M)$. Moreover, it is also clear that $\mathrm{cat}(A, M) \geq \mathrm{cat}(A, Y)$ provided $A \subset M \subset Y$. ∎

Example 9.4

(i) Let $S^{m-1} = \{x \in \mathbb{R}^m : |x| = 1\}$ denote the unit sphere in the Euclidean m dimensional space. Since S^{m-1} is not contractible in itself but can be covered by two closed hemispheres, then $\mathrm{cat}(S^{m-1}) = 2$. Remark that, obviously, $\mathrm{cat}(S^{m-1}, \mathbb{R}^m) = 1$.

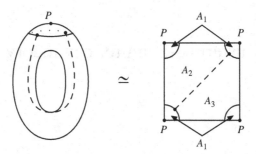

Figure 9.1

(ii) If S is the unit sphere in an infinite dimensional Hilbert space E then one has that $\mathrm{cat}(S) = 1$. Actually, S is contractible in itself, see [90].

(iii) If $T^2 = S^1 \times S^1$ denotes the two-dimensional torus in $\mathbb{R}^3$ then $\mathrm{cat}(T^2) = 3$. Actually, it is easy to see that $\mathrm{cat}(T^2) \leq 3$ because $T^2 \subset A_1 \cup A_2 \cup A_3$, each A_i being closed and contractible in T^2, see Figure 9.1 (the set A_1 is nothing but a closed neighbourhood of the point P).

It is possible to show (but it requires more work) that T^2 cannot be covered by two closed subsets of T^2 contractible in T^2. Thus $\mathrm{cat}(T^2) = 3$. In general, for the k dimensional torus $T^k = \mathbb{R}^k/\mathbb{Z}^k$ one has $\mathrm{cat}(T^k) = k + 1$. See also (9.2) at the end of this section.

(iv) Suppose that there is a topological group G which acts on E through isometries g. We say that $A \subset E$, respectively $J : E \to \mathbb{R}$, is *G-invariant* if $gu \in A$, respectively $J(gu) = J(u)$, for all $u \in A$, $g \in G$; a map $\eta : E \to E$ is said to be *G-equivariant* if $\eta(gu) = g\eta(u)$, for all $u \in A$, $g \in G$. If M and J are G-invariant we can consider J as defined on M/G. Usually, M/G has category greater than M since, roughly, the contractible sets in M/G are much fewer than those contractible in M. Actually, if $A \subset M$, the corresponding invariant set A/G might not be contractible in M/G even though A is contractible in M. The reason is that, working on M/G we can use G-equivariant deformations, only. In the next chapter we will investigate in detail the specific case $M = S$ (the unit sphere in E) and $G = \mathbb{Z}_2 \simeq \{\mathrm{Id}_E, -\mathrm{Id}_E\}$. In such a case, an invariant set is a set $A \subset S$ such that $u \in A$ if and only if $-u \in A$; J is invariant if and only if $J(-u) = J(u)$ and $\eta : E \to E$ is equivariant if and only if $\eta(-u) = -\eta(u)$. In other words A is *symmetric*, J is *even* and η is *odd*. In particular, we anticipate that, letting $\mathbb{P}^m = S^m/\mathbb{Z}_2$, then $\mathrm{cat}(\mathbb{P}^m) = m + 1$. See Theorem 10.7 later on. Similarly, if E is infinite dimensional and $\mathbb{P}^\infty = S/\mathbb{Z}_2$, then $\mathrm{cat}(\mathbb{P}^\infty) = +\infty$. ∎

Remark 9.5 Fournier and Willem introduced the definition of *relative* category. If $A, Y \subset M$ are closed $\text{cat}_{M,Y}(A)$ is the least integer k such that $A \subset A_0 \cup A_1 \cup \cdots \cup A_k$, with A_i $(i = 0, \ldots, k)$ closed, $A_1, \ldots, A_k$ contractible in M and A_0 satisfies the following property: there exists $h \in C([0,1] \times A, M)$ such that $h(0, u) = u, h(1, u) \in Y, \forall u \in A$ and $h(t, Y) \subset Y, \forall t \in [0, 1]$. When $Y = \emptyset$ we get $\text{cat}_{M,\emptyset}(A) = \text{cat}(A, M)$. For the properties of the relative category and its application to critical point theory, we refer to [171] which also contains several further references. Some properties of the relative category are also listed in Exercise 9.3(iv). ∎

Let us now prove the main properties of the category.

Lemma 9.6 *Let $A, B \subset M$.*

(i) *If $A \subset B$ then $\text{cat}(A, M) \leq \text{cat}(B, M)$;*
(ii) $\text{cat}(A \cup B, M) \leq \text{cat}(A, M) + \text{cat}(B, M)$;
(iii) *suppose that A is closed and let $\eta \in C(A, M)$ be a deformation, then*

$$\text{cat}(A, M) \leq \text{cat}(\overline{\eta(A)}, M). \tag{9.1}$$

Proof. The first two statements follow in a straightforward manner from the definition: (i) if $A \subset B$, any covering of B is a covering of A; (ii) if $A \subset \cup_1^h A_i$ and $B \subset \cup_1^k B_i$ (if one of them is infinite, the result is trivial) then $A \cup B \subset (\cup_1^h A_i) \cup (\cup_1^k B_i)$. For (iii), let $\text{cat}(\eta(A)) = k$ (again, if it is infinite, there is nothing to prove). Then $\eta(A) \subset C_1 \cup \cdots \cup C_k$, where C_i is closed and contractible in M. We set

$$A_i = \eta^{-1}(C_i) = \{x \in A : \eta(x) \in C_i\}.$$

Each A_i is closed in A and hence in M because A is closed. Moreover, each A_i is contractible in M because $C_i = \eta(A_i)$ is. Since, obviously, $A \subset A_1 \cup \cdots \cup A_k$ it follows, by definition, that $\text{cat}(A, M) \leq k(= \text{cat}(\eta(A), M))$. ∎

Remark 9.7 An example in which the strict inequality in (9.1) holds is the following: $M = S^1$ (the unit circle in $\mathbb{R}^2$), $A = S_+^1 : \{e^{i\theta} \in S^1 : \theta \in [0, \pi]\}$ and $\eta(e^{i\theta}) = H(1, \theta)$ where $H(t, \theta) = e^{i(t+1)\theta}$, $t \in [0, 1]$. Moreover, the assumption that A is closed can be eliminated in (iii) provided we suppose that the deformation η is defined on all M (the reader can give the proof for an exercise). Otherwise, if A is not closed and η is defined on A only, the claim can be false: it suffices to take $M = S^1$, $p \in S^1$ and $A = S^1 \setminus \{p\}$. ∎

In the next lemma we require that M satisfies the *extension property*:

- for any metric space Y, any closed subset S of Y and any map $f \in C(S, M)$, there is a neighbourhood $\mathcal{N}$ of S and a map $\widetilde{f} \in C(\mathcal{N}, M)$ which is an extension of f: $\widetilde{f}(u) = f(u)$ for all $u \in S$.

Let us point out that, as a consequence of the extension property, any point $p \in M$ has a neighbourhood which is contractible in M. It is possible to see that the Hilbert manifolds M we deal with have the above properties.

Lemma 9.8 *Suppose that* (**M**) *holds or, more in general, that* M *has the extension property and let* $A \subset M$ *be compact. Then*

(i) $\mathrm{cat}(A, M) < +\infty$,
(ii) *there is a neighbourhood* U_A *of* A *such that* $\mathrm{cat}(\overline{U}_A, M) = \mathrm{cat}(A, M)$.

Proof. Let us suppose that $\mathrm{cat}(A, M) = 1$. Then there exist $p \in M$ and an homotopy $H \in C([0, 1] \times A, M)$ such that $H(0, u) = u$ and $H(1, u) = p$ for all $u \in A$. First, we extend H to a map, still denoted by H, defined on

$$S = (\{0\} \times M) \cup ([0, 1] \times A) \cup (\{1\} \times M)$$

by setting

$$H(t, u) = \begin{cases} u & \text{if } t = 0, \quad u \in M \\ H(t, u) & \text{if } t \in [0, 1], \quad u \in A \\ p & \text{if } t = 1, \quad u \in M. \end{cases}$$

The set S is a closed subset of $Y = [0, 1] \times M$ and $H \in C(S, M)$. By the extension property there exist a neighbourhood $\mathcal{N}$ of S in Y and $\widetilde{H} \in C(\mathcal{N}, M)$ such that $\widetilde{H}(t, u) = H(t, u)$ for all $(t, u) \in S$. Since the set $[0, 1] \times A$ is compact and has empty intersection with the closed set $Y \setminus \mathcal{N}$, then the distance between these two sets is positive and thus we can find a neighbourhood U_A in M such that $[0, 1] \times \overline{U}_A$ is contained in $\mathcal{N}$, see Figure 9.2.
Since $\widetilde{H}(0, u) = H(0, u) = u$ and $\widetilde{H}(1, u) = H(1, p) = p$ then $\overline{U}_A$ is contractible in M and hence $\mathrm{cat}(\overline{U}_A, M) = 1$.

In particular, from the preceding proof it follows that any $q \in M$ has a contractible neighbourhood U_q, hence $\mathrm{cat}(\overline{U}_q, M) = 1$. Since A is compact, there is a finite number of points $q_1, \ldots, q_k \in A$ such that $A \subset \overline{U}_{q_1} \cup \cdots \cup \overline{U}_{q_k}$ and this implies that $\mathrm{cat}(A, M) \leq k < +\infty$, proving *(i)*.

To prove *(ii)* let $\mathrm{cat}(A, M) = k$. Then $A \subset \cup_1^k A_i$, with A_i closed and contractible. Substituting A_i with $A \cap A_i$, we can assume that the A_i are compact. For each A_i we find neighbourhoods U_i of A_i such that $\overline{U}_i$ is contractible. Let $U_A = \cup_1^k U_i$. Then $\overline{U}_A \subset \cup_1^k \overline{U}_i$ and hence $\mathrm{cat}(\overline{U}_A, M) \leq k = \mathrm{cat}(A, M)$. Conversely,

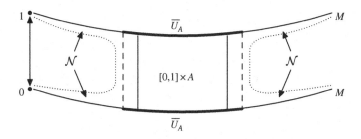

Figure 9.2

from $A \subset \overline{U}_A$ it follows (see Lemma 9.6(i)) that $\mathrm{cat}(A, M) \leq \mathrm{cat}(\overline{U}_A, M)$ and the proof is complete. ∎

Remark 9.9 In the preceding proof it has been shown that if M has the extension property then every homotopy defined on a closed subset A of M can be extended to a neighbourhood of A. ∎

Finally, it is worth mentioning an interesting lower bound of the category by means of the cup length. The *cup length* of M is defined by

$$\text{cup length}(M) = \sup\{k \in \mathbb{N} : \exists \alpha_1, \dots, \alpha_k \in \check{H}^*(M) \setminus 1, \ \alpha_1 \cup \cdots \cup \alpha_k \neq 0\}.$$

If no such class exists, we set cup length$(M) = 0$. Here $\check{H}^*(M)$ is the Alexander cohomology of M with real coefficients and $\cup$ denotes the cup product. It is possible to show, see for example [160], Theorem 5.14, that

$$\mathrm{cat}(M) \geq \text{cup length}(M) + 1. \tag{9.2}$$

For example, if M is the two dimensional torus T^2 it is easy to see that cup length$(T^2) = 2$ and then $\mathrm{cat}(T^2) \geq 3$. This, together with the discussion in Example 9.4(iii), proves that $\mathrm{cat}(T^2) = 3$.

We will not enter into more details, since these topics are beyond the scope of this book.

9.2 Lusternik–Schnirelman theorems

Let us reconsider the two dimensional torus $M = T^2$ in $\mathbb{R}^3$ and the functional $J(x, y, z) = z$, see Section 7.1. The sublevels $M^a = \{p \in M : J(p) \leq a\}$ are indicated in Figure 9.3. If $a < c_1$ then $M^a = \emptyset$ and $\mathrm{cat}(M^a, M) = 0$. If $a = a_1 \in [c_1, c_2)$, then M^a is a spherical cap which is contractible in M and hence $\mathrm{cat}(M^a, M) = 1$. If $a = a_2 \in [c_2, c_4)$, then M^a is not contractible

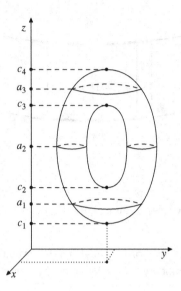

Figure 9.3

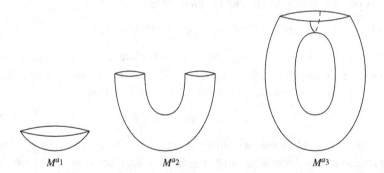

Figure 9.4

in M but can be covered by two contractible sets. Therefore, in this case one has $\mathrm{cat}(M^a, M) = 2$. Finally, if $a \geq c_4$ then $M^a = M$ and so $\mathrm{cat}(M^a, M) = \mathrm{cat}(M) = 3$. Summarizing, one has:

$$\mathrm{cat}(M^a, M) = \begin{cases} 0 & \text{if} \quad a < c_1 \\ 1 & \text{if} \quad c_1 \leq a < c_2 \\ 2 & \text{if} \quad c_2 \leq a < c_4 \\ 3 & \text{if} \quad c_4 \leq a. \end{cases}$$

Consider any closed subset A of M such that $\mathrm{cat}(A, M) \geq 2$. For example, any sublevel M^a with $a \geq c_2$ belongs to this class. Indeed, one has that

$$c_2 = \min\{\max_{u \in A} J(u) : \mathrm{cat}(A, M) \geq 2\}, \tag{9.3}$$

because if $\max_A J < c_2$ for some A, then there is $a < c_2$ such that $A \subset M^a$ and hence $\mathrm{cat}(A, M) \leq \mathrm{cat}(M^a, M) = 1$. In a similar way, $c_1 = \min\{\max_{u \in A} J(u) : \mathrm{cat}(A, M) \geq 1\}$ and $c_4 = \min\{\max_{u \in A} J(u) : \mathrm{cat}(A, M) \geq 3\}$. Actually, in the former case any point can be taken as A and $c_1 = \min_M J$; in the latter the only admissible set A is M itself and one finds that $c_4 = \max_M J$. Let us point out that one can find c_3 by reversing the procedure and taking max-min instead of min-max in (9.3). In general, there are functionals which possess exactly three critical points on the torus: the maximum, the minimum and one saddle point.

The preceding example is actually a particular case of a general fact: the L-S category allows us to define *min-max* levels which are critical. Let (M) hold and set

$$\mathrm{cat}_k(M) = \sup\{\mathrm{cat}(A, M) : A \subset M \ A \text{ compact}\}.$$

The notion of cat_k was introduced by F. Browder in [67]. Obviously, if M is compact then $\mathrm{cat}_k(M) = \mathrm{cat}(M)$. Moreover, this is also the case in all the applications we will deal with. Some specific properties of $\mathrm{cat}_k(M)$ are discussed in Exercise 9.3(ii) at the end of the chapter.

Next, let us consider the class

$$C_m = \{A \subset M : A \text{ is compact and } \mathrm{cat}(A, M) \geq m\}.$$

In the preceding definition, and always in the sequel, m is a positive integer such that $m \leq \mathrm{cat}_k(M)$, if $\mathrm{cat}_k(M) < +\infty$; otherwise, m can be any positive integer.

Obviously one has that $C_m \neq \emptyset$ for all such m. If $J \in C(M, \mathbb{R})$ we set:

$$c_m = \inf_{A \in C_m} \max_{u \in A} J(u). \tag{9.4}$$

The following properties are immediate consequences of the definition:

$$\left\{ \begin{array}{ll} \text{(i)} & c_1 = \inf_M J; \\ \text{(ii)} & c_1 \leq c_2 \leq \cdots \leq c_{m-1} \leq c_m \leq \cdots \text{ (because } C_m \subset C_{m-1}); \\ \text{(iii)} & c_m < +\infty, \quad \forall \, m \leq \mathrm{cat}_k(M); \\ \text{(iv)} & \text{if } \inf_M J > -\infty \text{ then any } c_m \text{ is finite: } -\infty < c_1 \leq c_m < +\infty. \end{array} \right. \tag{9.5}$$

Moreover, the following property will be frequently used in the sequel. Let η be a deformation in M. Then from Lemma 9.6(iii) it follows that

$$\mathrm{cat}(\eta(A), M) \geq \mathrm{cat}(A, M) \geq m$$

and hence (remark that $\eta(A)$ is compact because A is)

$$\eta(A) \in C_m, \quad \forall A \in C_m. \tag{9.6}$$

In Theorem 7.12, Section 7.5, we have shown that c_1 is a critical level of J provided (PS) holds. Let us recall the notation introduced before. Z denotes the set of critical points of J on M: $Z = \{z \in M : \nabla_M J(z) = 0\}$ and $Z_c = \{z \in Z : J(z) = c\}$. Moreover $c \in \mathbb{R}$ is a critical level of J on M if Z_c is not empty. Let us also point out that if $(PS)_c$ holds, then Z_c is compact.

Next, we are going to show that all the finite c_m are critical levels of J. In addition, if two of them coincide, then J has a continuum of critical points at that level.

Theorem 9.10 *Let (M) hold and let $J \in C^{1,1}(E, \mathbb{R})$ be bounded from below on M and let J satisfy (PS).*
Then J has at least $\operatorname{cat}_k(M)$ critical points on M. More precisely,

(i) *any c_m is a critical level of J,*
(ii) *let $c_m = c_{m+1} = \cdots = c_{m+q}$ for some integer $q \geq 1$ and let c denote this common value, then $\operatorname{cat}(Z_c, M) \geq q + 1$.*

Remark 9.11
(i) Since a discrete set has obviously category (with respect to M) equal to 1, statement (ii) implies that J has infinitely many critical points at level c.
(ii) Using slightly different arguments it has been proved by F. Browder [67] that J has at least $\operatorname{cat}_k(M)$ critical points on M, provided J is bounded from below on M and (PS) holds. ■

Proof of Theorem 9.10. We will prove separately the two statements. Actually, (i) would also follow from (ii) with $q = 0$. Let us recall that c_m is finite for all $m = 1, 2, \ldots$ because J is bounded from below on M, see Equation (9.5)(iv).

(i) Let us suppose, by contradiction, that $Z_{c_m} = \emptyset$. Since $\inf_M J > -\infty$ and (PS) holds, we can apply Lemma 7.10: there exists $\delta > 0$ and a deformation η such that

$$\eta(M^{c_m+\delta}) \subset M^{c_m-\delta}. \tag{9.7}$$

According to the definition of c_m there exists $A \in C_m$ such that $\max_A J \leq c_m+\delta$, namely $A \subset M^{c_m+\delta}$. From (9.7) it follows that $\eta(A) \subset M^{c_m-\delta}$, a contradiction because $\eta(A) \in C_m$, see (9.6).

(ii) We need to sharpen the deformation Lemma 7.10. Let $\alpha(t, u)$ denote the steepest descent flow corresponding to the Vector field $W(u) = -\nabla J(u)$.

Lemma 9.12 *For every neighbourhood U of Z_c there exists $\delta > 0$ and a deformation η in M such that*

$$\eta(M^{c+\delta} \setminus U) \subset M^{c-\delta}. \tag{9.8}$$

Proof. First let us show

(∗) for every neighbourhood U of Z_c there exists $\delta > 0$ such that:

$$u \notin U, \ |J(u) - c| \leq \delta \implies \|\nabla J(\alpha(t, u))\| \geq 2\delta, \quad \forall t \in [0, 1].$$

Arguing by contradiction, we can find sequences $t_k \in [0, 1]$ and $u_k \notin U$ such that

$$|J(u_k) - c| \leq \frac{1}{k}, \tag{9.9}$$

$$\|\nabla J(\alpha(t_k, u_k))\| \to 0. \tag{9.10}$$

Without relabelling, we can assume that $t_k \to \bar{t}$. Consider the sequence $v_k = \alpha(t_k, u_k)$. From (9.9) and (9.10) it follows that

$$J(v_k) \leq J(u_k) \leq c + \frac{1}{k} \quad \text{and} \quad \nabla J(v_k) \to 0.$$

Hence v_k is a PS sequence and therefore $v_k \to z$ (up to a subsequence). Obviously $\nabla J(z) = \lim \nabla J(v_k) = 0$ and thus the solution of the Cauchy problem that defines the steepest descent flow satisfies $\alpha(t, z) \equiv z$. Furthermore, one has $u_k = \alpha(-t_k, v_k)$ and hence $u_k \to \alpha(\bar{t}, z) = z$. As a consequence we infer $J(z) = \lim J(u_k) = c$. In other words, $z \in Z_c$, a contradiction to the fact that $u_k \to z$ and $u_k \notin U$, proving the claim. Now (∗) allows us to repeat the arguments of Lemma 7.6 yielding a deformation $\eta(u) = \alpha(1, u)$ such that $\eta(u) \in M^{c-\delta}$ provided $u \in M^{c+\delta} \setminus U$. ∎

Remark 9.13 The preceding proof makes it clear that if $M = E$, we can weaken the regularity assumption on J by requiring that $J \in C^1(E, \mathbb{R})$. It suffices to replace $-\nabla J(u)$ with a pseudogradient vector field, see Section 8.1.1, and to define the deformation η by using the corresponding flow. The proof requires changes similar to those made in the proof of Lemma 8.4. ∎

Proof of Theorem 9.10 completed. First of all, let us complete the proof of (ii). Suppose by contradiction that $\mathrm{cat}(Z_c, M) \leq q$. As remarked before, Z_c is compact and hence, by Lemma 9.8(ii), there exists a neighbourhood U of Z_c such that

$$\mathrm{cat}(\overline{U}, M) = \mathrm{cat}(Z_c, M) \leq q. \tag{9.11}$$

In correspondence to U we use Lemma 9.12 to find $\delta > 0$ and a deformation η satisfying (9.8). By the definition of $c = c_{m+q}$ there is $A \in C_{m+q}$ such that

$A \subset M^{c+\delta}$. Setting $A' = \overline{A \setminus U}$, the subadditivity property of the category, see Lemma 9.6(ii), yields

$$\text{cat}(A', M) \geq \text{cat}(A, M) - \text{cat}(\overline{U}, M) \geq m + q - q = m,$$

namely $A' \in C_m$. Then from (9.6) it follows that $\eta(A') \in C_m$, while (9.8) yields that $\eta(A') \subset M^{c-\delta}$. These two facts are in contradiction to the definition of c_m. The proof of (ii) is completed.   ■

Remark 9.14

(i) The proof of Theorem 9.10 highlights that we can weaken the assumption that J satisfies (PS). If $c_m < b$ for any m, it suffices that J satisfies $(PS)_c$ for all $c < b$. Indeed, to prove (i) one can merely suppose that $(PS)_{c_m}$ holds. See Theorem 10.12 below for a case when this weakening is essential.

(ii) The fact that c_m is a critical level is a particular case of a more general statement. Let $\mathcal{G}$ denote a class of subsets of M with the property that $\alpha(t, G) \in \mathcal{G}$ for all $G \in \mathcal{G}$ and all $t \geq 0$. Consider

$$b = \inf_{G \in \mathcal{G}} \sup_{u \in G} J(u),$$

suppose that b is finite and that $(PS)_b$ holds. Then b is a critical level of J. Each C_m defined above through the L-S category is a particular case of the class $\mathcal{G}$. Another example can be given considering a topological space T and a homotopy class $[h]$ of maps from T into M. Then $\mathcal{G} = \{h(T) : h \in [h]\}$ is a class which can be used for defining a min-max critical level. Let us point out that the main advantage one has using the classes C_m, is that one can handle the degenerate case $c_m = c_{m+q}$.

(iii) Another approach to find a multiplicity result like Theorem 9.10 is to consider the class

$$\widetilde{C}_m = \{A \subset M : A \text{ is closed and } \text{cat}(A, M) \geq m\}$$

instead of C_m. Statements (i) and (ii) can be proved in a quite similar manner. In addition, one could show that if $c_m = +\infty$ for some $m \in \mathbb{N}$, then $\sup_Z J = +\infty$ and hence J has infinitely many critical points on M. In this way one proves that J possesses on M at least $\text{cat}(M)$ critical points, see [159].

(iv) The critical points at level c_m can possibly be degenerate. For example, if M is the two dimensional torus, then any $J \in C^1(M, \mathbb{R})$ has at least three critical points. An example of a functional that has precisely three critical points on the torus is reported in Figure 9.5. The minimum of J is

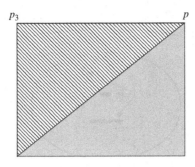

Figure 9.5 The dashed area represents the set where $J(p) < J(p_3)$. The grey area represents the set where $J(p) > J(p_3)$.

achieved at the point p_1 and the maximum at p_2. Moreover, J has a saddle point at p_3, which is degenerate. The degeneracy of p_3 is a general fact. Actually, if J has only non degenerate critical points z, namely $J''(z)$ has a trivial kernel, then Morse theory ensures that J has four critical points, see Chapter 12. We anticipate that, in general, Morse theory requires that the critical points are nondegenerate but gives rise to results more precise than those found by the L-S construction. Unfortunately, this a-priori assumption is not very appropriate for applications to differential equations. The fact that we do not need to assume any nondegeneracy hypothesis makes the L-S theory more suitable for applications to analysis.

(v) Using the relative category, see Remark 9.5, one can prove that the strip $M^b \setminus M^a$ contains at least $\mathrm{cat}_{M^b, M^a}(M^b)$ critical points of J constrained on M, provided that $(\mathbf{M})$ holds, $J \in C^{1,1}(E, \mathbb{R})$ is bounded from below on M and $(PS)_c$ is satisfied for all $c \in [a, b]$. See [171], Theorem 5.19. ∎

We end this section with a result that highlights a relationship between the category and the (PS) condition.

Theorem 9.15 *Let $(\mathbf{M})$ hold and let $J \in C^{1,1}(E, \mathbb{R})$ be bounded from below on M. Furthermore, suppose there exists $a \in \mathbb{R}$ such that $(PS)_c$ holds for all $c \leq a$. Then $\mathrm{cat}(M^a) < +\infty$.*

Proof. Step 1. From the assumptions it follows that $Z^* = Z \cap M^a$ is compact and there exists a neighbourhood U of Z^* such that

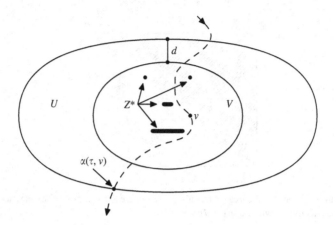

Figure 9.6 The dashed line represents the steepest descent flow entering in V.

$\mathrm{cat}(\overline{U}, M^a) = \mathrm{cat}(Z^*, M^a) < +\infty$. By continuity, we can also assume that

$$\|\nabla J(u)\| \le 1, \quad \forall u \in \overline{U}. \tag{9.12}$$

Step 2. We can also find a second neighbourhood V of Z^* such that $V \subset U$ and $d = \mathrm{dist}(\overline{V}, \partial U) > 0$. Let us show that the steepest descent flow entering in V can exit out of U only after a time greater than or equal to d. Precisely, one has

$$v \in \overline{V}, \qquad \alpha(\tau, v) \notin U \implies \tau > d. \tag{9.13}$$

Actually, let $\hat{t} = \sup\{t \ge 0 : \alpha(s, v) \in U, \ \forall s \in [0, t]\}$. Then (9.12) implies

$$d \le \|\alpha(\hat{t}, u) - u\| \le \int_0^{\hat{t}} \left\| \frac{d}{ds}\alpha \right\| \, dt \le \int_0^{\hat{t}} \|\nabla J(\alpha)\| \, dt \le \hat{t}.$$

Since $\tau \ge \hat{t}$, (9.13) follows.

Step 3. Since $(PS)_c$ holds for all $c \le a$ then there exists $\delta > 0$ such that

$$\|\nabla J(u)\| \ge \delta, \quad \forall u \in M^a \setminus V. \tag{9.14}$$

Let $a' = a - \inf_M J$, $T > a'/\delta^2$ and fix an arbitrary $p \in M^a$. If $\alpha(t, p)$ never enters in V then (9.14) yields

$$J(\alpha(T, u)) \le J(u) - T \cdot \delta^2 < a - a' = \inf_M J,$$

a contradiction that shows:

$$\forall p \in M^a \ \exists t \in [0, 1] : \alpha(t, p) \in V. \tag{9.15}$$

Step 4. Let $t_0 = 0 < t_1 < \cdots < t_n = T$ be such that $|t_i - t_{i-1}| \le d/2$. From (9.15) there exists $\bar{t} \in [0, T]$ such that $\alpha(\bar{t}, p) \in V$. Taken t_i such that

$|t_i - \bar{t}| \leq d/2$ and using (9.13) with $v = \alpha(\bar{t}, u)$ and $\tau = t_i$, we infer that $\alpha(t_i, p) \in U$. Setting $A_i = \{p \in M^a : \alpha(t_i, p) \in U\}$, we have shown that

$$M^a \subset \bigcup_{0 \leq i \leq n} A_i$$

and hence

$$\operatorname{cat}(M^a) \leq \sum_{1}^{n} \operatorname{cat}(A_i, M^a). \tag{9.16}$$

On the other hand, each A_i can be deformed into U, so that $\operatorname{cat}(A_i, M^a) \leq \operatorname{cat}(\overline{U}, M^a) < +\infty$. This and (9.16) prove the theorem. ∎

If $\sup_M J < +\infty$ we can take $a > \sup_M J$. Hence $M^a = M$ and Theorem 9.15 immediately yields the following corollary.

Corollary 9.16 *If J is also bounded from above on M, then* $\operatorname{cat}(M) < +\infty$.

Furthermore, we can also deduce the following.

Corollary 9.17 *Let J be bounded from below on M and suppose that* $\operatorname{cat}(M) = +\infty$ *and that* $(PS)_a$ *holds for all* $a < \sup_M J$.
Then $c_m \to \sup_M J$ and hence J has infinitely many critical points z_m such that $J(z_m) \to \sup_M J$.

Proof. For any $a < \sup_M J$ one has that $\operatorname{cat}(M^a, M) \leq \operatorname{cat}(M^a) < +\infty$. Taking $m > \operatorname{cat}(M^a, M)$ one finds $c_m \geq a$. ∎

9.3 Exercises

(i) Prove that if $A \cap B = \emptyset$ then $\operatorname{cat}(A \cup B, M) \leq \max\{\operatorname{cat}(A, M), \operatorname{cat}(B, M)\}$. Moreover, if M has the extension property and $A \cap B$ is compact, show that $\operatorname{cat}(A \cup B, M) \leq \max\{\operatorname{cat}(A, M), \operatorname{cat}(B, M)\} + \operatorname{cat}(A \cap B, M)$.

(ii) Let X be a topological space. Recall that $\operatorname{cat}_k(X)$ is the $\sup\{\operatorname{cat}(K, X) : K \subset X, K \text{ compact}\}$. Show that $\operatorname{cat}_k(X)$ satisfies the following properties.

(a) Let X_i be an increasing sequence of open subsets of X. If $X = \cup X_i$ then $\operatorname{cat}_k(X) \leq \liminf \operatorname{cat}_k(X_i)$.

(b) If $X = X_1 \cup X_2$m, with X_i closed in X, then $\operatorname{cat}_k(X) \leq \operatorname{cat}_k(X_1) + \operatorname{cat}_k(X_2)$.

(c) Let $Y \subseteq X$ and let $\eta \in C(X, X)$ be homotopic to the identity and such that $\eta(X) = Y$. Then $\text{cat}_k(X) = \text{cat}_k(Y)$.

(iii) Let $M \subset \mathbb{R}^n$ be smooth and compact. Suppose that $J \in C^{1,1}(\mathbb{R}^n, \mathbb{R})$ has only isolated critical points on M. For $a \in \mathbb{R}$ we set
$J_a = \{u \in M : J(u) = a\}$.

(a) Let c be an (isolated) critical level of J on M. Prove that

$$\text{cat}(J_{c-\varepsilon}, M) - 1 \leq \text{cat}(J_{c+\varepsilon}, M) \leq \text{cat}(J_{c-\varepsilon}, M) + 1.$$

(b) Using (a) prove that if the interval $[a, b]$ contains exactly one critical level, then $\text{cat}(M_a^b, M) \leq \text{cat}(J_a, M) + 1$, where
$M_a^b = \{u \in M : a \leq J(u) \leq b\}$.

(c) Take a finite sequence of noncritical levels $a_0 < a_1 < \cdots < a_m$ such that $a_0 < \min_M J$, $a_m > \max_M J$ and $[a, b]$ contains exactly one critical level of J on M. Prove that

$$\text{cat}(M) \leq \text{cat}\left(\bigcup J_{a_i}, M\right) \text{cat}\left(\bigcup M_{a_i+\varepsilon}^{a_{i+1}-\varepsilon}, M\right),$$

provided $\varepsilon > 0$ is small enough, such that $[a_i - \varepsilon, a_i + \varepsilon]$ does not contain any critical level.

(d) Prove that there exists a noncritical level a such that
$\text{cat}(J_a, M) \geq \frac{1}{2}(\text{cat}(M) - 1)$. [Hint: use (c).]

(iv) Prove the following properties of the relative category, defined in Remark 9.5:

(a) $\text{cat}_{M,Y}(Y) = 0$;

(b) $\text{cat}_{M,Y}(A \cup B) \leq \text{cat}_{M,Y}(A) + \text{cat}(B, M)$;

(c) if $Y \subset A \cap B$ and there exists $h \in C([0, 1] \times A, M)$ such that $h(0, u) = u$, $h(1, u) \in B$, $\forall u \in A$ and $h(t, Y) \subset Y$, $\forall t \in [0, 1]$, then $\text{cat}_{M,Y}(A) \leq \text{cat}_{M,Y}(B)$.

10

Critical points of even functionals on symmetric manifolds

In this chapter we will investigate the case that J is an even functional and M is a symmetric set. The abstract results will apply to a class of nonlinear elliptic problems with odd nonlinearities.

10.1 The Krasnoselski genus

Theorem 9.10 does not give any new result when $M = S$ is the unit sphere in a infinite dimensional Hilbert space E because $\mathrm{cat}(S) = 1$ in such a case. On the other hand, if A is a linear compact positive operator, the quadratic functional $J(u) = (Au \mid u)$ has infinitely many critical points z_k, corresponding to the solutions of the linear eigenvalue problem $Au = \lambda u$. We will show that, more in general, any *even* functional on an infinite dimensional Hilbert sphere, satisfying the (PS) condition on S and bounded from below on S, possesses infinitely many critical points on S. In order to obtain such a result one could consider the projective space $\mathbb{P}^\infty = S/\mathbb{Z}_2$ (see Example 9.4(iv)) and use the L-S theory. On the contrary, we will use a topological tool, the *genus*, which will substitute the category in the case of $\mathbb{Z}_2$ symmetry. The genus was introduced by M. A. Krasnoselski [110]; we will follow [149].

Let $\mathcal{A}$ denote the class of all closed subsets A of $E \setminus \{0\}$ which are symmetric, namely $u \in A \Longrightarrow -u \in A$.

Definition 10.1 *Let $A \in \mathcal{A}$. The genus $\gamma(A)$ of A is defined as the least integer n such that there exists $\phi \in C(E, \mathbb{R}^n)$ such that ϕ is odd and $\phi(x) \neq 0$ for all $x \in A$. We set $\gamma(A) = +\infty$ if there are no integers with the above property and $\gamma(\emptyset) = 0$.*

Remark 10.2 An equivalent way to define $\gamma(A)$ is to take the minimal integer n such that there exists an odd map $\phi \in C(A, \mathbb{R}^n \setminus \{0\})$. Actually, such a ϕ can be extended by the Dugundij theorem [90], to a map $\hat{\phi} \in C(E, \mathbb{R}^n)$. If ϕ^* is the odd part of $\hat{\phi}$, namely $\phi^*(u) = \frac{1}{2}(\hat{\phi}(u) - \hat{\phi}(-u))$, ϕ^* satisfies the properties required in the above definition. ∎

Example 10.3 Given any closed $A \subset E \setminus \{0\}$ we set $-A = \{u \in E : -u \in A\}$. If $A \cap (-A) = \emptyset$ then there exists an odd $\phi \in C(E, \mathbb{R})$ such that $\phi(u) = a \neq 0$ for all $u \in A$. Then $\phi(u) \neq 0$ for all $u \in A \cup (-A)$ and hence $\gamma(A \cup (-A)) = 1$. ∎

The following lemma shows that the genus verifies properties similar to those proved in Lemmas 9.6 and 9.8 for the category.

Lemma 10.4 *Let* $A_1, A_2 \in \mathcal{A}$.

  (i) *If* $A_1 \subset A_2$ *then* $\gamma(A_1) \leq \gamma(A_2)$;
  (ii) $\gamma(A_1 \cup A_2) \leq \gamma(A_1) + \gamma(A_2)$;
  (iii) *if* $\eta \in C(A, E)$ *is odd then* $\gamma(A) \leq \gamma(\eta(A))$;
  (iv) *if* A *is compact then* $\gamma(A) < +\infty$ *and there exists a symmetric neighbourhood* U_A *of* A *such that* $\gamma(\overline{U}_A) = \gamma(A)$.

Proof. Step 1. (i) is trivial.
Step 2. Let $\gamma(A_1) = n$ and $\gamma(A_2) = m$. Then there exist $\phi_1 \in C(E, \mathbb{R}^n)$, $\phi_2 \in C(E, \mathbb{R}^m)$ odd and such that $\phi_i(x) \neq 0$ for all $x \in A_i$. Define $\psi : E \to \mathbb{R}^n \times \mathbb{R}^m$ by setting

$$\psi(x) = (\phi_1(x), \phi_2(x)).$$

Since ψ is odd and $\psi(x) \neq 0$ for all $x \in A_1 \cup A_2$, then $\gamma(A_1 \cup A_2) \leq n + m$, proving (ii).

Step 3. If $\gamma(\eta(A)) = n$ then there exists an odd $\phi \in C(E, \mathbb{R}^n)$ such that $\phi(x) \neq 0$ for all $x \in \eta(A)$. Consider the composite map $\phi' = \phi \circ \eta$

$$A \quad \xrightarrow{\;\;\eta\;\;} \quad \eta(A)$$

$$\phi \circ \eta \searrow \qquad \downarrow \phi$$

$$\mathbb{R}^n \setminus \{0\}.$$

Since $\phi(x) \neq 0$ for all $x \in \eta(A)$ then $\phi' \in C(A, \mathbb{R}^n \setminus \{0\})$. Moreover, the map ϕ' is odd because η and ϕ are. This suffices (see Remark 10.2) to say that $\gamma(A) \leq n = \gamma(\eta(A))$ proving (iii).
Step 4. For all $x \in A$ let $\varepsilon > 0$ be such that $B_\varepsilon(x) \cap B_\varepsilon(-x) = \emptyset$. Then $C_x = B_\varepsilon(x) \cup B_\varepsilon(-x)$ is such that $\gamma(C_x) = 1$, see Example 10.3. If A is

compact we can find a finite set $\{x_1, \ldots, x_k\}$ such that $A \subset \cup_1^k C_{x_i}$. Then, using (i) above we get

$$\gamma(A) \le \sum_1^k \gamma(C_{x_i}) = k.$$

Furthermore, if $\gamma(A) = k$ there exists an odd $\phi \in C(E, \mathbb{R}^k)$ such that $\phi(x) \ne 0$ for all $x \in A$. By continuity, there is $\varepsilon > 0$ such that $\phi(x) \ne 0$ for all $x \in N_\varepsilon(A)$, the closed ε neighbourhood of A. Then $\gamma(N_\varepsilon(A)) \le k$. On the other hand $A \subset N_\varepsilon(A)$ implies by (i) that $\gamma(A) \le \gamma(N_\varepsilon(A))$, proving (iv). ∎

An important property of the genus is that $\gamma(S^{m-1}) = m$ (recall the notation S^{m-1} which stands for the unit sphere in R^m).

Theorem 10.5 *Let $E = \mathbb{R}^m$ and let $\partial\Omega$ denote the boundary of $\Omega \subset \mathbb{R}^m$, an open bounded symmetric neighbourhood of $x = 0$. Then $\gamma(\partial\Omega) = m$.*

Proof. Since the *identity* map can be used in the definition of γ then $\gamma(\partial\Omega) \le m$. Suppose, by contradiction, that $\gamma(\partial\Omega) = k < m$. Then there exists $\phi \in C(\mathbb{R}^m, \mathbb{R}^k)$ which is odd and such that $\phi(x) \ne 0$ for all $x \in \partial\Omega$. We can suppose that ϕ is a map from $\mathbb{R}^m$ into $\mathbb{R}^m$ by taking the last $m - k$ components of $\phi(x)$ equal to 0. Obviously, this extended map has still the property that $\phi(x) \ne 0$ for all $x \in \partial\Omega$ and hence the topological degree $\deg(\phi, \Omega, 0)$ is well defined. Since ϕ is odd, the Borsuk–Ulam theorem (see Theorem 3.9) applies and yields that $\deg(\phi, \Omega, 0) = 1 \pmod 2$. By the continuity of $\deg(\phi, \Omega, \cdot)$ it follows that there is $\varepsilon > 0$ such that

$$\deg(\phi, \Omega, y) = 1 \pmod 2 \quad \forall\, y \in B_\varepsilon$$

where $B_\varepsilon = \{y \in \mathbb{R}^m : |y| < \varepsilon\}$. This means that $\phi(x) = y$ has a solution in Ω for all $y \in B_\varepsilon$, namely that $B_\varepsilon \subset \phi(\Omega)$, while $\phi(\Omega) \subset \mathbb{R}^k$, with $k < m$. This contradiction proves the theorem. ∎

Corollary 10.6 $\gamma(S^{m-1}) = m$. *As a consequence, if E is infinite dimensional and separable and S denotes the unit sphere in E, then $\gamma(S) = +\infty$.*

We conclude this section with a result that links the genus and the L-S category in projective spaces, anticipated in Example 9.4(iv). If E is a Hilbert space, $S = S_E = \{u \in E : \|u\| = 1\}$, we set $\mathbb{P}S = S/\mathbb{Z}_2$, where $\mathbb{Z}_2 \simeq \{\mathrm{Id}_E, -\mathrm{Id}_E\}$. If $E = \mathbb{R}^{m+1}$ we find $\mathbb{P}S = \mathbb{P}^m$, while if E is an infinite dimensional space then $\mathbb{P}S = \mathbb{P}^\infty$. Moreover, if $A \subset S$, we set $A^* = A/\mathbb{Z}_2$.

Theorem 10.7 *Let $A \subset S$ be (symmetric and) such that $\gamma(A) < +\infty$. Then $\gamma(A) = \mathrm{cat}(A^*, \mathbb{P}S)$.*

As a consequence, we find that $\mathrm{cat}(\mathbb{P}^m, \mathbb{P}S) = \gamma(S^m) = m+1$ (more precisely, one can show that $\mathrm{cat}(\mathbb{P}^m) = \gamma(S^m) = m+1$), and when E is infinite dimensional and separable, this implies that $\mathrm{cat}(\mathbb{P}^\infty) = +\infty$. For more details and proofs, we refer to [148], Theorem 3.7.

10.2 Existence of critical points

The genus can be used to prove existence results of critical points similar to Theorem 9.10, provided J is even and $M \in \mathcal{A}$. More precisely, we set

$$\gamma_k(M) = \sup\{\gamma(K) : K \subset M, \ K \in \mathcal{A} \text{ and is compact}\}$$

and define, for all $m \le \gamma_k(M)$

$$\mathcal{A}_m = \{A \subset M : A \in \mathcal{A}, \ A \text{ is compact and } \gamma(A) \ge m\}$$

and

$$\sigma_m = \inf_{A \in \mathcal{A}_m} \max_{u \in A} J(u). \tag{10.1}$$

As for the critical levels defined through the category, here m is a positive integer such that $m \le \gamma_k(M)$, if $\gamma_k(M) < +\infty$; otherwise, m can be any positive integer. This will always be understood in the sequel.

Let us explicitly remark that $\sigma_m < +\infty$ and $\sigma_m \le \sigma_{m+1}$. Moreover, if J is bounded from below on M, then $\sigma_1 > -\infty$ and hence any σ_m is finite.

If we deal with problems without constraints, namely if we are looking for stationary points of $J \in C^1(E, \mathbb{R})$ on E, we understand that $\mathcal{A} = \{A \in E \setminus \{0\} : A \text{ is symmetric}\}$, that

$$\mathcal{A}_m = \{A \subset \mathcal{A}, \ A \text{ is compact and } \gamma(A) \ge m\}$$

and that σ_m is defined as before.

One can easily show the following general result which holds both in the case of critical points of J constrained on M and in the case without constraints.

Proposition 10.8 *Each finite σ_m is a critical level for $J \in C^1(E, \mathbb{R})$ (or a critical level for J on M) provided $(PS)_{\sigma_m}$ holds. Moreover, if $\sigma := \sigma_m = \sigma_{m+1} = \cdots = \sigma_{m+q} \in \mathbb{R}$ for some integer $q \ge 1$, then $\gamma(Z_\sigma) \ge q+1$.*

Proof. It suffices to take into account the following facts:

- the genus verifies properties similar to those of category; in particular compare Lemma 10.4 to Lemmas 9.6 and 9.10

- if J is even and M is symmetric then the steepest descent flow gives rise to odd deformations. Actually, in the present case the vector field $-\nabla J(u)$ is odd, or the pseudogradient vector field can be taken to be odd. Then $-\alpha(t, u)$ solves the Cauchy problem

$$y' = -\nabla J(y), \quad y(0) = -u$$

and thus $-\alpha(t, u) = \alpha(t, -u)$, namely α is odd in u.

In particular, the latter remark allows us to use Lemma 10.4(iii). ∎

An application of Proposition 10.8 will be given in Section 10.4 below.

More in general, taking into account the preceding observations, one can repeat all the arguments used in Section 9.2 to prove the following.

Theorem 10.9 *Let (M) hold and let $J \in C^{1,1}(E, \mathbb{R})$ be bounded from below on M and satisfy (PS). Furthermore, suppose that J is even and that $M \in \mathcal{A}$. Then J has at least $\gamma_k(M)$ critical points on M. More precisely, one has*

- *(i) σ_m is a critical level of J;*
- *(ii) let $\sigma_m = \sigma_{m+1} = \cdots = \sigma_{m+q}$ for some integer $q \geq 1$ and let σ denote this common value, then $\gamma(Z_\sigma) \geq q + 1$;*
- *(iii) if there is b such that $\sigma_m < b$ for any m, then it suffices to assume that J satisfies $(PS)_c$ for every $c < b$.*

As a direct application of Theorem 10.9 we can find a classical result due to Lusternik and Schnirelman: any even functional $J \in C^{1,1}(\mathbb{R}^m, \mathbb{R})$ has at least m pairs of critical points on the sphere $S^{m-1} = \{x \in \mathbb{R}^m : |x| = 1\}$. Actually, $\gamma_k(S^{m-1}) = \gamma(S^{m-1}) = m$, according to Corollary 10.6.

Furthermore, we can prove results similar to Theorem 9.15 and Corollaries 9.16 and 9.17.

Theorem 10.10 *Let (M) hold, with $M \in \mathcal{A}$ and let $J \in C^{1,1}(E, \mathbb{R})$ be bounded from below on M and even. Then*

- *(i) if $(PS)_c$ holds for all $c \leq a$ then $\gamma(M^a) < +\infty$;*
- *(ii) if J is also bounded from above on M, then $\gamma(M) < +\infty$;*
- *(iii) if $\gamma(M) = +\infty$ and $(PS)_a$ holds for all $a < \sup_M J$, then J (has infinitely many critical points and) $\sigma_m \to \sup_M J$.*

Remark 10.11 Using pseudogradient vector fields one can extend the above results to cover the case in which $J \in C^1(E, \mathbb{R})$ and M is a $C^{1,1}$ Hilbert (or Banach) manifold. Furthermore, it is worth mentioning that A. Szulkin [168] proved some L-S theorems for even C^1 functionals on a C^1 manifold. ∎

Theorems 10.9 and 10.10 together with Remark 10.11, apply to the nonlinear eigenvalue problem

$$J'(u) = \lambda u. \tag{10.2}$$

On the functional $J \in C^1(E, \mathbb{R})$ we assume the following conditions:

(a) $J(0) = 0$ and $J(u) < 0$, respectively $J(u) > 0$, for all $u \neq 0$;
(b) J is weakly continuous and J' is compact;
(c) $J'(u) \neq 0$ for all $u \neq 0$.

Theorem 10.12 *Let E be a separable infinite dimensional Hilbert space and suppose that $J \in C^1(E, \mathbb{R})$ satisfies (a)–(c) and is even.*
Then (10.2) has infinitely many solutions (μ_k, z_k) with $\|z_k\| = 1$ and $\mu_k \nearrow 0$, respectively $\mu_k \searrow 0$.

Proof. We will carry out the proof in the case that $J < 0$ (in the case that $J(u) > 0$ we take $-J$). We apply the preceding results with $M = S$. For this, we will show

(i) J is bounded from below on S,
(ii) condition $(PS)_c$ holds, for every $c < 0$.

If $\inf_S J = -\infty$ there exists a sequence $u_k \in S$ such that $J(u_k) \to -\infty$. Up to a subsequence, $u_k \to u$ weakly in E. Since J is weakly continuous, it follows that $J(u) \leq \liminf J(u_k) = -\infty$, a contradiction that proves (i).

To prove (ii), let $u_k \in S$ be $(PS)_c$ sequence, with $c < 0$. Again we can assume that $u_k \to u$ weakly in E. From the weak continuity, one has that $J(u) = c$. Since $c < 0$ then $u \neq 0$. For brevity we write ∇J to indicate the gradient of J on S. From

$$\nabla J(u_k) = J'(u_k) - (J'(u_k) \mid u_k) u_k, \tag{10.3}$$

we infer

$$(\nabla J(u_k) \mid J'(u_k)) = \|J'(u_k)\|^2 - (J'(u_k) \mid u_k)^2.$$

Recall that J' is compact and that $\|\nabla J(u_k)\| \to 0$ because u_k is a $(PS)_c$ sequence. Then taking the limit as $k \to \infty$ we get

$$0 = \|J'(u)\|^2 - (J'(u) \mid u)^2.$$

This and (c) imply $(J'(u) \mid u) \neq 0$. Then for n large we have that $(J'(u_k) \mid u_k) \neq 0$ and hence from (10.3) we obtain

$$u_k = \frac{1}{(J'(u_k) \mid u_k)}[J'(u_k) - \nabla J(u_k)].$$

Passing to the limit we finally find

$$u_k \to \frac{1}{(J'(u) \mid u)}[J'(u) - \nabla J(u)].$$

This proves that $(PS)_c$ holds.

Since $\gamma_k(S) = \gamma(S) = +\infty$ we find infinitely many critical points z_k that give rise to solutions of (10.2). Furthermore, one obviously has that $\sup_S J = 0$ and hence, applying Theorem 10.10 (iii) (if $J \in C^{1,1}(E, \mathbb{R})$, otherwise we use Remark 10.11) we find that $J(z_k) \nearrow 0$. Since J is weakly continuous, this implies that $z_k \to 0$ weakly and hence $\mu_k = (J'(z_k) \mid z_k) \nearrow 0$.   ∎

Remark 10.13 Let us emphasize that we have used the statement (iii) of Theorem 10.9. Actually, the $(PS)_c$ condition does not hold at the level $c = 0$; otherwise, Theorem 10.10 would imply that $\gamma(S) < +\infty$, a contradiction.   ∎

Theorem 10.12 applies to the nonlinear eigenvalue problem discussed in Proposition 5.13

$$\begin{cases} -\lambda \Delta u = f(x, u) & x \in \Omega \\ u(x) = 0 & x \in \partial\Omega, \end{cases} \qquad (EP_\lambda)$$

where $f(x, u) \in \mathbb{F}_p$, $1 < p < (n+2)/(n-2)$, and f is *odd* with respect to u. Take $J(u) = \Phi(u) = \int_\Omega F(x, u)\,dx$, where $F(x, u) = \int_0^u f(x, s)\,ds$ and remark that:

- f odd in $u \Longrightarrow \Phi$ is even on $E = H_0^1(\Omega)$;
- $uf(x, u) < 0$, respectively > 0, $\forall x \in \Omega$, $\forall u \neq 0 \Longrightarrow$ (J.1.1) holds.

A straight application of Theorem 10.12 yields $z_k \in S$ and $\mu_k \nearrow 0$, respectively $\mu_k \searrow 0$ such that $\Phi'(z_k) = \mu_k z_k$. This proves the following theorem.

Theorem 10.14 *Suppose that $f \in \mathbb{F}_p$, $1 < p < (n+2)/(n-2)$, and that f is odd with respect to u. Moreover, let $uf(x, u) < 0$, respectively > 0, $\forall x \in \Omega$, $u \neq 0$.*
Then (EP_λ) has infinitely many solutions (μ_k, z_k) with $\|z_k\| = 1$ and $\mu_k \nearrow 0$, respectively $\mu_k \searrow 0$.

The above theorem is the extension of Proposition 5.13 to odd nonlinearities.

Example 10.15 Let $f = |u|^{p-1}u$ with $1 \leq p < (n+2)/(n-2)$. Then the nonlinear eigenvalue problem

$$-\lambda \Delta u = |u|^{p-1}u \ (x \in \Omega), \quad u = 0 \ (x \in \partial\Omega),$$

has infinitely many solutions $(\mu_k, z_k) \in \mathbb{R} \times H_0^1(\Omega)$ such that $\|z_k\| = 1$ and $\mu_k \searrow 0$. By scaling, we also find infinitely many solutions of $-\Delta u = |u|^{p-1}u$,

$u \in H_0^1(\Omega)$, see Section 10.3. When $p = 1$ the preceding equation becomes the linear problem $-\lambda \Delta u = u$, $u \in H_0^1(\Omega)$ and $\mu_k = (\lambda_k)^{-1}$ are characteristic values of $-\Delta$ with Dirichlet boundary conditions on $\partial \Omega$. Actually, it is possible to show that the characteristic values found using the genus coincide with those found by the classical Courant–Fisher min-max procedure, see [82]. The reader can fill in the details as an exercise.   ■

10.3 Multiple critical points of even unbounded functionals

The genus can be used to find multiple critical points of functionals which are even and satisfy the assumptions of the mountain pass or linking theorem. First, we will discuss an extension of the MP Theorem 8.2. In the sequel, the Hilbert space E is assumed to be separable.

Let $J \in C^1(E, \mathbb{R})$ be even and set $E_+ = \{u \in E : J(u) \geq 0\}$. We suppose that

(MP-1) $J(0) = 0$ and $\exists\, r, \rho > 0$ such that

> (i) $J(u) > 0\ \forall u \in B_r \setminus \{0\}$ and
> (ii) $J(u) \geq \rho\ \forall u \in S_r$;

(MP-2′) for any m dimensional subspace $E^m \subset E$, $E^m \cap E_+$ is bounded.

Let us remark that (MP-2′) is nothing but the natural extension of (MP-2). As for (MP-1), see also Remark 8.3.

Let $\mathcal{H}^*$ denote the class of maps $h \in C(E, E)$ which are odd homeomorphisms and such that $h(\overline{B}_1) \subset E_+$. $\mathcal{H}^*$ is not empty because the map $h_r : u \to r\,u$ belongs to $\mathcal{H}^*$.

We set

$$\Gamma_m = \{A \subset \mathcal{A} : A \text{ is compact, and } \gamma(A \cap h(S)) \geq m,\ \forall h \in \mathcal{H}^*\}$$

where, as usual, $S = \partial B_1$, $\mathcal{A} = \{A \subset E \setminus \{0\} : A \text{ is closed and symmetric}\}$ and γ denotes the *genus*, see Section 10.1. The following lemma describes the properties of Γ_m.

Lemma 10.16 *Let $J \in C^1(E, \mathbb{R})$ be even and satisfy* (MP-1) *and* (MP-2′). *Then*

> (i) $\Gamma_m \neq \emptyset$ *for all m;*
> (ii) $\Gamma_{m+1} \subset \Gamma_m$;
> (iii) *if $A \in \Gamma_m$ and $U \in \mathcal{A}$, with $\gamma(U) \leq q < m$, then $\overline{A \setminus U} \in \Gamma_{m-q}$;*
> (iv) *if η is an odd homeomorphism in E such that $\eta^{-1}(E_+) \subset E_+$, then $\eta(A) \in \Gamma_m$ provided $A \in \Gamma_m$.*

Proof. (i) By (MP-2′) there exists $R > 0$ such that

$$E^m \cap E_+ \subset \overline{B_R} \cap E^m.$$

We will show that $B_R^m := \overline{B_R} \cap E^m \in \Gamma_m$. For any $h \in \mathcal{H}^*$ one has that $h(B_1) \subset E_+$ and then $E^m \cap h(B_1) \subset E^m \cap E_+ \subset B_R^m$. This readily implies that $E^m \cap h(S) \subset B_R^m \cap h(S)$. Since, obviously, one also has that $B_R^m \cap h(S) \subset E^m \cap h(S)$, we infer

$$B_R^m \cap h(S) = E^m \cap h(S). \tag{10.4}$$

Since h is an odd homeomorphism, then $E^m \cap h(B_1)$ is a symmetric neighbourhood Ω of 0. It is also easy to check that $\partial\Omega = \partial(E^m \cap h(B_1))$ is contained in $E^m \cap h(S)$. Then Theorem 10.5 implies that

$$\gamma(B_R^m \cap h(S)) = \gamma(E^m \cap h(S)) \geq \gamma(\partial\Omega) = m.$$

This shows that $B_R^m \in \Gamma_m$.

(ii) This is a trivial consequence of the monotonicity property of the genus.

(iii) Obviously, $\overline{A \setminus U} \in \mathcal{A}$ and is compact. Furthermore one has

$$[\overline{A \setminus U}] \cap h(S) = \overline{[A \cap h(S)] \setminus U}.$$

Then Lemma 10.4(ii) yields

$$\gamma([\overline{A \setminus U}] \cap h(S)) = \gamma(\overline{[A \cap h(S)] \setminus U}) \geq \gamma(A \cap h(S)) - \gamma(\overline{U}) \geq m - q,$$

whence $\overline{A \setminus U} \in \Gamma_{m-q}$.

(iv) The set $A' = \eta(A)$ is compact and belongs to $\mathcal{A}$ because η is an odd homeomorphism. To prove that $A' \in \Gamma_m$ we have to show that $\gamma(A' \cap h(S)) \geq m$ for any $h \in \mathcal{H}^*$. There holds

$$A' \cap h(S) = \eta[A \cap \eta^{-1}(h(S))]. \tag{10.5}$$

Moreover, if η satisfies the condition $\eta^{-1}(E_+) \subset E_+$ then $\eta^{-1} \circ h \in \mathcal{H}^*$ for all $h \in \mathcal{H}^*$. Thus, if $A \in \Gamma_m$ one has

$$\gamma(A \cap \eta^{-1}(h(S))) \geq m. \tag{10.6}$$

Since η is odd, Lemma 10.4(iii) jointly with (10.5) and (10.6) imply

$$\gamma(A' \cap h(S)) = \gamma(\eta[A \cap \eta^{-1}(h(S))]) \geq \gamma(A \cap \eta^{-1}(h(S))) \geq m,$$

and hence $A' = \eta(A) \in \Gamma_m$. ∎

Remark 10.17 The condition required in (iv) is equivalent to asking that $\eta(u) \in E_+ \implies u \in E_+$, namely

$$J(\eta(u)) \geq 0 \implies J(u) \geq 0.$$

In particular, if η is a steepest descent flow, then $J(\eta(u)) \leq J(u)$ and hence $J(u) \geq 0$ provided $J(\eta(u)) \geq 0$. ∎

The preceding properties of Γ_n allow us to define min-max levels of J by setting

$$b_m = \inf_{A \in \Gamma_m} \max\{J(u) : u \in A\}. \tag{10.7}$$

Let us point out that each b_m is finite because the sets $A \in \Gamma_m$ are compact. Moreover one has

Theorem 10.18 *Let $J \in C^1(E, \mathbb{R})$ be even and satisfy* (MP-1) *and* (MP-2'). *Then*

  (i) *for every positive integer m one has $b_{m+1} \geq b_m \geq \rho > 0$;*
 (ii) *each b_m is a critical level of J, provided* $(PS)_{b_m}$ *holds;*
(iii) *if $b = b_m = b_{m+1} = \cdots = b_{m+q}$, then $\gamma(Z_b) \geq q + 1$, provided* $(PS)_b$
      holds.

As a consequence if, in addition to the preceding assumptions, J satisfies $(PS)_c$ *for every $c > 0$, J possesses infinitely many critical points.*

Proof. Recalling that $h_r \in \mathcal{H}^*$ and $h_r(B_1) = S_r$ then $A \cap S_r \neq \emptyset$ for any $A \in \Gamma_m$. Therefore

$$b_m \geq \inf\{J(u) : u \in S_r\} \geq \rho > 0.$$

Moreover, Lemma 10.16(ii) implies that $b_m \leq b_{m+1}$ and (i) follows.

Next, it suffices to prove the stronger statement (iii). By contradiction, suppose that $\gamma(Z_b) \leq q$. Using Lemma 10.4(iv) there is a neighbourhood U of Z_b such that $\overline{U} \in \mathcal{A}$ and

$$\gamma(\overline{U}) = \gamma(Z_b) \leq q.$$

Applying the deformation Lemma 9.12, jointly with Remark 9.13 to take into account that $J \in C^1(E, \mathbb{R})$, we find a homeomorphism η and a positive δ such that (see Remark 10.17)

$$\eta^{-1}(E_+) \subset E_+ \tag{10.8}$$

and

$$J(\eta(u)) < b - \delta, \quad \forall u \in J^{b+\delta} \setminus U. \tag{10.9}$$

Moreover η is odd because J is even. By the definition of $b = b_{m+q}$, there exists $A \in \Gamma_{m+q}$ such that $A \subset J^{b+\delta}$. By Lemma 10.4(ii) it follows that $\tilde{A} := \overline{A \setminus U} \in \Gamma_m$. In view of (10.8), Lemma 10.16(iv) applies yielding $\eta(\tilde{A}) \in \Gamma_m$. Finally (10.9) implies

$$\eta(\tilde{A}) \subset J^{b-\delta},$$

a contradiction with the definition of $b = b_m$. ∎

We will now deal with the case in which J is even and has the geometry of the linking theorem 8.24. We assume that $E = V \oplus W$, with $d = \dim V < +\infty$, $W = V^\perp$ and let $J \in C^1(E, \mathbb{R})$ satisfy (MP-2′) and

(L-4′) $J(0) = 0$ and $\exists\, r, \rho > 0$ such that

    (i) $J(u) > 0, \quad \forall u \in (B_r \setminus \{0\}) \cap W$, and

    (ii) $J(u) \geq \rho, \quad \forall u \in S_r \cap W$.

Let us define the counterparts of $\mathcal{H}^*$ and Γ_m by setting

$$\widetilde{\mathcal{H}} = \{h \in C(E, E) : h \text{ is an odd homeomorphism, and } h(B_1) \subset E_+ \cup \overline{B_r}\}$$

and

$$\widetilde{\Gamma}_m = \{A \subset \mathcal{A} : A \text{ is compact, and } \gamma(A \cap h(S)) \geq m \;\forall\, h \in \widetilde{\mathcal{H}}\}.$$

As before, $\widetilde{\mathcal{H}} \neq \emptyset$ because $h_r \in \widetilde{\mathcal{H}}$.

Lemma 10.19 *If J is even and satisfies* (L-4′) *and* (MP-2′) *then*

    (i) $\widetilde{\Gamma}_m \neq \emptyset$ for all m;

    (ii) $\widetilde{\Gamma}_{m+1} \subseteq \widetilde{\Gamma}_m$;

    (iii) if $A \in \widetilde{\Gamma}_m$ and $U \in \mathcal{A}$, with $\gamma(U) \leq q < m$, then $\overline{A \setminus U} \in \widetilde{\Gamma}_{m-q}$;

    (iv) if η is an odd homeomorphism in E such that $\eta(u) = u$ for all u with $J(u) < 0$ and $\eta^{-1}(E_+) \subset E_+$, then $\eta(A) \in \widetilde{\Gamma}_m$ provided $A \in \widetilde{\Gamma}_m$.

Proof. (i) As in Lemma 10.16, there exists $R > 0$, such that $B_R^m := \overline{B_R} \cap E^m \supset E_+ \cap E^m$. Taking R possibly larger, one also has that

$$B_R^m \supset (E_+ \cup \overline{B_r}) \cap E^m.$$

Then, according to the definition of $\widetilde{\mathcal{H}}$, one has

$$B_R^m \supset h(B_1) \cap E^m, \qquad \forall\, h \in \widetilde{\mathcal{H}}.$$

The rest of the proof is the same as that carried out in Lemma 10.16(i).

    (ii)–(iii) These properties also follow reasoning as for the corresponding claims in Lemma 10.16.

    (iv) Let us show that $\eta^{-1} \circ h(B_1) \subset E_+ \cup \overline{B_r}$. For any $u \in B_1$ let v be such that $\eta(v) = h(u)$. Since $h \in \widetilde{\mathcal{H}}$ it follows that $\eta(v) = h(u) \in E_+ \cup \overline{B_r}$. Either $\eta(v) \in E_+$ and then $v \in E_+$ because $\eta^{-1}(E_+) \subset E_+$ or $\eta(v) = v$. In any case one has that $v = \eta^{-1} \circ h(u) \in E_+ \cup \overline{B_r}$. Since $\eta^{-1} \circ h$ is an odd homeomorphism, it follows that $\eta^{-1} \circ h \in \widetilde{\mathcal{H}}$, whenever $h \in \widetilde{\mathcal{H}}$. Repeating the arguments used to prove Lemma 10.16(iv), the result follows. ∎

Let us define

$$\tilde{b}_m = \inf_{A \in \tilde{\Gamma}_m} \{J(u) : u \in A\}.$$

Theorem 10.20 *Let* $J \in C^1(E, \mathbb{R})$ *be even and satisfy* (L-4') *and* (MP-2'). *Then, for all* $m > d$, *one has:*

(i) $\tilde{b}_{m+1} \geq \tilde{b}_m \geq \rho > 0$;
(ii) *each* $\tilde{b}_m$ *is a critical level for* J, *provided* $(PS)_{\tilde{b}_m}$ *holds;*
(iii) *if* $\tilde{b} = \tilde{b}_m = \cdots = \tilde{b}_{m+q}$, *then* $\gamma(Z_{\tilde{b}}) \geq q + 1$, *provided* $(PS)_{\tilde{b}}$ *holds.*

As a consequence, if, in addition to the preceding assumptions, J *satisfies* $(PS)_c$ *for every* $c > 0$, J *possesses infinitely many critical points.*

Proof. Let $A \in \tilde{\Gamma}_m$ with $m > d$. Then, taking $h = h_r$ we find

$$\gamma(A \cap h_r(S)) = \gamma(A \cap S_r) \geq m > d. \qquad (10.10)$$

This implies that

$$(A \cap S_r) \cap W \neq \emptyset. \qquad (10.11)$$

To see this, we can argue by contradiction: if $(A \cap S_r) \cap W = \emptyset$, then, denoted by P the canonical projection onto V, one has that

$$P(A \cap \partial B_r) \subset V \setminus \{0\}.$$

Thus, such a P is an odd continuous map from $A \cap \partial B_r$ to V, that does not vanish on $A \cap \partial B_r$. Since $\dim V = d$, then the definition of genus implies that $\gamma(A \cap S_r) \leq d$, a contradiction to (10.10).
From (10.11) and (L-4') it follows that

$$\tilde{b}_m \geq \max\{J(u) : u \in A \cap S_r\} \geq \rho > 0.$$

The rest of the theorem is proved as Theorem 10.18, noticing that Lemma 10.19(iv) applies because the map η given by the deformation lemma can obviously be taken to satisfy $\eta(u) = u$ for all u such that $J(u) < 0$. ∎

The next result deals, roughly, with a functional J which has a strict local minimum at 0, is bounded from below and has a negative global minimum. According to the mountain pass theorem, such an f possesses a second, non-trivial critical point at a positive level. We will show that if f is even and $\{u : f(u) < 0\}$ has genus $\tilde{d}$, then f has $2\tilde{d}$ pairs of nontrivial critical points.
Precisely, let us substitute (MP-2') with the following statement.

(MP-2'') There exist a subspace $\tilde{V}$ of E with $\dim(\tilde{V}) = \tilde{d}$ and a compact, symmetric set $\mathcal{K} \subset \tilde{V}$ such that $J < 0$ on $\mathcal{K}$ and 0 lies in a bounded component in $\tilde{V}$ of $\tilde{V} - \mathcal{K}$.

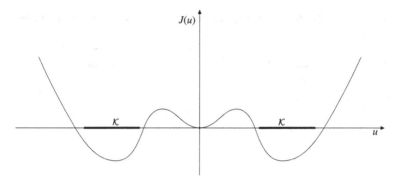

Figure 10.1 Behaviour of J satisfying (MP-2″). Notice that this is the symmetric counterpart of Figure 8.2.

Theorem 10.21 *Let $J \in C^1(E, \mathbb{R})$ be even and satisfy* (MP-1), (MP-2″) *and* (PS). *Then each $\tilde{b}_m$, $1 \leq m \leq \tilde{d}$, is a positive critical level for J, and J possesses at least $\tilde{d}$ pairs of nontrivial critical points $\pm u_m$, with $J(\pm u_m) > 0$.*
If, in addition, J is bounded below on E, then J possesses at least other $\tilde{d}$ pairs of nontrivial critical points $\pm v_k$, $1 \leq k \leq \tilde{d}$, with $J(\pm v_k) < 0$.

Proof. To prove the first statement, let us remark that the only role played by (MP-2′) was to show that $\Gamma_m \neq \emptyset$. We shall prove that this is still the case for $1 \leq m \leq \tilde{d}$, whenever (MP-2″) holds. Let again $A = \overline{B}_R \cap E^m$. For R large and $1 \leq n \leq \tilde{d}$, (MP-2″) implies that $A \supset \mathcal{K} \cap E^m$. Therefore the component $\mathcal{Q}$ of $E_+ \cap E^m$ containing 0 lies in A. Thus, for all $h \in \mathcal{H}^*$ one has that $A \cap h(\partial B_1) \supset \mathcal{Q} \cap h(\partial B_1)$ and hence

$$\gamma(A \cap h(\partial B_1)) \geq \gamma(\mathcal{Q} \cap h(\partial B_1)) \geq m.$$

The last inequality is due to the fact that $\mathcal{Q} \cap h(\partial B_1)$ contains the boundary of a symmetric, bounded, neighbourhood of 0 in E. Then, repeating the arguments of Theorem 10.18, the result follows.

Let J be, in addition, bounded from below on E and consider the inf-sup level

$$\tilde{c}_k = \inf_{\gamma(A) \geq k} \max[J(u) : u \in A].$$

Since $\mathcal{K}$ contains the boundary of a symmetric, bounded neighbourhood of 0 in $\tilde{V}$, then $\gamma(\mathcal{K}) = \tilde{d}$ and there holds

$$\tilde{c}_{\tilde{d}} \leq \max[J(u) : u \in \mathcal{K}] < 0.$$

As a consequence, for all $1 \leq k \leq \tilde{d}$ one has $\tilde{c}_k \leq \tilde{c}_{\tilde{d}} < 0$ and each $\tilde{c}_k$ carries a pair of nontrivial critical points. This completes the proof of the theorem. ∎

10.4 Applications to Dirichlet boundary value problems

In this final section, we will apply the preceding abstract results to find multiple solutions of semilinear Dirichlet BVPs with odd nonlinearities. To avoid technicalities we will focus on model problems, only.

We first consider the sublinear problem

$$\begin{cases} -\Delta u = \lambda u - |u|^{p-1}u & x \in \Omega \\ u = 0 & x \in \partial\Omega, \end{cases} \tag{10.12}$$

where $p > 1$. We have shown, see Example 5.11, that (10.12) possesses at least a (positive) solution for all $\lambda > \lambda_1$, λ_1 being the first eigenvalue of $-\Delta$ on E. We want to improve this result in the case of an odd nonlinearity. As in Example 5.11, we perform a truncation, by setting

$$f_\lambda(u) = \begin{cases} \lambda u - |u|^{p-1}u & \text{if } -\lambda^{1/(p-1)} \le u \le \lambda^{1/(p-1)} \\ 0 & \text{otherwise,} \end{cases}$$

and consider the auxiliary boundary value problem

$$\begin{cases} -\Delta u(x) = f_\lambda(u) & x \in \Omega \\ u(x) = 0 & x \in \partial\Omega. \end{cases} \tag{10.13}$$

The same arguments used in Example 5.11 show that any solution u of (10.13) satisfies $|u(x)| \le \lambda^{1/(p-1)}$ for all $x \in \Omega$, and hence is a solution of (5.5). For $u \in E = H_0^1(\Omega)$, let

$$J_\lambda(u) = \frac{1}{2}\|u\|^2 - \int_\Omega F_\lambda(u)\,dx, \qquad F_\lambda(u) = \int_0^u f_\lambda(s)\,ds.$$

Obviously, J_λ is even and of class C^1. Moreover, we know that J_λ is bounded from below on E, is coercive and that every (PS) sequence is bounded. This immediately implies that (PS) holds. As in (10.1), we set

$$\sigma_{\lambda,m} = \inf_{\gamma(A) \ge m} \sup_{u \in A} J_\lambda(u), \qquad A \in \mathcal{A}.$$

We claim that if $\lambda > \lambda_k$, where λ_k is the kth eigenvalue of $-\Delta$ on E, then $\sigma_{\lambda,k} < 0$. Actually, let φ_i denote an eigenfunction corresponding to λ_i, with $\|\varphi_i\| = 1$ and $(\varphi_i \mid \varphi_j) = 0$ for $i \ne j$, and let us consider the $k - 1$ dimensional sphere

$$S_{k,\varepsilon} = \left\{ v = \sum_1^k a_i\varphi_i : \sum_1^k a_i^2 = \varepsilon^2 \right\}.$$

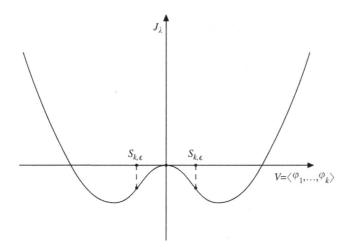

Figure 10.2 Behaviour of $J_\lambda(u)$ with $u \in V = \langle \varphi_1, \ldots, \varphi_k \rangle$ and $\lambda > \lambda_k$.

From Corollary 10.6 one infers that $\gamma(S_{k,\varepsilon}) = k$. Furthermore, for $v \in S_{k,\varepsilon}$ and ε small, one has that $F_\lambda(v) = \frac{1}{2}\lambda v^2 - 1/(p+1)|v|^{p+1}$ and hence

$$J_\lambda(v) = \frac{1}{2}\varepsilon^2 - \frac{1}{2}\lambda \sum_1^k a_i^2 \int_\Omega |\varphi_i|^2 + o(\varepsilon^2)$$

$$= \frac{1}{2}\varepsilon^2 - \frac{1}{2}\lambda \sum_1^k a_i^2 \lambda_i^{-1} + o(\varepsilon^2)$$

$$< \frac{1}{2}\varepsilon^2 - \frac{1}{2}\frac{\lambda}{\lambda_k}\varepsilon^2 + o(\varepsilon^2).$$

Thus if $\lambda > \lambda_k$ we deduce for $\varepsilon \ll 1$

$$\sup_{v \in S_{k,\varepsilon}} J_\lambda(v) < 0,$$

and this implies that $\sigma_{\lambda,k} < 0$ provided $\lambda > \lambda_k$. A straight application of Proposition 10.8 yields that J_λ has at least k pairs of nontrivial critical points which give rise to k nontrivial solutions of (10.12). More in general one can show

Theorem 10.22 *Consider the problem*

$$\begin{cases} -\Delta u = \lambda u - f(u) & x \in \Omega \\ u = 0 & x \in \partial\Omega, \end{cases} \tag{D_λ^-}$$

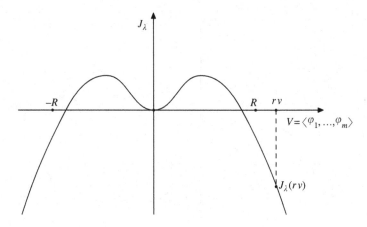

Figure 10.3 Behaviour of $J_\lambda(u)$ with $u \in V = \langle \varphi_1, \ldots, \varphi_m \rangle$.

where f is locally Hölder continuous, odd and satisfies:

$$\lim_{u \to 0} \frac{f(u)}{u} = 0, \qquad \lim_{|u| \to +\infty} \frac{f(u)}{u} = +\infty, \tag{10.14}$$

uniformly with respect to $x \in \Omega$. Then (D_λ^-) has at least k pairs of nontrivial solutions, provided $\lambda > \lambda_k$.

Next, let us consider the model problem

$$\begin{cases} -\Delta u = \lambda u + |u|^{p-1} u & x \in \Omega \\ u = 0 & x \in \partial\Omega, \end{cases} \tag{10.15}$$

where $1 < p < (n+2)/(n-2)$. This is a superlinear problem like those discussed in Theorems 8.11 and 8.28.

Now the corresponding functional J_λ is defined on E by

$$J_\lambda(u) = \frac{1}{2}\|u\|^2 - \frac{1}{2}\lambda \int_\Omega |u|^2 \, dx - \frac{1}{p+1} \int_\Omega |u|^{p+1} \, dx.$$

As before, J_λ is even. Moreover J_λ satisfies $(PS)_c$ for every $c > 0$ and (MP-1), respectively (L-4'), if $\lambda < \lambda_1$, respectively $\lambda \geq \lambda_1$.
We claim that J_λ satisfies the assumption (MP-2'). If this is true, then Theorem 10.18, respectively Theorem 10.20, applies and hence (10.15) has infinitely many solutions for every $\lambda \in \mathbb{R}$. To show that (MP-2') holds, it suffices to take any $v \in \mathrm{span}\{\varphi_1, \ldots, \varphi_m\}$ (we use the same notation employed before, when we discussed the problem (10.12)), $v = \sum_1^m a_i \varphi_i$, with $\|v\| = 1$ and consider $J_\lambda(rv)$, $r > 0$. It is clear that

$$J_\lambda(rv) = \frac{1}{2}r^2 - \frac{1}{2}\lambda r^2 \|v\|_{L^2}^2 - \frac{1}{p+1} r^{p+1} \|v\|_{L^{p+1}}^{p+1}.$$

This immediately implies that there exists $R > 0$ such that $J_\lambda(rv) < 0$ for every $r \geq R$, proving that (MP-2′) holds.

More in general, quite similar arguments lead to the following theorem.

Theorem 10.23 *Consider the problem*

$$\begin{cases} -\Delta u = \lambda u + f(x, u) & x \in \Omega \\ u = 0 & x \in \partial\Omega, \end{cases} \qquad (D_\lambda^+)$$

where $f \in \mathbb{F}_p$, $1 < p < (n+2)/(n-2)$, *is odd and satisfies (8.9)–(8.10), namely*

$$\lim_{u \to 0^+} \frac{f(x, u)}{u} = \lambda \in \mathbb{R}, \qquad \text{for a.e. } x \in \Omega;$$

$$\exists\, r > 0,\ \theta \in (0, \tfrac{1}{2})\ :\ F(x, u) \leq \theta u f(x, u), \quad \forall x \in \Omega,\ |u| \geq r.$$

Then, for every $\lambda \in \mathbb{R}$, (D_λ^+) *has infinitely many pairs of (nontrivial) solutions.*

The above theorem is the counterpart, for odd nonlinearities, of Theorem 7.14 (if $\lambda < \lambda_1$) or Theorem 8.28 (if $\lambda \geq \lambda_1$).

Remark 10.24 (i) It is possible to show that the min-max critical levels b_m, defined in (10.7), or the corresponding $\tilde{b}_m$ when $\lambda \geq \lambda_2$, diverge at $+\infty$ as $m \to +\infty$. Moreover, from the equation it follows that

$$\|u_m\|^2 = \int_\Omega [\lambda u_m^2 + f(x, u_m) u_m]\, dx.$$

This and (8.10) imply

$$J(u_m) = \int_\Omega \left[\tfrac{1}{2} f(x, u_m) u_m - F(x, u_m) \right] dx \leq c_1 \|u_m\|^2 + c_2,$$

and hence $\|u_m\| \to \infty$.

(ii) A similar result can be proved for problems where the nonlinearity is odd and satisfies the same assumptions as the nonlinearity discussed in Theorem 7.14. ∎

Our last application deals with problems like

$$\begin{cases} -\Delta u = \mu(|u|^{s-1} u - |u|^{p-1} u) & x \in \Omega \\ u = 0 & x \in \partial\Omega, \end{cases} \qquad (10.16)$$

with $1 < s < p$. Suppose first that $p < (n+2)/(n-2)$. The solutions of (10.16) are critical points of

$$J_\mu(u) = \frac{1}{2}\|u\|^2 - \mu \int_\Omega \left[\frac{1}{s+1}|u|^{s+1} - \frac{1}{p+1}|u|^{p+1} \right] dx, \qquad u \in E = H_0^1(\Omega). \qquad (10.17)$$

Let us show that Theorem 10.21 applies. Actually, $u = 0$ is a strict local minimum of J_μ because $1 < s < p$ and thus (MP-1) holds. Take once more any $v = \sum_1^m a_i \varphi_i$, with $\|v\| = 1$. It is easy to check that there exists $\bar{r} > 0$ such that taking $0 < r < \bar{r}$, one has that

$$\int_\Omega \left[\frac{1}{s+1} |rv|^{s+1} - \frac{1}{p+1} |rv|^{p+1} \right] dx > 0.$$

Then there exists $\mu_m > 0$ large enough such that for all $\mu > \mu_m$ one has that $J_\mu(rv) < 0$. This implies that (MP-2$''$) holds with $\tilde{V} = \text{span}\{\varphi_1, \ldots, \varphi_m\}$ and $\mathcal{K} = \{u = r \sum_1^m a_i \varphi_i \; : \; 0 < r < \bar{r}\}$. Finally, it is easy to check as for (D_λ^-), that J_μ is bounded from below on E and that (PS) holds. In conclusion, an application of Theorem 10.21 yields, for any integer $m > 0$, the existence of $\mu_m > 0$ such that (10.16) possesses at least $2m$ pairs of nontrivial solutions provided $\mu > \mu_m$. If $p \geq (n+2)/(n-2)$, we perform a truncation by setting

$$\tilde{f}(u) = \begin{cases} |u|^{s-1} u - |u|^{p-1} u & \text{if } |u| \leq 1 \\ 0 & \text{otherwise,} \end{cases}$$

and consider the auxiliary functional

$$\tilde{J}_\mu(u) = \frac{1}{2}\|u\|^2 - \mu \int_\Omega \tilde{F}(u)\, dx, \qquad \tilde{F}(u) = \int_0^u \tilde{f}(s)\, ds.$$

Repeating the preceding arguments one shows that there exists $\mu_m > 0$ such that $\tilde{J}_\mu$ possesses at least $2m$ pairs of critical points provided $\mu > \mu_m$ which give rise to solutions of $-\Delta u = \mu \tilde{f}(u)$, $u \in E$. By the maximum principle, these solutions satisfy (10.16).

More in general, one can use similar arguments to handle a problem such as

$$\begin{cases} -\Delta u = \mu f(x, u) & x \in \Omega \\ u = 0 & x \in \partial\Omega. \end{cases} \tag{10.18}$$

Let $F(x, u) = \int_0^u f(x, t)\, dt$ and set

$$J_\mu(u) = \frac{1}{2}\|u\|^2 - \mu \int_\Omega F(x, u)\, dx, \qquad u \in E.$$

Theorem 10.25 *Suppose that f is locally Hölder continuous, is odd and satisfies:*

(a_1) $\lim_{u \to 0} f(x, u)/u = 0$, *uniformly with respect to $x \in \Omega$,*
(a_2) $f(x, u) > 0$ *in a deleted neighbourhood of $u = 0$,*
(a_3) *there exists $\bar{u} > 0$ such that $f(x, \bar{v}) < 0$, for all $x \in \overline{\Omega}$.*

Then for any integer $m > 0$ there exists $\mu_m > 0$ such that for all $\mu > \mu_m$, (10.18) has at least 2m distinct pairs of solutions $\pm\underline{u}_j$, $\pm\bar{u}_j$ with $J_\mu(\pm\underline{u}_j) < 0 < J_\mu(\pm\bar{u}_j)$.

Of course, the preceding theorem is the counterpart of Theorem 8.14 for odd functionals.

We conclude this section by summarizing, with some bifurcation diagrams, the results found above. As usual, the figures are intended only to be suggestive.

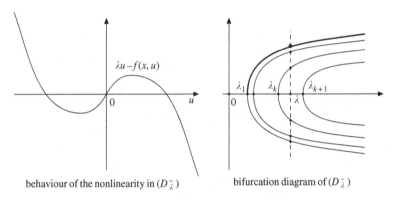

behaviour of the nonlinearity in (D_λ^-) bifurcation diagram of (D_λ^-)

Figure 10.4 Nonlinearity and solutions of $(\boldsymbol{D_\lambda^-})$. The bold line denotes the branch of positive solutions.

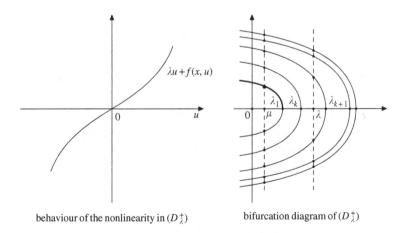

behaviour of the nonlinearity in (D_λ^+) bifurcation diagram of (D_λ^+)

Figure 10.5 Nonlinearity and solutions of (D_λ^+). The bold line denotes the branch of positive solutions.

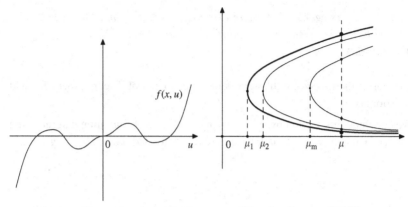

behaviour of the nonlinearity in (10.18)   bifurcation diagram of (10.18)

Figure 10.6 Nonlinearity and solutions of (10.18). The bold line denotes the branch of positive solutions.

10.5 Exercises

(i) Consider the superlinear BVP (D) discussed in Section 7.6 and suppose that, in addition to the assumptions of Theorem 7.14, f is odd. Prove that (D) has infinitely many pairs of solutions.

(ii) Consider a functional $J \in C^1(E, \mathbb{R})$ which satisfies the assumptions (a)–(c) stated before Theorem 10.12 and let M be the unit sphere S in E. Let $S^m \subset S$ denote the m dimensional unit sphere and set
$\tilde{\mathcal{A}}_m = \{A \subset S : A = \phi(S^m),\ \phi \in C(S^m, S),\ \phi \text{ odd}\}$. Define

$$\tilde{\sigma}_m = \inf_{A \in \tilde{\mathcal{A}}_m} \max_{u \in A} J(u).$$

(a) Show that $\tilde{\sigma}_m$ is a critical level for J on S.
(b) Show that $\tilde{\sigma}_m \uparrow 0$. [Hint: prove that $\tilde{\sigma}_m \geq \sigma_m$, where σ_m is defined in (10.1).]

(iii) Let $E_m \subset E$ be an m-dimensional subspace and suppose that $A \in \mathcal{A}$ be such that $\gamma(A) > m$. Prove that $A \cap E_m \neq \emptyset$. (See the first part of the proof of Theorem 10.20).

11

Further results on elliptic Dirichlet problems

In this chapter we collect some further results on elliptic Dirichlet BVPs. Section 11.1 is devoted to the existence of solitary waves. Section 11.2 deals with an elliptic problem with critical nonlinearity, following a celebrated paper by H. Brezis and L. Nirenberg [64]. In Section 11.3 equations with discontinuous nonlinearities are studied and the final Section 11.4 is concerned with problems with concave-convex nonlinearities.

11.1 Radial solutions of semilinear elliptic equation on $\mathbb{R}^n$

In this section we will deal with the following semilinear elliptic equation in $\mathbb{R}^n$

$$-\Delta u + u = u^p, \quad u > 0, \quad u \in W^{1,2}(\mathbb{R}^n), \tag{11.1}$$

where $n \geq 3$ and $1 < p < (n+2)/(n-2)$. Similar results can be proved when $n = 2$ and $p > 1$. Equation (11.1) arises, for example, when we look for a solitary wave of the nonlinear Klein–Gordon equation

$$\frac{\partial^2 \psi}{\partial t^2} - \Delta \psi + a\psi = |\psi|^{p-1}\psi, \quad (a > 0). \tag{11.2}$$

Above, $(t, x) \in \mathbb{R} \times \mathbb{R}^n$ and $\psi = \psi(t, x)$ is a complex valued function. Making the *Ansatz* $\psi(t, x) = e^{i\omega t}u(x)$ with $0 < \omega < a$, we find for u an equation like (11.1) (to simplify the notation, we set $a^2 - \omega^2 = 1$), where the condition $u \in W^{1,2}(\mathbb{R}^n)$ is required to obtain solutions with physical interest. We will look for *radial* solutions of (11.1).

Let $W^{1,2}$ denote the Sobolev space $W^{1,2}(\mathbb{R}^n)$ endowed with scalar product

$$(u \mid v) = \int_{\mathbb{R}^n} [\nabla u \cdot \nabla u + uv] \, dx$$

and consider the functional $J \in C^2(W^{1,2}, \mathbb{R})$,

$$J(u) = \frac{1}{2}\|u\|^2 - \frac{1}{p+1}\int_{\mathbb{R}^n} |u|^{p+1} \, dx.$$

Let us point out that the last integral makes sense because $W^{1,2} \subset L^{p+1}(\mathbb{R}^n)$, for $1 < p < (n+2)/(n-2)$. We shall show that J has a MP critical point which gives rise to a radial solution of (11.1). Since it is clear that J has the MP geometry, the main difficulty relies in the PS condition because, in general, $W^{1,2}$ is not compactly embedded in $L^{p+1}(\mathbb{R}^n)$ even if $1 < p < (n + 2)/(n - 2)$ (to check this lack of compactness it suffices to take any function $u \in W^{1,2}$, any sequence $\xi_k \in \mathbb{R}^n$ such that $|\xi_k| \to \infty$, and consider the translates $u_k(x) = u(x + \xi_k)$). This problem will be bypassed working with radial functions. Let $W_r^{1,2}$ denote the space of functions in $W^{1,2}$ which are radial. If $u \in W_r^{1,2}$ then $u(x) = \tilde{u}(|x|)$ for some $\tilde{u} : \mathbb{R}_+ \mapsto \mathbb{R}$. In the sequel, to simplify the notation, we will identify $\tilde{u}$ with u, using the symbol $u(r)$ for functions belonging to $W_r^{1,2}$.

Below we collect some results dealing with properties of $W_r^{1,2}$ which are interesting in themselves. We first prove the following lemma.

Lemma 11.1 *Let $n \geq 3$. There exists $c_n > 0$, depending only on n, such that for all $u \in W_r^{1,2}$ there holds*

$$|u(r)| \leq c_n r^{(1-n)/2}\|u\|, \qquad \forall\, r \geq 1. \tag{11.3}$$

Proof. By density, we can suppose that $u \in C_0^\infty(\mathbb{R}^n)$. If the prime symbol denotes the derivative with respect to r, we have $(r^{n-1}u^2)' = 2r^{n-1}uu' + (n-1)r^{n-2}u^2$, whence $(r^{n-1}u^2)' \geq 2r^{n-1}uu'$. Integrating over $[r, \infty)$ we find

$$r^{n-1}u^2(r) \leq -2 \int_r^\infty r^{n-1}uu'\, dr \leq c\|u\|^2,$$

where c depends on n, only, proving (11.3). ∎

We can now state the key result for proving the PS condition.

Theorem 11.2 [164] *The embedding of $W_r^{1,2}$ in $L^q(\mathbb{R}^n)$, $n \geq 3$, is compact for all $2 < q < 2^*$.*

Proof. Let $u_k \in W_r^{1,2}$ be such that $u_k \rightharpoonup 0$. From (11.3) it follows that

$$|u_k(r)| \leq C_1 r^{(1-n)/2}\|u_k\| \leq C_2 r^{(1-n)/2}.$$

Since $q > 2$ we deduce that, given $\varepsilon > 0$, there exists $C_3 > 0$ and $\overline{R} > 0$ such that $|u_k(r)|^q \leq C_3\varepsilon |u_k(r)|^2$, for all $r \geq \overline{R}$. This implies

$$\int_{|x| \geq \overline{R}} |u_k(x)|^q\, dx \leq C_3\varepsilon \int_{|x| \geq \overline{R}} |u_k(x)|^2\, dx \leq C_3\varepsilon \|u_k\|^2 \leq C_4\varepsilon. \tag{11.4}$$

Moreover, from the standard Sobolev embedding theorem, we have that $u_k \to 0$, strongly in $L^q(B_{\bar{R}})$, for every $2 \le q < 2^*$. Thus there exists $k_0 > 0$ such that for all $k \ge k_0$ one has

$$\int_{|x| \le \bar{R}} |u_k(x)|^q \, dx \le \varepsilon.$$

This and (11.4) imply that $\int_{\mathbb{R}^n} |u_k(x)|^q \, dx \le C_5 \varepsilon$ for $k \ge k_0$, proving that $u_k \to 0$, strongly in $L^q(\mathbb{R}^n)$, for every $2 < q < 2^*$. ∎

Following W. Strauss [164], we are now in a position to prove the following theorem.

Theorem 11.3 *If* $1 < p < (n+2)/(n-2)$, *(11.1) has a classical radial solution.*

Proof. We will work in $W_r^{1,2}$. Let $c > 0$ denote the MP level of J on $W_r^{1,2}$. Using Theorem 11.2 it is easy to check that c is a critical level which carries a critical point $u \in W_r^{1,2}$ such that $J'(u) \perp W_r^{1,2}$. Let us show that u indeed satisfies $J'(u) = 0$. Let $\sigma \in O(n)$ denote a generic rotation in $\mathbb{R}^n$ and define, for all $v \in W^{1,2}$, $v_\sigma(x) := v(\sigma^{-1}x)$. One immediately verifies (that $J(v_\sigma) = J(v)$ and) that $J'(v_\sigma) = (J'(v))_\sigma$. Since u is radial, then $u = u_\sigma$ and hence $J'(u) = (J'(u))_\sigma$, namely $J'(u) \in W_r^{1,2}$. Since one also has $J'(u) \perp W_r^{1,2}$, we infer that $J'(u) = 0$. By elliptic regularity, $u \in C^2$ and then the usual argument implies that $u > 0$ on $\mathbb{R}^n$. This completes the proof. ∎

Remark 11.4

(a) The fact, proved above, that u is radial is a particular case of a more general abstract result. More precisely, suppose that the topological group G acts on the Hilbert space E through isometries and let $J \in C^1(E, \mathbb{R})$ be G-invariant. Then any critical point of J on $\mathrm{Fix}(G) = \{u \in E : gu = u, \forall g \in G\}$ is a critical point of J on E. This is called the *symmetric criticality principle* by R. Palais [142].

(b) It is possible to show that any C^2 radial solution u of (11.1) has an exponential decay at infinity: namely there are $C, \delta > 0$ such that

$$|u(r)| \le C r^{(1-n)/2} e^{-\delta r}, \qquad r \gg 1.$$

We will be sketchy, leaving the details to the reader. Since u verifies $-(r^{n-1}u')' = r^{n-1}u^p$ then, setting $w(r) = r^{n-1}u^2(r)$, one finds that $\frac{1}{2}w'' \ge (1 + br^{-2} - u^{p-1})w$, where $b = (n-1)(n-2)/2$. Using (11.3), it follows that $w'' \ge mw$, for some $m > 0$, for all $r \gg 1$. From this, an elementary argument implies that $w \le Ce^{-\sqrt{m}r}$, whence $|u(r)| \le C r^{(1-n)/2} e^{-\sqrt{m}r/2}$.

(c) For further results dealing with more general equations like
$-\Delta u + u = f(u)$, we refer to [53]. ∎

11.2 Boundary value problems with critical exponent

In this section we will study the BVP with critical exponent

$$-\Delta u = \lambda u + u^{(n+2)/(n-2)}, \quad u > 0, \quad u \in H_0^1, \tag{11.5}$$

where, as usual, $n \geq 3$, Ω is a bounded domain in $\mathbb{R}^n$ with smooth boundary $\partial\Omega$ and $H_0^1 = H_0^1(\Omega)$ is endowed with the norm $\|u\|^2 = \int_\Omega |\nabla u|^2 \, dx$.
Solutions of (11.5) are critical points of $J_\lambda \in C^\infty(H_0^1, \mathbb{R})$,

$$J_\lambda(u) = \frac{1}{2}\|u\|^2 - \frac{1}{2}\lambda|u|_2^2 - \frac{1}{2^*}|u|_{2^*}^{2^*},$$

where $|u|_p$ denotes the norm in $L^p(\Omega)$. Let us make a couple of remarks.

Remark 11.5
(a) For $\lambda < \lambda_1$ (the first the eigenvalue of $-\Delta$ on H_0^1) the functional J_λ
satisfies (MP-1). Moreover, for any fixed $\bar{u} \in E \setminus \{0\}$, one has that
$J_\lambda(t\bar{u}) \to -\infty$ at $t \to \infty$ and then (MP-2) holds with, say, $e = t\bar{u}, t \gg 1$.
As a consequence, if $\lambda < \lambda_1$, then J_λ has the MP geometry.
(b) On the other hand, if Ω is star shaped, (11.5) has no solution for $\lambda \leq 0$ (by
the Pohozaev identity). ∎

From these two remarks we deduce that J_λ does not satisfy the *(PS)* condition
for $\lambda \leq 0$. Of course, the failure of the *(PS)* condition is related to the fact that
the embedding of $L^{2^*}(\Omega)$ into $H_0^1(\Omega)$ is not compact. We want to investigate
what happens for $\lambda > 0$. The main result we are going to discuss is the following
one, contained in the aforementioned paper by Brezis and Nirenberg.

Theorem 11.6 *If $n \geq 4$ problem (11.5) has a solution for all $\lambda \in \,]0, \lambda_1[$.*
If $n = 3$ there exists $\lambda^* \in [0, \lambda_1[$ such that (11.5) has a solution if and only if
$\lambda \in]\lambda^*, \lambda_1[$.

Remark 11.7 It might happen that $\lambda^* > 0$. For example, if Ω is the unit ball
and $n = 3$, then one can show that $\lambda^* = \lambda_1/4$, see the paper by Brezis and
Nirenberg [64]. ∎

Remark 11.8 Let $\lambda^* = 0$ and let us consider solutions u_k of (11.5) corresponding to a sequence $\lambda_k \downarrow 0$. If Ω is star shaped, one has that $\|u_k\| \to +\infty$.

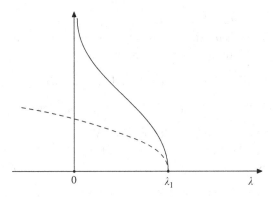

Figure 11.1 Bifurcation diagram of positive solutions of (11.5) when $\lambda^* = 0$ and Ω is star shaped (solid curve). The dashed line represents the branch of positive solutions of the subcritical equation $-\Delta u = \lambda u + u^p$, $u \in H_0^1(\Omega)$, $1 < p < (n+2)/(n-2)$.

Otherwise one finds that, up to a subsequence, $u_k \to u_0$, which is a solution of (11.5) with $\lambda = 0$, in contradiction with the Pohozaev identity. ∎

Roughly, we will show two facts:

(i) the $(PS)_c$ condition holds for c smaller than a certain threshold involving the best Sobolev constant, see below;
(ii) the MP critical level of J_λ lies below such a threshold, provided $\lambda \in]0, \lambda_1[$, if $n \geq 4$, respectively $\lambda \in]\lambda^*, \lambda_1[$, if $n = 3$.

To carry out this program, we first need some preliminaries. We denote by S the best Sobolev constant for the embedding of $H_0^1(\Omega)$ into $L^*(\Omega)$, namely

$$S = \inf\{\|u\|^2 \; : \; u \in H_0^1, \; |u|_{2*} = 1\}.$$

Recall that S does not depend on Ω and is never achieved, unless $\Omega = \mathbb{R}^n$ (otherwise there would be a positive solution $u \in H_0^1$ of $-\Delta u = Su^{(n+2)/(n-2)}$, in contradiction with the Pohozaev identity). If $\Omega = \mathbb{R}^n$ the best constant S is achieved by the function

$$U(x) = \frac{c}{(1 + |x|^2)^{(n-2)/2}}, \quad c = (n(n-2))^{(n-2)/4},$$

that satisfies

$$-\Delta U = S\, U^{(n+2)/(n-2)}, \quad x \in \mathbb{R}^n.$$

Since this problem is dilation invariant, it turns out that also

$$U_\varepsilon(x) := \varepsilon^{-(n-2)/2} U(x/\varepsilon), \quad \varepsilon > 0$$

as well as the translated $U_\varepsilon(x - \xi)$, $\xi \in \mathbb{R}^n$, satisfy $-\Delta U_\varepsilon = S U_\varepsilon^{(n+2)/(n-2)}$ ($x \in \mathbb{R}^n$) (actually it is possible to show that they are the only regular positive solutions of this equation). Moreover, one has that

$$U_\varepsilon \in L^{2^*}(\mathbb{R}^n), \quad \nabla U_\varepsilon \in L^2(\mathbb{R}^n), \text{ and } \|U_\varepsilon\|^2 = |U_\varepsilon|_{2^*}^{2^*} = S^{n/2}.$$

As anticipated in the items (i) and (ii) before, we shall prove:

- J_λ satisfies the $(PS)_c$ condition for all $c < \frac{1}{n}S^{n/2}$;
- if c_λ denotes the MP critical level, one has that $c_\lambda < \frac{1}{n}S^{n/2}$ provided $\lambda \in]0, \lambda_1[$ (respectively $\lambda \in]\lambda^*, \lambda_1[$) if $n \geq 4$ (respectively if $n = 3$).

Let us begin with the (PS) condition. First we show a lemma, due to Brezis and Lieb [63]. Let us point out explicitly that in this lemma Ω is any domain, possibly unbounded, in $\mathbb{R}^n$.

Lemma 11.9 *Let* $u_m \in L^p(\Omega)$, $1 \leq p < \infty$ *be such that*

(a) $|u_m|_p \leq$ constant
(b) $u_m \to u$ *a.e. in* Ω.

Then

$$\lim |u_m - u|_p^p = \lim |u_m|_p^p - |u|_p^p. \tag{11.6}$$

Remark 11.10
(a) If, in addition to (a) and (b), one has also that $|u_m|_p \to |u|_p$ then u_m converges to u strongly in $L^p(\Omega)$.
(b) If we replace L^p with any Hilbert space, one has that

$$\|u_m - u\|^2 = \|u_m\|^2 + \|u\|^2 - 2(u_m|u). \tag{11.7}$$

Hence if u_m converges weakly to u then (11.6) holds. ∎

Proof of the lemma. By the Fatou lemma,

$$|u|_p \leq \liminf_{m \to \infty} |u_m|_p \leq \text{constant}$$

and this implies that $u \in L^p(\Omega)$. Now we use the following inequality: for all $\varepsilon > 0$ there exists $c(\varepsilon) > 0$ such that

$$\left| |a + b|^p - |a|^p - |b|^p \right| \leq \varepsilon |a|^p + c(\varepsilon)|b|^p, \quad a, b \in \mathbb{R} \tag{11.8}$$

which, in turn, follows easily from

$$\lim_{t \to +\infty} \frac{|t + 1|^p - |t|^p - 1}{|t|^p} = 0.$$

Using (11.8) with $a = u_m - u$, and $b = u$ we get

$$\left| \underbrace{|u_m|^p - |u_m - u|^p - |u|^p}_{v_m} \right| \leq \varepsilon |u_m - u|^p + c(\varepsilon)|u|^p,$$

namely

$$|v_m| - \varepsilon |u_m - u|^p \leq c(\varepsilon)|u|^p.$$

By the dominated convergence theorem we infer

$$\lim_{m \to \infty} \int_\Omega \left(|v_m| - \varepsilon |u_m - u|^p \right) = 0. \tag{11.9}$$

Moreover, from

$$|v_m| \leq \left(|v_m| - \varepsilon |u_m - u|^p \right) + \varepsilon |u_m - u|^p$$

we get

$$\int_\Omega |v_m| \leq \int_\Omega \left(|v_m| - \varepsilon |u_m - u|^p \right) + \varepsilon \int_\Omega |u_m - u|^p.$$

Using (11.9) it follows that

$$\int_\Omega |v_m| \leq o(1) + c_1 \varepsilon, \quad c_1 = \sup_m |u_m - u|_p^p < \infty.$$

In conclusion, we find

$$\int_\Omega |v_m| = \int_\Omega \left| |u_m|^p - |u_m - u|^p - |u|^p \right| \to 0,$$

proving the lemma.

We can now prove the next lemma.

Lemma 11.11 J_λ *satisfies the* $(PS)_c$ *condition for all* $c < \frac{1}{n} S^{n/2}$.

Proof. Let $J_\lambda(u_m) \to c$ and $J'_\lambda(u_m) \to 0$. As in the subcritical case one has that $\|u_m\|$ is bounded and hence we can assume that $u_m \rightharpoonup u$ in H_0^1 and in $L^{2^*}(\Omega)$, $u_m \to u$ in $L^2(\Omega)$ and $u_m \to u$ a.e. in Ω. From $J'_\lambda(u_m) \to 0$ it follows that

$$(J'_\lambda(u_m)|\varphi) = (u_m|\varphi) - \lambda \int u_m \varphi - \int |u_m|^{2^*-2} u_m \varphi = o(1), \quad \forall \varphi \in H_0^1.$$

Then

$$(u|\varphi) - \lambda \int u\varphi - \int |u|^{2^*-2} u\varphi = 0, \quad \forall \varphi \in H_0^1.$$

This means that u is a weak solution of (11.5). Taking $\varphi = u$ we find

$$\|u\|^2 - \lambda|u|_2^2 = |u|_{2^*}^{2^*} \ (\geq 0) \tag{11.10}$$

and hence

$$J_\lambda(u) = \frac{1}{2}\|u\|^2 - \frac{1}{2}\lambda|u|_2^2 - \frac{1}{2^*}|u|_{2^*}^{2^*} = \left(\frac{1}{2} - \frac{1}{2^*}\right)|u|_{2^*}^{2^*} = \frac{1}{n}|u|_{2^*}^{2^*}. \tag{11.11}$$

After these preliminaries we can now prove that $w_m := u_m - u$ converges to 0 strongly in H_0^1. Using (11.7) we infer:

$$\|u_m\|^2 = \|w_m\|^2 + \|u\|^2 + o(1); \tag{11.12}$$

$$|u_m|_2^2 = |w_m|_2^2 + |u|_2^2 + o(1) = |u|_2^2 + o(1). \tag{11.13}$$

Furthermore, using the Brezis–Lieb lemma we find

$$|u_m|_{2^*}^{2^*} = |w_m|_{2^*}^{2^*} + |u|_{2^*}^{2^*} + o(1). \tag{11.14}$$

From (11.12), (11.13), (11.14) it follows that

$$J_\lambda(u_m) = \frac{1}{2}\|u_m\|^2 - \frac{1}{2}\lambda|u_m|_2^2 - \frac{1}{2^*}|u_m|_{2^*}^{2^*}$$

$$= J_\lambda(u) + \frac{1}{2}\|w_m\|^2 - \frac{1}{2^*}|w_m|_{2^*}^{2^*} + o(1). \tag{11.15}$$

Moreover, $(J_\lambda'(u_m)|u_m) \to 0$ implies

$$\|u_m\|^2 - \lambda|u_m|_2^2 - |u_m|_{2^*}^{2^*} = o(1).$$

Then, using again (11.12), (11.13), (11.14) we find

$$\|w_m\|^2 + \|u\|^2 - \lambda|u|_2^2 - \left(|w_m|_{2^*}^{2^*} + |u|_{2^*}^{2^*}\right) = o(1)$$

and hence

$$\|w_m\|^2 - |w_m|_{2^*}^{2^*} = -\left(\|u\|^2 - \lambda|u|_2^2 - |u|_{2^*}^{2^*}\right) + o(1) = o(1),$$

because $\|u\|^2 - \lambda|u|_2^2 - |u|_{2^*}^{2^*} = 0$, see (11.10). It follows that

$$\lim \|w_m\|^2 = \lim |w_m|_{2^*}^{2^*} \ (\equiv \alpha). \tag{11.16}$$

By the Sobolev embeddings we know that $S|w_m|_{2^*}^2 \leq \|w_m\|^2$ and hence

$$S\alpha^{2/2^*} \leq \alpha.$$

If $\alpha = 0$ we are done. If $\alpha \neq 0$ then $\alpha \geq S^{n/2}$. In such a case, from $J_\lambda(u_m) \to c$ and using (11.15) and (11.16) we get

$$c = J_\lambda(u) + \frac{1}{2}\|w_m\|^2 - \frac{1}{2^*}|w_m|_{2^*}^{2^*} + o(1) = J_\lambda(u) + \left(\frac{1}{2} - \frac{1}{2^*}\right)\alpha + o(1).$$

Since $J_\lambda(u) \geq 0$, see (11.11),

$$c \geq \frac{1}{n}\alpha + o(1) \geq \frac{1}{n}S^{n/2} + o(1),$$

a contradiction, because $c < \frac{1}{n}S^{n/2}$, by our assumption. So $\alpha = 0$. ∎

To complete the proof of Theorem 11.6 it remains to show that the MP level c_λ is smaller than $\frac{1}{n}S^{n/2}$ for $\lambda \in]0, \lambda_1[$ if $n \geq 4$, respectively $\lambda \in]\lambda^*, \lambda_1[$ if $n = 3$. We will prove this fact in the former case, only. The case $n = 3$ requires some more technicalities and we refer to the original paper by Brezis and Nirenberg.

Without loss of generality, we assume that $0 \in \Omega$. Let $\varphi \in C_0^\infty(\Omega)$, $\varphi(x) \equiv 1$ in the ball $|x| < \rho$. We set

$$u_\varepsilon(x) = \varphi(x)U_\varepsilon(x).$$

Lemma 11.12 *For $\varepsilon \to 0$ one has that $\|u_\varepsilon\|^2 - \lambda|u_\varepsilon|_2^2 < S|u_\varepsilon|_{2*}^2$.*

Proof. We claim:

$$\|u_\varepsilon\|^2 = \int_{\mathbb{R}^n} |\nabla U_\varepsilon|^2 + O(\varepsilon^{n-2}) = S^{n/2} + O(\varepsilon^{n-2}); \qquad (11.17)$$

$$|u_\varepsilon|_{2*}^{2*} = \int_{\mathbb{R}^n} U_\varepsilon^{2*} + O(\varepsilon^{n-2}) = S^{n/2} + O(\varepsilon^n); \qquad (11.18)$$

$$|u_\varepsilon|_2^2 = \begin{cases} \text{constant } \varepsilon^2|\log \varepsilon| + O(\varepsilon^2) & \text{if } n = 4 \\ \text{constant } \varepsilon^2 + O(\varepsilon^{n-2}) & \text{if } n \geq 5. \end{cases} \qquad (11.19)$$

If the claim holds true, then for $n \geq 5$,

$$\frac{\|u_\varepsilon\|^2 - \lambda|u_\varepsilon|_2^2}{|u_\varepsilon|_{2*}^2} = \frac{S^{n/2} - \lambda\varepsilon^2 + O(\varepsilon^{n-2})}{(S^{n/2} + O(\varepsilon^n))^{(n-2)/n}} = S - \lambda\varepsilon^2 + O(\varepsilon^{n-2}) < S,$$

provided $|\varepsilon|$ is small enough. Similarly, for $n = 4$,

$$\frac{\|u_\varepsilon\|^2 - \lambda|u_\varepsilon|_2^2}{|u_\varepsilon|_{2*}^2} = S - \lambda\varepsilon^2|\log \varepsilon| + O(\varepsilon^2) < S \quad \text{as } \varepsilon \to 0,$$

and the lemma follows. It remains to prove the claim.

Proof of (11.17). Since, up to a constant,

$$u_\varepsilon(x) = \varphi(x) \cdot \varepsilon^{(n-2)/2}\left(\varepsilon^2 + |x|^2\right)^{-(n-2)/2},$$

we find

$$\nabla u_\varepsilon(x) = \frac{\varepsilon^{(n-2)/2}\nabla\varphi(x)}{\left(\varepsilon^2 + |x|^2\right)^{(n-2)/2}} - (n-2)\frac{\varepsilon^{(n-2)/2}x\varphi(x)}{\left(\varepsilon^2 + |x|^2\right)^{n/2}}.$$

Since $\varphi(x) \equiv 1$ on $|x| < \rho$, it follows that

$$\begin{aligned}
\|u_\varepsilon\|^2 &= (n-2)^2\varepsilon^{n-2}\int_\Omega \frac{|x|^2}{\left(\varepsilon^2 + |x|^2\right)^n} + O(\varepsilon^{n-2}) \\
&= (n-2)^2\varepsilon^{n-2}\int_{\mathbb{R}^n} \frac{|x|^2}{\left(\varepsilon^2 + |x|^2\right)^n} + O(\varepsilon^{n-2}).
\end{aligned}$$

Performing the change of variable $x = \varepsilon y$ we get

$$\begin{aligned}
\|u_\varepsilon\|^2 &= (n-2)^2\varepsilon^{n-2}\int_{\mathbb{R}^n} \frac{\varepsilon^{n+2}|y|^2}{\varepsilon^{2n}\left(1 + |y|^2\right)^n} + O(\varepsilon^{n-2}) \\
&= (n-2)^2\int_{\mathbb{R}^n} \frac{|y|^2}{\left(1 + |y|^2\right)^n} + O(\varepsilon^{n-2}).
\end{aligned}$$

Now, the constant in the definition of U is such that

$$(n-2)^2\int_{\mathbb{R}^n} \frac{|y|^2}{\left(1 + |y|^2\right)^n} = \int_{\mathbb{R}^n} |\nabla U(y)|^2.$$

In conclusion we find:

$$\|u_\varepsilon\|^2 = \int_{\mathbb{R}^n} |\nabla U(y)|^2 + O(\varepsilon^{n-2}) = S^{n/2} + O(\varepsilon^{n-2}),$$

proving (11.17).

Proof of (11.18). There holds

$$\begin{aligned}
|u_\varepsilon|_{2^*}^{2^*} &= \varepsilon^n\int_\Omega \frac{\varphi^{2^*}(x)}{\left(\varepsilon^2 + |x|^2\right)^n} \\
&= \varepsilon^n\int_\Omega \frac{\varphi^{2^*}(x) - 1}{\left(\varepsilon^2 + |x|^2\right)^n} + \varepsilon^n\int_\Omega \frac{1}{\left(\varepsilon^2 + |x|^2\right)^n} \\
&= \int_{\mathbb{R}^n} \frac{1}{\left(1 + |y|^2\right)^n} + O(\varepsilon^n) = S^{n/2} + O(\varepsilon^n).
\end{aligned}$$

Proof of (11.19). One has

$$|u_\varepsilon|_2^2 = \varepsilon^{n-2}\int_\Omega \frac{1}{\left(\varepsilon^2 + |x|^2\right)^{n-2}} + O(\varepsilon^{n-2}). \tag{11.20}$$

Consider first the case $n \geq 5$: one finds that

$$\int_\Omega \frac{1}{\left(\varepsilon^2 + |x|^2\right)^{n-2}} = \int_{\mathbb{R}^n} \frac{1}{\left(\varepsilon^2 + |x|^2\right)^{n-2}} + O(1)$$

$$= \varepsilon^{4-n} \int_\Omega \frac{1}{\left(1 + |y|^2\right)^{n-2}} + O(1) = c_1 \varepsilon^{4-n} + O(1).$$

Substituting into (11.20) we find

$$|u_\varepsilon|_2^2 = c_1 \varepsilon^2 + O(\varepsilon^{n-2}).$$

If $n = 4$ we get

$$|u_\varepsilon|_2^2 = \varepsilon^2 \int_\Omega \frac{1}{\left(\varepsilon^2 + |x|^2\right)^2} + O(\varepsilon^2).$$

We can choose $R_1 < R_2$ such that

$$\int_{|x|<R_1} \frac{1}{\left(\varepsilon^2 + |x|^2\right)^2} \leq \int_\Omega \frac{1}{\left(\varepsilon^2 + |x|^2\right)^2} \leq \int_{|x|<R_2} \frac{1}{\left(\varepsilon^2 + |x|^2\right)^2}.$$

Since (below ω denotes the measure of the unit sphere in $\mathbb{R}^3$)

$$\int_{|x|<R} \frac{1}{\left(\varepsilon^2 + |x|^2\right)^2} = \omega \int_0^R \frac{r^3 dr}{\left(\varepsilon^2 + r^2\right)^2} = \omega |\log \varepsilon| + O(1),$$

we get

$$|u_\varepsilon|_2^2 = \text{constant}.\varepsilon^2 |\log \varepsilon| + O(\varepsilon^2).$$

This completes the proof of the lemma. ∎

We can finally complete the proof of Theorem 11.6 by showing the following.

Lemma 11.13 *If c_λ denotes the MP critical level, one has that $c_\lambda < \frac{1}{n} S^{n/2}$ provided $\lambda \in]0, \lambda_1[$ (respectively $\lambda \in]\lambda^*, \lambda_1[$) if $n \geq 4$ (respectively if $n = 3$).*

Proof. First, let us take in Remark 11.5(a) $\bar{u} = u_\varepsilon$, with a fixed $\varepsilon > 0$ such that Lemma 11.12 holds. Taking the half-line $\{tu_\varepsilon : t \geq 0\}$ as test curve for the MP level, we have that

$$c_\lambda \leq \max_{t \geq 0} J_\lambda(tu_\varepsilon).$$

Since

$$J_\lambda(tu_\varepsilon) = \frac{1}{2} t^2 \left(\|u_\varepsilon\|^2 - \lambda |u_\varepsilon|_2^2 \right) - \frac{1}{2^*} t^{2^*} |u_\varepsilon|_{2^*}^{2^*},$$

then $\max_{t \geq 0} J_\lambda(tu_\varepsilon)$ is achieved at

$$\tau = \left(\frac{\|u_\varepsilon\|^2 - \lambda|u_\varepsilon|_2^2}{|u_\varepsilon|_{2^*}^{2^*}} \right)^{(n-2)/4},$$

and

$$J_\lambda(\tau u_\varepsilon) = \frac{1}{n} \left(\frac{\|u_\varepsilon\|^2 - \lambda|u_\varepsilon|_2^2}{|u_\varepsilon|_{2^*}^{2^*}} \right)^{n/2}.$$

Then, using also Lemma 11.12, we find

$$c_\lambda \leq \max_{t \geq 0} J_\lambda(tu_\varepsilon) = J_\lambda(\tau u_\varepsilon) = \frac{1}{n} \left(\frac{\|u_\varepsilon\|^2 - \lambda|u_\varepsilon|_2^2}{|u_\varepsilon|_{2^*}^{2^*}} \right)^{n/2} < \frac{1}{n} S^{n/2},$$

and this completes the proof. ∎

Remark 11.14 It is easy to see that $c_\lambda < \frac{1}{n} S^{n/2}$ is equivalent to the condition $S_\lambda < S$, where $S_\lambda := \inf\{\|u_\varepsilon\|^2 - \lambda|u_\varepsilon|_2^2 : u \in H_2^1, |u_\varepsilon|_{2^*} = 1\}$. ∎

11.3 Discontinuous nonlinearities

Here we consider the case in which the nonlinearity f has a discontinuity with respect to u. Problems with this feature arise, for example, in plasma physics.

11.3.1 A general result

Following [12], let us consider the following Dirichlet boundary value problem

$$\begin{cases} -\Delta u = f(u) + h(x) & x \in \Omega \\ u = 0 & x \in \partial\Omega, \end{cases} \tag{11.21}$$

where $h \in L^2(\Omega)$ and f satisfies:

(a) there is $a \in \mathbb{R}$ such that $f \in C(\mathbb{R} \setminus \{a\}, \mathbb{R})$ and $f(a-) = \lim_{u \to a-} f(u)$, $f(a+) = \lim_{u \to a+} f(u)$ exists and $f(a-) < f(a+)$. Moreover $f(a) \in T_a$, where $T_a = [f(a-), f(a+)]$;

(b) $\liminf_{u \to +\infty} f(u) > -\infty$ and $\limsup_{u \to -\infty} f(u) < +\infty$;

(c) there is $m \geq 0$ such that $u \mapsto mu + f(u)$ is strictly increasing.

Remark 11.15 To avoid technicalities, we assumed that f has a discontinuity at a single point a. More in general, the arguments that we will carry out can

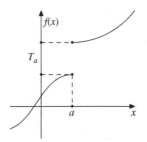 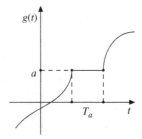

Figure 11.2 The functions f and g.

be easily extended to cover the case in which $f \in C(\mathbb{R} \setminus A, \mathbb{R})$, where A is a set with no accumulation points. ∎

We use the same symbol f to denote the multivalued map which is equal to $f(u)$ for $u \neq a$ and such that $f(a) = T_a$. With this notation, by solution of (11.21) we mean a function $u \in H_0^1(\Omega) \cap H^{2,2}(\Omega)$ such that

$$-\Delta u(x) - h(x) \in f(u(x)), \qquad \text{a.e. in } \Omega.$$

Of course, defining

$$\Omega_a(u) = \{x \in \Omega : u(x) = a\},$$

if $|\Omega_a(u)| = 0$ ($|A|$ stands for the measure of the set $A \subset \mathbb{R}^n$) then u satisfies $-\Delta u(x) = f(u(x)) + h(x)$, for a.e. $x \in \Omega$.

In order to find solutions of (11.21) we will transform such an equation in an equivalent problem, involving a C^1 functional.

From (b) and (c) it follows that there exists $M > 0$ such that $u \mapsto Mu + f(u)$ is strictly increasing and such that $Mu + f(u) \to \pm\infty$ as $u \to \pm\infty$. Then the function g defined by setting $g(t) = u$ if and only if $t \in Mu + f(u)$, namely

$$g(t) = \begin{cases} a & \text{if } t - Ma \in T_a \\ u \quad \text{such that } Mu + f(u) = t & \text{if } t - Ma \notin T_a, \end{cases}$$

is defined on all of $\mathbb{R}$, is continuous and $G(t) = \int_0^t g(s)\,ds$ is of class C^1. Let $E = L^2(\Omega)$, let $K : E \mapsto E$ be defined by

$$u = K(v) \quad \Longleftrightarrow \quad -\Delta u + Mu = v, \quad u \in H_0^1(\Omega),$$

and set $J : E \mapsto \mathbb{R}$,

$$J(v) = \int_\Omega \left[G(v) - \tfrac{1}{2}vK(v) - vK(h) \right] dx.$$

The following theorem shows the links between J and (11.21).

Theorem 11.16 *Let $h \in L^2(\Omega)$ and let f satisfy (a) and (b).*

(i) *$J \in C^1(E, \mathbb{R})$; moreover, if $J'(w) = 0$ then $z = K(w + h)$ is a solution of (11.21), in the sense specified above.*

(ii) *If either $-h(x) \notin T_a$ a.e. in Ω_a or if w is a local minimum of J, then $|\Omega_a(z)| = 0$ and thus z satisfies (11.21) a.e. in Ω.*

Proof. From (b) it follows that there is $c > 0$ such that $|Mu + f(u)| \geq M|u| - c$. Setting $t = Mu + f(u)$ we get immediately that $|g(t)| \leq \frac{1}{M}|t| + c_1$ and $|G(t)| \leq c_2 t^2 + c_3$. Using Theorem 1.7, it follows that $J \in C^1(E, \mathbb{R})$.

Let $w \in E$ be such that $J'(w) = 0$ and set $z = K(w + h)$. Then $z \in H_0^1(\Omega) \cap H^{2,2}(\Omega)$ and $G'(w) = K(w) + K(h)$. This implies $g(w) = z$ or else

$$w - Mz \in f(z). \tag{11.22}$$

On the other hand, from $z = K(w + h)$ it also follows that $-\Delta z + Mz = w + h$. This and (11.22) imply that $-\Delta z - h \in f(z)$, a.e. in Ω, proving (i).

Next, a theorem by Stampacchia yields $-\Delta z(x) = 0$ a.e. on Ω_a and thus $-h(x) \in T_a$ a.e. on $\Omega_a(z)$. This implies that $|\Omega_a(z)| = 0$ provided $-h \notin T_a$. Finally, suppose that ($-h \in T_a$ a.e. on $\Omega_a(z)$, and) w is a local minimum of J and set

$$T_a^+ = [f(a-), \tfrac{1}{2}(f(a-) + f(a+)], \quad T_a^- = T \setminus T^+,$$
$$\Omega^\pm = \{x \in \Omega : -h(x) \in T^\pm\}.$$

Define

$$\chi(x) = \begin{cases} 1 & x \in \Omega^+ \\ -1 & x \in \Omega^- \\ 0 & x \in \Omega \setminus \Omega_a. \end{cases}$$

For $\varepsilon > 0$ small enough one has that

$$-h(x) + \varepsilon \chi(x) \in T_a, \qquad \text{a.e. on } \Omega_a. \tag{11.23}$$

There holds

$$\frac{d}{d\varepsilon} J(w + \varepsilon \chi) = \int_\Omega g(w + \varepsilon \chi) \chi \, dx - \varepsilon \int_\Omega \chi K(\chi) \, dx - \int_\Omega \chi K(w + h) \, dx.$$

From $w + h = -\Delta z + Mz$ it follows that $w + h = Ma$ a.e. on $\Omega_a(z)$. This and the definition of χ imply

$$\int_\Omega g(w + \varepsilon \chi) \chi \, dx = \int_{\Omega_a(z)} g(w + \varepsilon \chi) \chi \, dx = \int_{\Omega_a(z)} g(Ma - h + \varepsilon \chi) \chi \, dx.$$

Set $t = Ma - h + \varepsilon\chi$. Using (11.23) we infer that $t - Ma \in T_a$ and therefore, by the definition of g we have $g(t) = a$. This and the preceding equation yield

$$\int_\Omega g(w + \varepsilon\chi)\chi \, dx = a \int_{\Omega_a(z)} \chi \, dx.$$

One also has

$$\int_\Omega \chi K(w + h) \, dx = \int_\Omega \chi z \, dx = \int_{\Omega_a(z)} \chi z \, dx = a \int_{\Omega_a(z)} \chi \, dx.$$

Moreover, setting $K(\chi) = \eta$, we find $-\Delta\eta + M\eta = \chi$ and hence

$$\int_\Omega \chi K(\chi) \, dx = \int_\Omega (|\nabla\eta|^2 + M\eta^2) \, dx.$$

Putting together these calculations we deduce

$$\frac{d}{d\varepsilon} J(w + \varepsilon\chi) = -\varepsilon \int_\Omega \chi K(\chi) \, dx = -\varepsilon \int_\Omega (|\nabla\eta|^2 + M\eta^2) \, dx. \qquad (11.24)$$

Since w is a local minimum of J, then $\eta \equiv 0$, which implies that $\chi \equiv 0$. This is equivalent to say that $|\Omega_a(z)| = 0$, and the proof is complete. ∎

Remark 11.17 The arguments above are inspired by Clarke's *dual variational principle* [80], see also the book by Ekeland [93] for applications to hamiltonian systems. The smoothing effect of this principle was first highlighted in [12]. ∎

Let us indicate a simple case in which J has indeed a minimum on E and Theorem 11.16 applies. Suppose that f satisfies

$$|f(u)| \leq \alpha|u| + k, \qquad \alpha < \lambda_1, \quad k > 0. \qquad (11.25)$$

Lemma 11.18 *If, in addition to (a)–(c), (11.25) holds, then J is bounded from below and coercive.*

Proof. If (11.25) holds, then repeating the previous argument, one finds

$$G(t) \geq \frac{1}{2} \frac{t^2}{\alpha + M} - c_0|t|.$$

This yields

$$\int_\Omega G(v) \, dx \geq \frac{1}{2} \frac{1}{\alpha + M} \|v\|_{L^2}^2 - c_1 \|v\|_{L^2}. \qquad (11.26)$$

On the other hand, the definition of K easily implies

$$\int_\Omega v K(v) \, dx \leq \frac{1}{\lambda_1 + M} \|v\|_{L^2}^2,$$

and we find

$$J(v) \geq \frac{1}{2} \frac{1}{\alpha + M} \|v\|_{L^2}^2 - \frac{1}{2} \frac{1}{\lambda_1 + M} \|v\|_{L^2}^2 - c_2 \|v\|_{L^2}.$$

Since $\alpha < \lambda_1$ it follows that J is bounded below and coercive. ∎

Next, we prove the following form of the (PS) condition (see Remark 8.7)

Lemma 11.19 J *satisfies* $(\widetilde{PS})_c$ *for all* $c \in \mathbb{R}$, *namely if* $v_j \in E$ *is such that* $J(v_j) \to c$ *and* $J'(v_j) \to 0$, *then there exists* $v^* \in E$ *such that* $J(v^*) = c$ *and* $J'(v^*) = 0$.

Proof. Here to simplify the notation we put $M = 0$ and $h = 0$. From $J(v_j) \to c$ it follows that $\|v_j\|_{L^2}$ is bounded and hence $v_j \rightharpoonup v^*$ (up to a subsequence). Set $u^* = K(v^*)$. From $J'(v_j) \to 0$ and the compactness of K, we infer that

$$g(v_j) \to K(v^*) = u^*, \qquad \text{strongly in } E \text{ and a.e. in } \Omega. \tag{11.27}$$

Let us first prove that $v_j \to v^*$ strongly in $L^2(\Omega')$, where $\Omega' = \Omega \setminus \Omega_a(u^*)$. Since $u^*(x) \neq a$ in Ω' and since f is continuous out of $u = a$, we get from (11.27)

$$v_j \to f(u^*), \qquad \text{a.e. in } \Omega'. \tag{11.28}$$

From (11.25) it follows that $|v_j| \leq c_1 |g(v_j)| + c_2$ and this, jointly with (11.27), implies that there exists $\phi \in L^2(\Omega)$ such that $|v_j| \leq \phi$. Therefore, using also (11.28), we find that $v_j \to f(u^*)$ in $L^2(\Omega')$. Since $v_j \rightharpoonup v^*$, we deduce

$$v_j \to v^*, \qquad \text{strongly in } L^2(\Omega'). \tag{11.29}$$

Since g is asymptotically linear, it follows that

$$g(v_j) \to g(v^*), \quad \text{in } L^2(\Omega'), \qquad \int_{\Omega'} G(v_j)\,dx \to \int_{\Omega'} G(v^*)\,dx. \tag{11.30}$$

To complete the proof, it is necessary to analyse separately the cases $0 \in T_a$ and $0 \notin T_a$.

First we claim that

$$0 \notin T_a, \quad \Longrightarrow \quad |\Omega_a(u^*)| = 0. \tag{11.31}$$

To prove the claim, let $T_a = [b_1, b_2]$ with $b_1 > 0$ (if $T_a \subset (-\infty, 0[$ the argument is quite similar), and let χ_a denote the characteristic function of $\Omega_a(u^*)$. Since $v_j \rightharpoonup v^*$ we find

$$\int_{\Omega_a(u^*)} v_j\,dx = \int_\Omega v_j \chi_a\,dx \to \int_{\Omega_a(u^*)} u^*\,dx.$$

As remarked before, one has that $v^* = -\Delta u^* = 0$ a.e. in $\Omega_a(u^*)$, and thus

$$\int_{\Omega_a(u^*)} v_j \, dx \to 0. \tag{11.32}$$

We can also use (11.27) to find that $g(v_j(x)) \to a$ for a.e. $x \in \Omega_a(u^*)$. Since g is continuous and increasing, it follows that $\lim \inf v_j(x) \geq b_1$. Recalling that $|v_j| \leq \phi \in L^2(\Omega)$, we use Fatou's lemma and (11.32) to get

$$b_1 |\Omega_a(u^*)| \leq \lim \inf \int_{\Omega_a(u^*)} v_j \, dx = 0,$$

proving the claim.

We are now ready to complete the proof of the lemma. If $0 \notin T_a$, then $|\Omega_a(u^*)| = 0$ and this together with (11.30) immediately implies that $g(v_j) \to g(v^*)$ in $L^2(\Omega)$ as well as $\int_\Omega G(v_j) \, dx \to \int_\Omega G(v^*)$. Therefore

$$\begin{aligned}
J(v^*) &= \int_\Omega \left[G(v^*) - \tfrac{1}{2} v^* K(v^*) \right] dx \\
&= \lim_j \int_\Omega \left[G(v_j) - \tfrac{1}{2} v_j K(v_j) \right] dx = \lim_j J(v_j) = c,
\end{aligned}$$

and

$$J'(v^*) = g(v^*) - K(v^*) = \lim_j \left[g(v_j) - K(v_j) \right] = \lim_j J'(v_j) = 0.$$

If $0 \in T_a$ then $g(v^*(x)) = g(0) = a = v^*(x)$ for a.e. $x \in \Omega_a(u^*)$ and once more $J'(v^*) = 0$. Moreover, $G(t) = at$ for $t \in T_a$ and one readily finds $|G(v_j) - av_j| \to 0$ a.e. in $\Omega_a(u^*)$. Thus

$$\int_{\Omega_a(u^*)} |G(v_j) - av_j| \, dx \to 0.$$

This, (11.30) and (11.32) imply

$$\int_\Omega G(v_j) \, dx = \int_{\Omega'} G(v_j) \, dx + \int_{\Omega_a(u^*)} G(v_j) \, dx \to \int_{\Omega'} G(v^*) \, dx.$$

Finally, $v^* = 0$ in $\Omega_a(u^*)$ and $G(0) = 0$ yield $\int_{\Omega'} G(v^*) \, dx = \int_\Omega G(v^*) \, dx$, and one finds that $J(v^*) = c$, as before. ∎

As pointed out in Remark 8.7 the condition $(\widetilde{PS})_c$ can replace the usual $(PS)_c$ condition to find a critical point of a functional which is coercive and bounded from below. Therefore we conclude with the following theorem.

Theorem 11.20 *Suppose that, in addition to (a)–(c), f verifes (11.25). Then J has a minimum on E which corresponds, through Theorem 11.16, to a function $z \in H_0^1(\Omega) \cap H^{2,2}(\Omega)$, satisfying (11.21) a.e. in Ω.*

Remark 11.21 Dealing with other specific nonlinearities, it could be conveni-
ent to work on a different space E. For example, if $f(u) \sim |u|^{p-1}u$ as $|u|$ tends to
0 and to ∞, with $1 < p < (n+2)/(n-2)$, it is convenient to take $E = L^\beta(\Omega)$,
where β is the conjugate exponent of $p+1$. Actually, here $G(t) \sim |t|^\beta$ and

$$J(v) \sim \|v\|_{L^\beta}^\beta - \int_\Omega vK(v)\,dx.$$

Note that $\beta < 2$ and thus J has the mountain-pass geometry. For some further examples in which Theorem 11.16 applies, we refer to [12]. See also
Exercises 8.5(iii) and 11.5(i). ∎

11.3.2 A problem with multiple solutions

Following [23], we now consider the problem

$$\begin{cases} -\Delta u = h(u-a)f_0(u) & x \in \Omega \\ u = 0 & x \in \partial\Omega, \end{cases} \tag{11.33}$$

where $a > 0$, f_0 is continuous and satisfies appropriate conditions at infinity
and h is the Heaviside function

$$h(t) = \begin{cases} 0 & t \le 0 \\ 1 & t > 0. \end{cases}$$

The choice of this specific nonlinearity is motivated by applications arising
in plasma physics in which one deals with quantities v that satisfy $-\Delta v = 0$
below a threshold a, and $-\Delta v = \phi(v)$ for $v > a$. Setting $u = v - u_{|\partial\Omega}$ and
$f_0(u) = \phi(u + u_{|\partial\Omega})$, we get exactly (11.33).

We assume that f_0 satisfies

(a′) $f_0 \in C(\mathbb{R})$, $f_0(u) \ge 0$, and is nondecreasing;
(b′) $|f_0(u)| \le \alpha|u| + k$, with $\alpha < \lambda_1$ and $k > 0$.

Note that (b′) is nothing but (11.25). We set $f(u) = h(u-a)f_0(u)$ and $b = f_0(a)$:
$b = \lim_{u \downarrow a} f(u)$ is the size of the jump of $f(u)$ at $u = a$.

Using the same arguments carried out in the preceding subsection, we take
$m > 0$, so that $f_m(u) := mu + f(u)$ is strictly increasing. As before, f_m also
denotes the multivalued function obtained by filling up the jump at $u = a$.
Define g_m and $G_m(t)$ by

$$g_m(t) = u \quad \Longleftrightarrow \quad t \in f_m(u), \qquad G_m(t) = \int_0^t g_m(s)\,ds.$$

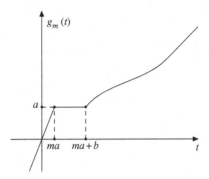

Figure 11.3 The function g_m.

Note that, from (a') and (b') it follows that

$$\frac{1}{2} \frac{t^2}{\alpha + m} - c_0|t| \leq G_m(t) \leq \frac{t^2}{2m}. \tag{11.34}$$

Finally, let $E = L^2(\Omega)$ and let $K_m : E \mapsto E$ be such that

$$K_m(v) = u \quad \Longleftrightarrow \quad -\Delta u + mu = v, \qquad u \in H_0^1(\Omega) \cap H^{2,2}(\Omega).$$

According to Theorems 11.16 and 11.20, the functional $J : E \mapsto \mathbb{R}$,

$$J(u) = \int_\Omega G_m(u)\,\mathrm{d}x - \frac{1}{2}\int_\Omega u K_m(u)\,\mathrm{d}x$$

is of class C^1 and if u is a critical point of J then $v = K_m(u)$ is a solution of (11.33). Moreover, from (11.34) it follows that J is bounded from below and coercive. As in Lemma 11.19, it is also easy to see that J satisfies the $(\widetilde{PS})_c$ condition, for all $c \in \mathbb{R}$.

We are going to show that, for suitable values of the parameter a, b, J has the same geometry as the functional studied in Theorem 8.14, see Figure 11.4, and hence it possesses two nontrivial critical points: one minimum w_1 with $J(w_1) < 0$ and a MP w_2 such that $J(w_2) > 0$. Let

$$\theta = \|\varphi_1\|_{L^1} \, \|\varphi_1\|_{L^2}^{-2},$$

where φ_1 is such that $-\Delta\varphi_1 = \lambda_1\varphi_1$ in Ω, $\varphi_1 = 0$ on $\partial\Omega$, and is normalized by taking $\|\varphi_1\|_\infty = 1$. Let us point out that, in general, θ depends upon Ω.

Lemma 11.22

(i) *If* $b/a > 2\lambda_1\theta$, *then for all* $m \ll 1$, $J(e) < 0$, *with* $e = b\varphi_1$.

(ii) *There are* $r, \rho > 0$, $r < \|e\|_{L^2}$, *such that, for all* $m \ll 1$, $J(v) \geq \rho$ *for all* $\|v\|_{L^2} = r$.

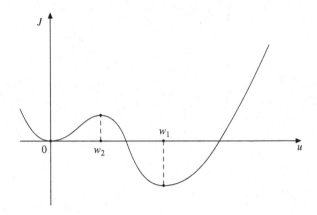

Figure 11.4 Behaviour of J for $b/a > 2\lambda_1\theta$.

Proof. (i) From $0 < b\varphi_1 \le b$ it follows that $g(b\varphi_1) \le a$. Then

$$J(b\varphi_1) \le ab \int_\Omega \varphi_1 \, dx - \tfrac{1}{2}b^2 \int_\Omega \varphi_1 K_m(\varphi_1) \, dx$$

$$\le ab\|\varphi_1\|_{L^1} - \tfrac{1}{2}b^2(\lambda_1 + m)^{-1}\|\varphi_1\|_{L^2},$$

proving (i).

(ii) Take $q = 2n/(n+2) < 2$, and let r be its conjugate exponent. Using the Hölder inequality, the Sobolev embedding theorem and elliptic regularity, we get

$$\int_\Omega v K_m(v) \, dx \le \|K_m(v)\|_{L^r}\|v\|_{L^q} \le c_1\|K_m(v)\|_{H^{2,q}}\|v\|_{L^q} \le c_2\|v\|_{L^q}^2. \tag{11.35}$$

Let us point out that the constants c_1, c_2 can be chosen independent of m, for m small. Next we can choose $\beta > 0$ such that, for m small,

$$g(v) \ge \min\left\{\frac{v}{m}, \beta q v^{q-1}\right\}, \qquad \forall \, v > 0.$$

Then

$$\int_\Omega G(v) \, dx \ge \int_{|v| \ge \varepsilon_m} \left[\beta|v|^q - \gamma_m\right] \, dx,$$

where $\varepsilon_m > 0$ is the solution of $t/m = \beta q t^{q-1}$ and $\gamma_m \to 0$ as $m \to 0$. Since $\varepsilon_m \to 0$ as $m \to 0$, we can find $\gamma'_m \to 0$ as $m \to 0$ such that

$$\int_\Omega G(v) \, dx \ge \beta\|v\|_{L^q}^q - \gamma'_m.$$

This and (11.35) imply

$$J(v) \geq \beta \|v\|_{L^q}^q - \gamma'_m - c_2 \|v\|_{L^q}^2.$$

Since $q < 2$ and $\gamma'_m \to 0$, (ii) follows. ∎

Theorem 11.23 *Suppose that (a')–(b') hold and that $b/a > 2\lambda_1\theta$. Then J has two nontrivial critical points w_1, w_2. The former is a minimum of J and the latter is a MP critical point. The functions $z_i = K_m w_i$ are positive solutions of (11.33) such that $\|z_i\|_\infty > a$. Moreover, $|\Omega_a(z_1)| = 0$.*

Proof. The existence of w_i follows immediately from the properties of J listed before and from Lemma 11.22. By the maximum principle it follows that w_i exceed a in Ω. The property that $|\Omega_a(z_1)| = 0$, has been proved in Theorem 11.16(ii). ∎

Remark 11.24 It is possible to prove that, if Ω is a ball, w_1, w_2 are radially symmetric and for the corresponding z_i there holds

$$\frac{\partial z_i}{\partial r} < 0, \qquad \forall\, r > 0, \qquad i = 1, 2. \tag{11.36}$$

Let us give an outline of the proof. If Ω is a ball, the *Schwarz symmetrization* of $v \in L^2(\Omega)$, $v \geq 0$, is the radially symmetric, radially nonincreasing function v^* such that

$$\text{meas}\{x \in \Omega : v^*(x) > s\} = \text{meas}\{x \in \Omega : v^*(x) > s\}, \qquad \forall\, s > 0.$$

There holds, see for example [45],

$$\int_\Omega G_m(v)\, dx = \int_\Omega G_m(v^*)\, dx, \qquad \int_\Omega v K_m(v)\, dx \leq \int_\Omega v^* K_m(v^*)\, dx.$$

From these inequalities it readily follows that the minimum is symmetric. As for the MP critical point w_2, one can use the fact that the map $v \mapsto v^*$ is continuous in $L^2(\Omega)$, to substitute any path in the MP min-max scheme by a path constituted by symmetric functions. This leads to a proof that J has a symmetric MP critical point. Property (11.36) follows by applying a weak form of the maximum principle ([101], Theorem 8.19) to $\partial z_i/\partial r$.

Similar arguments can be carried out if Ω is for example convex and symmetric with respect to the plane $x_j = 0$, for some $j = 1, \ldots, n$. In this case one has that $\partial z_i/\partial x_j < 0$.

From (11.36) it follows that $|\Omega_a(z_2)| = 0$. We have seen before, see Theorem 11.23(ii), that $|\Omega_a(z_1)| = 0$, for any domain Ω. We do not know whether, in this generality, such a property holds for z_2 as well. ∎

Another problem with a discontinuous nonlinearity arising in fluiddynamics will be discussed in Appendix 4.

11.4 Problems with concave-convex nonlinearities

Even though, for expository reasons, we have discussed elliptic problems either with topological theoretic methods or by means of variational tools, there are circumstances in which it is convenient to study a specific problem using both the abstract settings. For example, we have seen that, using topological methods, a Dirichlet problem like

$$-\Delta u = \lambda u^q, \qquad u \in H_0^1(\Omega),$$

has a unique positive solution for all $\lambda > 0$ provided that $0 < q < 1$, see Section 3.6.1. On the other hand, we also know that a superlinear problem like

$$-\Delta u = u^p, \qquad u \in H_0^1(\Omega),$$

can be handled by variational methods, yielding a positive solution for all $1 < p < (n+2)/(n-2)$, see Theorems 7.14 and 8.11. Below, we treat a case involving a nonlinearity which is the sum of a concave term and a convex one. We will see how combining topological and variational tools, it is possible to prove existence and multiplicity results for such a class of equations. For simplicity, we deal with the existence of solutions of the following model problem

$$\begin{cases} -\Delta u = \lambda u^q + u^p & x \in \Omega \\ u > 0 & x \in \Omega \\ u = 0 & x \in \partial\Omega, \end{cases} \qquad (D_\lambda)$$

where $0 < q < 1 < p$.

We first prove a result in which the concave term plays the main role.

Theorem 11.25 *Let* $0 < q < 1 < p$. *Then there exists* $\Lambda > 0$ *such that:*

  (i) for all $\lambda \in]0, \Lambda[$, (D_λ) *has a solution;*
  (ii) for $\lambda = \Lambda$, (D_λ) *has at least a weak solution;*
  (iii) for all $\lambda > \Lambda$, (D_λ) *has no solutions.*

Proof. The proof is based on the method of sub- and super-solutions, see Section 3.6.1. We will be sketchy. The details are easy and left to the reader. By simple modifications of the arguments carried out to solve (3.24), one shows that:

(a) for all $\varepsilon > 0$ small, $\varepsilon\varphi_1$ is a sub-solution of (D_λ) for all $\lambda > 0$;
(b) if $-\Delta\psi = 1$ in Ω and $\psi = 0$ on $\partial\Omega$, then there exists $M > 0$ such that
    $M\psi$ is a super-solution of (D_λ) for all $\lambda > 0$ small enough;
(c) $\varepsilon\varphi_1 \leq M\psi$ provided $\varepsilon \ll 1$.

From (a)–(c) it follows that for all $\lambda > 0$ small enough, (D_λ) has a solution u such that $\varepsilon\varphi_1 \le u \le M\psi$. Then, letting

$$\Lambda = \sup\{\lambda > 0 : (D_\lambda) \text{ has a solution}\},$$

it follows that $\Lambda > 0$. Next, let $0 < \lambda < \Lambda$. Then there exists μ, with $\lambda < \mu < \Lambda$, such that (D_μ) has a solution u_μ, which is a super-solution for (D_λ) for all $\lambda < \mu$. Taking ε such that $\varepsilon\varphi_1 \le u_\mu$, we find a solution of (D_λ). This proves (i). Statement (ii) follows immediately by a limiting procedure. Finally, let $\bar\lambda > 0$ be such that $\lambda t^q + t^p > \lambda_1 t$, for all $t > 0$ and all $\lambda \ge \bar\lambda$. Then (D_λ) has no solution for $\lambda \ge \bar\lambda$, otherwise if $\lambda \ge \bar\lambda$ we get

$$\lambda_1 \int_\Omega u\varphi_1 \, dx = - \int_\Omega \varphi_1 \Delta u \, dx = \int_\Omega \left(\lambda u^q + u^p\right) \varphi_1 \, dx > \lambda_1 \int_\Omega u\varphi_1 \, dx,$$

a contradiction. This implies that $\Lambda < \bar\lambda$, proving (iii). ∎

Remark 11.26 Theorem 11.25 is a particular case of a more general result proved in [24], where we also refer for more details. In particular, it is shown that for all $\lambda \in (0, \Lambda)$, (D_λ) possesses a *minimal solution* u_λ. Moreover, this minimal solution is increasing with respect to λ, in the sense that

$$0 < \lambda_1 < \lambda_2 < \Lambda \quad \Longrightarrow \quad u_{\lambda_1} \le u_{\lambda_2}, \; u_{\lambda_1} \not\equiv u_{\lambda_2}.$$

This will be used in Theorem 11.27 below. ∎

Next, we prove the existence of a second solution. Here the main role is played by the convex term u^p.

Theorem 11.27 *Let $0 < q < 1 < p \le (n+2)/(n-2)$. Then for all $\lambda \in \,]0, \Lambda[$ (D_λ) has at least two positive solutions.*

We shall use variational tools. Set

$$f_\lambda(u) = \begin{cases} 0 & \text{if } u \le 0 \\ \lambda u^q + u^p & \text{if } u > 0 \end{cases} \qquad F_\lambda(u) = \int_0^u f_\lambda(s) \, ds,$$

and denote by J_λ the functional on $E = H_0^1(\Omega)$ defined by

$$J_\lambda(u) = \tfrac{1}{2}\|u\|^2 - \int_\Omega F_\lambda(u) \, dx.$$

One has that $J_\lambda \in C^1(E, \mathbb{R})$ and its critical points are solutions of (D_λ). Roughly, we will first show that the minimal solution u_λ found before (see Remark 11.26) is a local minimum of J_λ; next, we will apply the MP theorem to find a second solution.

We need a preliminary result.

Lemma 11.28 *For all* $0 < \lambda < \Lambda$, J_λ *has a local minimum in the* C^1 *topology.*

Proof. According to Theorem 11.25(i) and Remark 11.26, if $0 < \lambda_1 < \lambda < \lambda_2 < \Lambda$ (D_λ) has minimal solutions $u_i = u_{\lambda_i}$ ($i = 1, 2$) such that $u_1 \leq u_2$ and $u_1 \not\equiv u_2$. Moreover,

$$-\Delta(u_2 - u_1) > \lambda_1(u_2^q - u_1^q) + (u_2^p - u_1^p), \qquad x \in \Omega,$$

and thus the Hopf maximum principle yields $u_1 < u_2$ in Ω and $\partial(u_1 - u_2)/\partial\nu < 0$ on $\partial\Omega$ (ν denotes the unit outer normal at $\partial\Omega$). We set

$$\tilde{f}_\lambda(u) = \begin{cases} f_\lambda(u_1) & \text{if } u \leq u_1 \\ f_\lambda(u) & \text{if } u_1 < u < u_2 \\ f_\lambda(u_2) & \text{if } u \geq u_2, \end{cases}$$

$\tilde{F}_\lambda(u) = \int_0^u \tilde{f}_\lambda(s)\,ds$ and $\tilde{J}_\lambda(u) = \frac{1}{2}\|u\|^2 - \int_\Omega \tilde{F}_\lambda(u)\,dx$. It is clear that $\tilde{J}_\lambda$ achieves the global minimum on E at some u satisfying $-\Delta u = \tilde{f}_\lambda(u)$. Using again the Hopf maximum principle, we get that $u_1 < u < u_2$ in Ω, as well as

$$\frac{\partial}{\partial\nu}(u - u_1) < 0, \quad \frac{\partial}{\partial\nu}(u - u_2) > 0, \quad x \in \partial\Omega.$$

If $\|v - u\|_{C^1} \leq \varepsilon$ then $u_1 \leq v \leq u_2$. Since $\tilde{J}_\lambda(v) = J_\lambda(v)$, the result follows. ∎

For fixed $\lambda \in {]0, \Lambda[}$, we look for a second solution in the form $u = \bar{u} + v$, where $v > 0$ and $\bar{u} = \bar{u}_\lambda$ is the solution found in the preceding lemma. A straight calculation shows that v satisfies

$$-\Delta v = \lambda(\bar{u} + v)^q - \lambda\bar{u}^q + (\bar{u} + v)^p - \bar{u}^p.$$

Denote by $g_\lambda(v)$ the right hand side of the preceding equation (with $g_\lambda(z) \equiv 0$ for $z \leq 0$) and set

$$\mathcal{J}_\lambda(v) = \frac{1}{2}\|v\|^2 - \int_\Omega G_\lambda(v)\,dx, \qquad G_\lambda(v) = \int_0^v g_\lambda(s)\,ds.$$

Lemma 11.29 $v = 0$ *is a local minimum of* $\mathcal{J}_\lambda$ *in* E.

Proof. If v^+ denotes the positive part of v, a straight calculation yields

$$\mathcal{J}_\lambda(v) = \frac{1}{2}\|v^-\|^2 + J_\lambda(\bar{u} + v^+) - J_\lambda(\bar{u}) \geq J_\lambda(\bar{u} + v^+) - J_\lambda(\bar{u}).$$

This and Lemma 11.28 imply that $\mathcal{J}_\lambda(v) \geq 0$, provided $\|v\|_{C^1}$ is sufficiently small. We shall now prove that $\mathcal{J}_\lambda(v) \geq 0$, provided $\|v\|_E \ll 1$. We will use an argument by Brezis and Nirenberg [65]. By contradiction, let us suppose that there exists a sequence $w_\varepsilon \to 0$ in E such that $\mathcal{J}_\lambda(w_\varepsilon) < \mathcal{J}_\lambda(0) = 0$. Without loss of generality, we can take w_ε to be such that

$$\mathcal{J}_\lambda(w_\varepsilon) = \min_{\|w\|\le\varepsilon} \mathcal{J}_\lambda(w).$$

It follows that there exists $\mu_\varepsilon \le 0$ such that $\mathcal{J}'_\lambda(w_\varepsilon) = \mu_\varepsilon w_\varepsilon$. This readily implies that $w_\varepsilon \in E$ is a weak solution of

$$-\Delta w_\varepsilon = \frac{1}{1-\mu_\varepsilon} g_\lambda(w_\varepsilon), \quad \text{with} \quad 0 < \frac{1}{1-\mu_\varepsilon} < 1.$$

Since the right hand side converges weakly to 0 in E, by elliptic regularity it follows that $w_\varepsilon \to 0$ in C^1 and this is not possible since we have shown that $v = 0$ is a local minimum of $\mathcal{J}_\lambda$ in the C^1 topology. ∎

Proof of Theorem 11.27. For fixed $v_1 \in E$, $v_1 > 0$, one easily checks that $\mathcal{J}_\lambda(tv_1) \to -\infty$ as $t \to +\infty$. Thus $\mathcal{J}_\lambda$ has the MP geometry. It remains to prove the (PS) condition. If $p < (n+2)/(n-2)$ one uses standard arguments. Let us consider the case $p = (n+2)/(n-2)$. As before, one shows that $(PS)_c$ holds for all $c < \frac{1}{n}S^{n/2}$. Let us prove that the MP level c_λ of $\mathcal{J}_\lambda$ is smaller than $\frac{1}{n}S^{n/2}$. We will modify the arguments used in Section 11.2, where we refer for more details and notation. We evaluate $\mathcal{J}_\lambda$ on $t\,\varphi\,U_\varepsilon$. Let $n \ge 4$. From $(a+b)^p \ge a^p + b^p + \alpha a^{p-1}b$ $(a, b \ge 0)$, we deduce that $g_\lambda(v) \ge v^p + \alpha v \bar{u}^{p-1}$. This implies

$$G_\lambda(v) \ge \tfrac{1}{p+1}v^{p+1} + \tfrac{1}{2}\alpha v^2 \bar{u}^{p-1},$$

and hence

$$\mathcal{J}_\lambda(t\varphi U_\varepsilon) \le \tfrac{1}{2}t^2\|\varphi U_\varepsilon\|^2 - \tfrac{1}{p+1}t^{p+1}\int_\Omega (\varphi U_\varepsilon)^{p+1}\,dx - \tfrac{1}{2}\alpha t^2 \int_\Omega \bar{u}^{p-1}\varphi^2 U_\varepsilon^2\,dx.$$

Since on the support of φU_ε, $\bar{u} \ge$ constant > 0 we find

$$\mathcal{J}_\lambda(t\varphi U_\varepsilon) \le \tfrac{1}{2}t^2\|\varphi U_\varepsilon\|^2 - \tfrac{1}{p+1}t^{p+1}\int_\Omega (\varphi U_\varepsilon)^{p+1}\,dx - \tfrac{1}{2}\alpha c_1 t^2 \int_\Omega \varphi^2 U_\varepsilon^2\,dx,$$

for some $c_1 > 0$. We can now repeat the arguments carried out in Section 11.2 proving that for the MP level c_λ there holds $c_\lambda < \frac{1}{n}S^{n/2}$. In the case $n = 3$ we argue as follows. We use that $(a + b)^5 \ge a^5 + b^5 + 5ab^4$ to infer, for some constant $c_2 > 0$,

$$\mathcal{J}_\lambda(t\varphi U_\varepsilon) \le \tfrac{1}{2}t^2\|\varphi U_\varepsilon\|^2 - \tfrac{1}{6}t^6 \int_\Omega (\varphi U_\varepsilon)^6\,dx - c_2 t^5 \int_\Omega (\varphi_\varepsilon U)^5\,dx,$$

From

$$\begin{cases} A_\varepsilon := \|\varphi U_\varepsilon\|^2 = S^{3/2} + O(\varepsilon) \\ B_\varepsilon := \int_\Omega (\varphi U_\varepsilon)^6\,dx = S^{3/2} + O(\varepsilon^3) \\ C_\varepsilon := \tfrac{1}{5}c_2 \int_\Omega (\varphi U_\varepsilon)^5\,dx = k\varepsilon^{1/2} + O(\varepsilon^{5/2}), \quad k > 0, \end{cases}$$

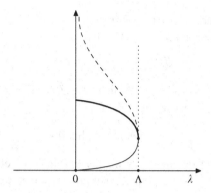

Figure 11.5 Bifurcation diagram of positive solutions of (D_λ). The bold, respectively dashed, line represents the subcritical case $1 < p < (n+2)/(n-2)$, respectively critical case $p = (n+2)/(n-2)$. The lower part of the branch exists for all $0 < q < 1 < p$.

we deduce that $\max_{t>0} \mathcal{J}_\lambda(t\varphi U_\varepsilon)$ is achieved at $t = t(\varepsilon)$ such that $A = Bt^4 + Ct^3$. Taking into account the values of A, B, C we find that $t = t(\varepsilon)$ satisfies

$$S^{3/2} + O(\varepsilon) = t^4(S^{3/2} + O(\varepsilon^3)) + t^3(k\varepsilon^{1/2} + O(\varepsilon^{5/2})).$$

From this it readily follows that

$$t(\varepsilon) = 1 - \frac{k}{4S^{3/2}}\varepsilon^{1/2} + o(\varepsilon^{1/2}).$$

A straight calculation shows that

$$\max_{t>0} \mathcal{J}_\lambda(t\varphi U_\varepsilon) = \mathcal{J}_\lambda(t(\varepsilon)\varphi U_\varepsilon) = \tfrac{1}{3}S^{3/2} - \tfrac{7}{10}k\varepsilon^{1/3} + o(\varepsilon^{1/2}),$$

and this implies $c_\lambda \leq \max_{t>0} \mathcal{J}_\lambda(t\varphi U_\varepsilon) < \tfrac{1}{3}S^{3/2}$, provided $\varepsilon >$ is sufficiently small. Finally, an application of the MP theorem yields a second solution of (D_λ). ∎

Remark 11.30 In [24] the problem

$$\begin{cases} -\Delta u = \lambda|u|^{q-1}u + |u|^{p-1}u & x \in \Omega \\ u = 0 & x \in \partial\Omega, \end{cases}$$

is also considered. Taking advantage of the oddness of the nonlinearity, one can show that $\exists\, \lambda^* > 0$ such that for all $\lambda \in (0, \lambda^*)$ one has:

(i) if $0 < q < 1 < p \leq (n+2)/(n-2)$ the preceding problem has infinitely many solutions with negative energy;

(ii) if $0 < q < 1 < p < (n+2)/(n-2)$ the preceding problem has infinitely many solutions with positive energy. ∎

11.5 Exercises

(i) Prove that the problem

$$-\Delta u = \mathrm{sgn}(u)|u|^q, \quad x \in \Omega, \qquad u = 0, \quad x \in \partial\Omega, \quad 0 < q < 1,$$

has infinitely many solutions. [Hint: use the preceding method to deal with a C^1 functional on a suitable $L^r(\Omega)$, and apply the Lusternik–Schnirelman theory.]

(ii) Consider the problem

$$\begin{cases} -\Delta u = 0 & \text{if } u \leq \delta \\ -\Delta u = \alpha u & \text{if } u > \delta \\ u = d & \text{on } \partial\Omega, \end{cases}$$

and prove that there exists $a^* > 0$ such that the problem has two nontrivial solutions for all $0 < \delta - d < a^*$. [Hint: perform the change of variable $u \mapsto u - d$ and apply Theorem 11.23 with $a = \delta - d$.] If z_1, z_2 denote the solutions corresponding to the minimum and the MP, respectively, show that as $a \to 0+$, $z_2 \to 0$ while $z_1 \to z^*$ in $H_0^1(\Omega)$, where z^* denotes the positive solution of $-\Delta u = \alpha(u + d)$ in Ω, $u = 0$ on $\partial\Omega$. Moreover, prove that for $\delta > d + \alpha(\lambda_1 - \alpha)^{-1}$ there are no nontrivial solutions.

(iii) Referring to Remark 11.30 in the case $1 < q < p < (n+2)/(n-2)$, prove the following facts:

(a) J_λ has infinitely many critical points with negative energy, provided $\lambda > 0$ is sufficiently small. [Hint: Show that there is a ball $B_r \subset E$ where $J_\lambda < 0$ and is bounded from below and apply Theorem 10.10 with $M = B_r$.]

(b) J_λ has infinitely many critical points with positive energy, provided $\lambda > 0$ is sufficiently small. [Hint: Apply Theorem 10.20.]

(c) Give another proof of the same results by using the natural constraint $M_\lambda = \{u \in H_0^1(\Omega) \setminus \{0\} : \|u\|^2 = \lambda \int_\Omega |u|^{q+1} + \int_\Omega |u|^{p+1}\}$ introduced in the Exercise 6 – (iii). [Hint: Use the fact that M_λ is the disjoint union of two manifolds.]

12

Morse theory

This chapter is devoted to the basic theory developed by M. Morse, which relates the structure of the critical points of a regular function on a manifold to its topology.

12.1 A short review of basic facts in algebraic topology

In this section we recall some basic notions and results in algebraic topology. To keep the presentation short, we will introduce the concepts in an axiomatic way, referring the interested reader to more complete treatments, like the books [131, 162]. Then we will review briefly the explicit construction of the singular homology theory, omitting most of the proofs.

12.1.1 The axiomatic construction

Definition 12.1 *Let $(G_i)_i$ be a sequence of Abelian groups, and let $(\varphi_i)_i$ be a sequence of homomorphisms*

$$\cdots \longrightarrow G_i \xrightarrow{\varphi_i} G_{i+1} \xrightarrow{\varphi_{i+1}} G_{i+2} \longrightarrow \cdots$$

We say that the sequence is exact *if for every i there holds* $\mathrm{im}(\varphi_i) = \ker(\varphi_{i+1})$.

Example 12.2 (i) Let G_1, G_2 be Abelian groups, and consider the following part of a sequence

$$0 \longrightarrow G_1 \xrightarrow{\varphi} G_2 \longrightarrow 0.$$

Then we have exactness if and only if φ is an isomorphism.

(ii) Let G_1, G_2, G_3 be Abelian groups, and suppose the following part of the sequence is exact

$$0 \longrightarrow G_1 \xrightarrow{\varphi_1} G_2 \xrightarrow{\varphi_2} G_3 \longrightarrow 0.$$

Then from the exactness it follows that φ_1 is injective, and that the image of φ_1 (which is isomorphic to G_1) is equal to the kernel of φ_2. Still from exactness, we have that φ_2 is surjective. Therefore, by the fundamental theorem of the homomorphism, we derive that G_2 is isomorphic to $\ker(\varphi_2) \oplus \mathrm{im}(\varphi_2) \simeq G_1 \oplus G_3$. ∎

Definition 12.3 *A pair of spaces* (X, A) *is a topological space* X *together with a subset* $A \subseteq X$. *We will write that* $(X, A) \subseteq (Y, B)$ *if* $X \subseteq Y$ *and if* $A \subseteq B$. *A map of pairs* (X, A), (Y, B) *is a continuous map* f *from* X *into* Y *such that* $f(A) \subseteq B$. *Two maps of pairs* $f_0, f_1 : (X, A) \to (Y, B)$ *are homotopic if there exists a map* $h : [0, 1] \times (X, A) \to (Y, B)$ *such that* $h(0, \cdot) = f_0$ *and* $h(1, \cdot) = f_1$.

The homology groups. (a) To every relative integer $q \in \mathbb{Z}$ and for every pair (X, A) is associated a group $H_q(X, A)$, denoted by $H_q(X)$ if $A = \emptyset$.

(b) To every map of pairs $f : (X, A) \to (Y, B)$ is associated a homomorphism $f_* : H_q(X, A) \to H_q(Y, B)$.

(c) To every $q \in \mathbb{Z}$ and every pair (X, A) is associated a homomorphism $\partial : H_q(X, A) \to H_{q-1}(A)$.

The following axioms are required.

Axiom 1 If $f = \mathrm{Id}|_X$, then $f_* = \mathrm{Id}|_{H_q(X,A)}$.

Axiom 2 If $f : (X, A) \to (Y, B)$ and $g : (Y, B) \to (Z, C)$ are maps of pairs, then $(g \circ f)_* = g_* \circ f_*$.

Axiom 3 If $f : (X, A) \to (Y, B)$ is a map of pairs, then $\partial \circ f_* = (f|_A)_* \circ \partial$.

Axiom 4 Let $i : A \to X$ and $j : (X, \emptyset) \to (X, A)$ be inclusions, then the following sequence is exact

$$\cdots \xrightarrow{\partial} H_q(A) \xrightarrow{i_*} H_q(X) \xrightarrow{j_*} H_q(X, A) \xrightarrow{\partial} H_{q-1}(A) \longrightarrow \cdots .$$

Axiom 5 If $f, g : (X, A) \to (Y, B)$ are homotopic maps of pairs, then $f_* = g_*$.

Axiom 6 (Excision) If U is an open set of X with $\overline{U} \subseteq \mathrm{int}(A)$, and if $i : (X \setminus U, A \setminus U) \to (X, A)$ denotes the inclusion, then i_* is an isomorphism

Axiom 7 If X consists of a single point p, then

$$H_q(\{p\}) = \begin{cases} \mathbb{Z} & \text{if } q = 0 \\ 0 & \text{if } q \neq 0. \end{cases}$$

We list next some of the basic properties which can be derived from the above axioms.

Property 1 If $(X,A) = \cup_{i=1}^{k}(X_i,A_i)$ with $(X_i)_i$ closed and disjoint, then $H_q(X,A) = \oplus_{i=1}^{k} H_q(X_i,A_i)$.

Property 2 (Mayer–Vietoris sequence) Let $X = X_1 \cup X_2$, with X_1, X_2 open sets, and let $A_1 \subseteq X_1, A_2 \subseteq X_2$ also be open sets. Then, if $X_1 \cap X_2 \neq \emptyset$, letting $A = A_1 \cup A_2$ and denoting by $i : (X_1,A_1) \to (X,A)$, $j : (X_2,A_2) \to (X,A)$, $l : (X_1 \cap X_2, A_1 \cap A_2) \to (X_1,A_1)$, $k : (X_1 \cap X_2, A_1 \cap A_2) \to (X_2,A_2)$ the natural inclusions, there exists an exact sequence

$$\cdots \longrightarrow \quad H_q(X_1,A_1) \oplus H_q(X_2,A_2)$$

$$\overset{\Psi}{\longrightarrow} \quad H_q(X,A) \overset{\Delta}{\longrightarrow} H_{q-1}(X_1 \cap X_2, A_1 \cap A_2)$$

$$\overset{\Phi}{\longrightarrow} \quad H_{q-1}(X_1,A_1) \oplus H_{q-1}(X_2,A_2) \longrightarrow \cdots,$$

where $\Psi = i_* - j_*$ and $\Phi = (k_*, l_*)$.

Property 3 If $H_q(X) = H_q(A)$ for every q, then $H_q(X,A) = 0$ for every $q \in \mathbb{Z}$.

Property 4 If A is a *deformation retract* of X (namely if there exists $h : [0,1] \times X$ such that $h(0,\cdot) = \mathrm{Id}|_X$, $h(1,\cdot) \subseteq A$ and $h(t,\cdot)|_A = \mathrm{Id}|_A$ for every t), then $H_q(X,A) = 0$ for every $q \in \mathbb{Z}$.

Property 5 Suppose that $A' \subseteq A \subseteq X$, and that A' is a deformation retract of A. Then $H_q(X,A)$ is isomorphic to $H_q(X,A')$ for every $q \in \mathbb{Z}$.

Property 6 If X is arcwise connected, then $H_0(X)$ is isomorphic to $\mathbb{Z}$.

Property 7 Suppose that $X_1, X_2 \subseteq X$, and that (X,X_1,X_2) are a *proper triad*, namely that the inclusions $k_1 : (X_1, X_1 \cap X_2) \to (X_1 \cup X_2, X_2)$, $k_2 : (X_2, X_1 \cap X_2) \to (X_1 \cup X_2, X_1)$ induce an isomorphism between the relative homology groups in any dimension. Then the sequence of the triad

$$\cdots \longrightarrow \quad H_q(X_1, X_1 \cap X_2) \longrightarrow H_q(X,X_2) \longrightarrow H_q(X, X_1 \cup X_2)$$

$$\overset{\partial}{\longrightarrow} \quad H_{q-1}(X_1, X_1 \cap X_2) \longrightarrow \cdots$$

is exact.

Property 8 (Generalized excision property) Suppose that $A \subseteq X$ and that U, V are open sets with $\overline{V} \subseteq U \subseteq A$ and that $(X\backslash U, A\backslash U)$ is a deformation retract of $(X\backslash V, A\backslash V)$. Then the inclusion $(X\backslash U, A\backslash U) \to (X, A)$ induces an isomorphism between the relative homology groups in any dimension.

We treat next some basic examples of computations of homology (and relative homology) groups of simple sets.

Example 12.4 *Homology of S^n.* We consider first the case $n = 0$, for which the sphere consists of two points, denoted by p_1 and p_2. Using Property 1 and Axiom 7 we find immediately that

$$H_q(S^0) \simeq \begin{cases} \mathbb{Z}^2 & \text{for } q = 0 \\ 0 & \text{for } q \neq 0. \end{cases} \tag{12.1}$$

We are going to prove next, by induction on the dimension n, that

$$H_q(S^n) \simeq \begin{cases} \mathbb{Z} & \text{for } q = 0, n \\ 0 & \text{otherwise.} \end{cases} \tag{12.2}$$

To show this, we embed S^n into $\mathbb{R}^{n+1}$ as $S^n = \{(x_1, \ldots x_{n+1}) : \sum_{i=1}^{n+1} x_i^2 = 1\}$ and use the Mayer–Vietoris sequence with $X = S^n, X_1 = S^n \setminus \{P_N\}$ and $X_2 = S^n \backslash \{P_S\}$, where $P_N = (0, \ldots, 0, 1)$ and $P_S = (0, \ldots, 0, -1)$ denote respectively the north and the south poles of the sphere. Clearly X_1, X_2 are open in X and $X_1 \cup X_2 = X$, so we have the exactness of the following sequence

$$\cdots \longrightarrow \quad H_q(X_1) \oplus H_q(X_2) \longrightarrow H_q(X)$$
$$\overset{\partial}{\longrightarrow} \quad H_{q-1}(X_1 \cap X_2) \longrightarrow H_{q-1}(X_1) \oplus H_{q-1}(X_2) \longrightarrow \cdots .$$

We notice at this point that $X_1 \cap X_2$ can be deformed to the equator of S^n, namely $S^n \cap \{x_{n+1} = 0\} \simeq S^{n-1}$, and that both X_1 and X_2 are contractible to a point. Therefore by Property 4 the above exact sequence becomes

$$\cdots \longrightarrow \quad H_q(\{p\}) \oplus H_q(\{p\}) \longrightarrow H_q(S^n)$$
$$\overset{\partial}{\longrightarrow} \quad H_{q-1}(S^{n-1}) \longrightarrow H_{q-1}(\{p\}) \oplus H_{q-1}(\{p\}) \longrightarrow \cdots .$$

For $q = 1$, we have that $H_q(\{p\}) = H_{q-1}(\{p\}) = 0$ so we get the exactness of

$$0 \longrightarrow H_q(S^n) \overset{\partial}{\longrightarrow} H_{q-1}(S^{n-1}) \longrightarrow 0,$$

which implies

$$H_q(S^n) \simeq H_{q-1}(S^{n-1}), \qquad n \geq 1, q > 1. \tag{12.3}$$

On the other hand, for $q = 1$ we have the exactness of

$$0 \longrightarrow H_1(S^n) \overset{\partial}{\longrightarrow} H_0(S^{n-1}) \overset{\Phi_0}{\longrightarrow} \mathbb{Z}^2, \tag{12.4}$$

from which we get that

$$H_1(S^n) \simeq \ker(\Phi_0), \qquad n \geq 1. \tag{12.5}$$

Taking $n = 1$ in (12.4), we have from (12.5) that $H_1(S_1) \simeq \ker(\Phi_0)$. By construction, $\Phi_0(\alpha_1, \alpha_2) = (\alpha_1 + \alpha_2, \alpha_1 + \alpha_2)$, where $(\alpha_1, \alpha_2) \in H_0(S^0) \simeq \mathbb{Z}^2$. Therefore we deduce that $H_1(S^1) \simeq \mathbb{Z}$. On the other hand, from (12.3) and Axiom 7 it follows that $H_q(S^1) \simeq H_{q-1}(S^0) = 0$ for $q \geq 2$.

Proceeding by induction, let us assume that (12.2) holds, and let us prove the analogous formula for S^{n+1}. For $q > 1$ we can apply (12.3) to get

$$H_q(S^{n+1}) \simeq H_{q-1}(S^n) \simeq \begin{cases} \mathbb{Z} & \text{if } q - 1 = n \\ 0 & \text{otherwise.} \end{cases}$$

For $q = 1$ and $n \geq 1$, we notice that S^n is arcwise connected, and hence the action of Φ_0 is $\Phi_0(\alpha) = (\alpha, \alpha)$, for $\alpha \in H_0(S^n) \simeq \mathbb{Z}$. Hence by (12.5) we finally get that $H^1(S^{n+1}) \simeq 0$. This concludes the proof. ∎

Example 12.5 *Relative homology of Euclidean balls relative to their boundary.*
Let $B^n = \{x \in \mathbb{R}^n \ : \ |x| \leq 1\}$, and let us denote by S^{n-1} the boundary of B^n, by $S^{n-1}_\pm$ the sets $\{x \in S^{n-1} \ : \ \pm x_n \geq 0\}$, and by S^{n-2} the equator $\{x \in S^{n-1} \ : \ x_n = 0\}$.

We observe first that B^n, $S^{n-1}_\pm$ and $(B^n, S^{n-1}_\pm)$ have trivial homology groups. In fact, all the three sets are contractible to a point. To prove this fact for the pair, it is sufficient to use Property 3 above.

We claim next that the triad $(B^n, S^{n-1}_+, S^{n-1}_-)$ is proper. To see this, we consider for example the inclusion $k_1 : (S^{n-1}_+, S^{n-2}) \to (S^{n-1}, S^{n-1}_-)$ and show that it induces an isomorphism between the corresponding relative homology groups. The same will hold true for the other inclusion $k_2 : (S^{n-1}_-, S^{n-2}) \to (S^{n-1}, S^{n-1}_+)$. This fact indeed follows from Property 8, taking $X = S^{n-1}$, $A = S^{n-1}_-$, $U = S^{n-1} \setminus S^{n-2}$, and $V = \{x \in S^{n-1} \ : \ x_n \leq -\frac{1}{2}\}$.

We are now in a position to compute the relative groups $H_q(B^n, S^{n-1})$. By the first observation we have that $H_q(B^n, S^{n-1}) = 0$ for all q. From the claim and Property 7, taking $X = B^n, X_1 = S^{n-1}_+$ and $X_2 = S^{n-1}_-$, we then deduce the exactness of the sequence

$$\cdots \longrightarrow \quad H_q(B^n, S^{n-1}_-) \longrightarrow H_q(B^n, S^{n-1}) \longrightarrow H_{q-1}(S^{n-1}_+, S^{n-2})$$
$$\longrightarrow \quad H_{q-1}(B^{n-1}, S^{n-2}) \longrightarrow \cdots,$$

from which we find that $H_q(B^n, S^{n-1}) \simeq H_{q-1}(S^{n-1}_+, S^{n-2})$. On the other hand, the pair (S^{n-1}_+, S^{n-2}) is homeomorphic to (B^{n-1}, S^{n-2}), so one has the isomorphism of the relative homology groups $H_q(B^n, S^{n-1}) \simeq H_{q-1}(B^{n-1}, S^{n-2})$.

Iterating this procedure we find that

$$H_q(B^n, S^{n-1}) \simeq H_{q-n}(B^0, \emptyset) \simeq \begin{cases} \mathbb{Z} & \text{for } q = n \\ 0 & \text{for } q \neq n. \end{cases}$$ ∎

12.1.2 Singular homology groups

Singular homology groups provide an explicit and intuitive construction of a homology theory. This is based on the notion of *singular simplex, singular chain* and the *boundary operator*.

Definition 12.6 *For r non-negative integer, we define the simplex $s_r \subseteq \mathbb{R}^{r+1}$ as*

$$s_r = \left\{ t_0 \mathbf{e}_0 + \cdots + t_r \mathbf{e}_r \; : \; t_0, t_i \geq 0, \sum_{i=0}^{r} t_i = 1 \right\},$$

where $\mathbf{e}_0 = (1, 0, \ldots, 0), \ldots, \mathbf{e}_r = (0, \ldots, 0, 1)$. A singular r-simplex of a topological space X is a continuous map $\sigma_r : s_r \to X$.

We also denote by $C_r(X)$ the Abelian group generated by formal linear combinations (with relative integer coefficients) of singular r-simplexes of X. The elements of $C_r(X)$ are called *singular r dimensional chains of X*.

On $C_r(X)$ is defined naturally a *boundary operator* ∂_r which is a linear map into $C_{r-1}(X)$. Given a singular r-simplex $\sigma_r : s_r \to X$, we define the jth face of σ_r by $\sigma_{r-1}^j := \sigma_r|_{s_{r-1}^j}$, namely the restriction of σ_r to the set

$$s_{r-1}^j = \{ t_0 \mathbf{e}_0 + \cdots + t_r \mathbf{e}_r \in s_r \; : \; t_j = 0 \}.$$

Then ∂_r is defined through the linear extension on the singular r-chains of the map

$$\partial_r \sigma_r = \sum_{j=0}^{r} (-1)^j \sigma_{r-1}^j.$$ (12.6)

As one can easily verify, the boundary operator satisfies

$$\partial_{r-1} \circ \partial_r = 0.$$ (12.7)

The latter property turns out to be very important, indeed in this way the singular chains form a *homological complex*

$$\cdots \longrightarrow C_{r+1} \xrightarrow{\partial_{r+1}} C_r \xrightarrow{\partial_r} C_{r-1} \longrightarrow \cdots C_1 \xrightarrow{\partial_1} C_0 \xrightarrow{\partial_0} 0,$$

in the sense that every map is a homomorphism, and that the composition of two consecutive maps is the trivial one.

The elements of the subgroup $Z_r(X) = \ker(\partial_r) \subseteq C_r$ are called *cycles*, while those of the subgroup $B_r(X) = \operatorname{im}(\partial_{r+1})$ are called *boundaries*. By (12.7), we have clearly that $B_r(X) \subseteq Z_r(X)$. The *rth singular homology group of X* is defined as the quotient

$$H_r(X) = Z_r(X)/B_r(X).$$

The whole sequence $(H_r(X))_r$ is denoted by $H_*(X)$, and is usually extended to be zero for $r < 0$.

Example 12.7 We consider the simple case in which X consists of a single point. For $q \geq 0$ there exists a unique simplex $\sigma_q : s_q \to X$, which is a constant map. For q even, using (12.6), we get that $\partial_q \sigma_q = \sigma_{q-1}$ and hence $Z_q = \ker(\partial_q) = 0$, from which we find $H_q(X) = 0$. For $q \geq 3$ odd, and for a singular q-simplex σ_q, we have from the previous case that $\sigma_q = \partial_{q+1}\sigma_{q+1}$. Therefore $\sigma_q \in \operatorname{im}(\partial_{q+1})$ and hence $Z_q(X) = B_q(X)$, so $H_q(X) \simeq 0$. Finally, for $q = 1$ we still have $\partial_1\sigma_1 = 0$, so $B_0(X) \simeq 0$. It then follows that $H_0(X) \simeq \mathbb{Z}$. ∎

Let us now consider the case of a continuous map f from a topological space X into a second space Y. Then f induces naturally a homomorphism $f_* : C_r(X) \to C_r(Y)$ via the composition $f \circ \sigma_r$. One can check that the map f_* commutes with the boundary operator, and therefore $f_*(Z_r(X)) \subseteq Z_r(Y)$, with also $f_*(B_r(X)) \subseteq B_r(Y)$. Passing to quotients, it follows that f induces also a homomorphism, still denoted by f_*, from $H_r(X)$ into $H_r(Y)$.

We list now some of the properties of this class of homomorphisms. If $f : X \to Y$ and $g : Y \to Z$ are continuous, then $(g \circ f)_* = g_* \circ f_*$. If $f = \operatorname{Id}$ on X, then also f_* is the identity on $H_*(X)$. If two maps f and g are homotopic, then $f_* = g_*$. Moreover if two spaces X, Y are *homotopically equivalent*, namely if there exist $f : X \to Y$ and $g : Y \to X$ for which $g \circ f$ is homotopically equivalent to $\operatorname{Id}|_X$ and $f \circ g$ is homotopically equivalent to $\operatorname{Id}|_Y$, then $H_r(X)$ is isomorphic to $H_r(Y)$ for every r. This applies in particular to the case in which Y is a deformation retract of X.

12.1.3 Singular relative homology groups

Let X be a topological space and A a subset of X. The inclusion $i : A \to X$ induces a homomorphism $i_* : C_q(A) \to C_q(X)$ which is clearly injective. Therefore we can consider the quotient group

$$C_q(X, A) = C_q(X)/C_q(A).$$

Since i_* commutes with ∂, then we have an induced homomorphism, still denoted by ∂, from $C_q(X,A)$ into $C_{q-1}(X,A)$. Then, letting $Z_q(X,A) = \ker(\partial_q)$ denote the *relative cycles* and $B_q(X,A) = \mathrm{im}(\partial_{q+1})$ denote the *relative boundaries*, we define the *relative homology group*

$$H_q(X,A) = Z_q(X,A)/B_q(X,A).$$

Using the equivalence classes in $C_q(X)$, a relative cycle is a family of elements of the form $(z+w)_{w \in C_q(A)}$, with z such that $\partial z \in C_{q-1}(A)$. A relative boundary is a class $(u+w)_{w \in C_q(A)}$ with $u = \partial_{q+1} s$, $s \in C_{q+1}(X)$.

Example 12.8 Consider a torus $X \simeq \mathbb{T}^2$ embedded in $\mathbb{R}^3$ as in Figure 12.1 and let $A = \{x_3 < \frac{1}{2}\} \cap X$. The singular 1-simplex z depicted on the right is an element of $Z_1(X,A)$ since its boundary lies in $C_0(A)$. The 1-cycle C depicted on the left is homologous to zero because it is the relative boundary of a 2-chain in X. In fact, considering a chain b with image $X \cap \{x_2 > 0\}$, we easily see that $\partial b = c + c_1$, where $c_1 \in C_1(A)$. ∎

Consider a homology class of X, namely an element of $H_q(X)$. This can be represented as $(z + \partial u)_{u \in C_{q+1}(X)}$, where $z \in C_q(X)$ is such that $\partial z = 0$. This is indeed also a relative homology class, if $A \subseteq X$, induced by the inclusion $j :$ $(X, \emptyset) \to (X, A)$. Let us apply the operator ∂ to a class of $H_q(X, A)$, represented as $(z + \partial u + w)_{w \in C_q(A)}$. Then

$$\partial(z + \partial u + w)_{w \in C_q(A)} = (\partial z + \partial w)_{w \in C_q(A)}.$$

Since $\partial \partial z = 0$, using representatives of the type $Z_{q-1}(X) + C_{q-1}(A)$ in $H_{q-1}(X,A)$, we can assume $\partial z \in Z_{q-1}(A)$ and $\partial w \in C_{q-1}(A)$. Therefore $(\partial z + \partial w)_{w \in C_q(A)}$ is a homology class in $H_{q-1}(A)$, and we have a homomorphism (still denoted by ∂)

$$\partial : H_q(X,A) \to H_{q-1}(A).$$

In this way, considering this operator and the natural inclusions $i : A \to X$, $j : (X, \emptyset) \to (X, A)$ we obtain a sequence of homomorphisms

$$\cdots \longrightarrow H_q(A) \xrightarrow{i_*} H_q(X) \xrightarrow{j_*} H_q(X,A) \xrightarrow{\partial} H_{q-1}(A) \longrightarrow \cdots$$

which turns out to be exact. Moreover, with these definitions of homology and relative homology, all the above axioms are verified.

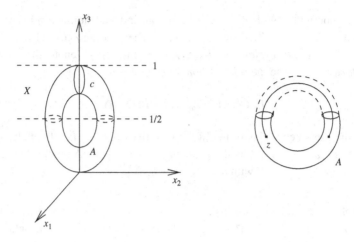

Figure 12.1

12.2 The Morse inequalities

We are now in a position to introduce the Morse inequalities, and we will mainly follow the approach in [132]. We consider a smooth compact finite dimensional manifold M (modelled on $\mathbb{R}^n$ according to the definition of Section 6.1). It is well known (see for example [163]) that using a partition of unity it is possible to construct on every such manifold a *Riemannian metric*, namely a smooth positive-definite symmetric bilinear form on tangent vectors. This clearly induces a scalar product on T_pM for every $p \in M$, and hence by duality also an identification between tangent vector fields and linear one-forms on M. Below, we will simply use the notation $(\cdot, \cdot)$ to denote the scalar product of two tangent vectors at a given point of M.

As we have seen in Section 5.2, the differential of a regular function f on M generates the gradient vector field ∇f on M, and if f is of class C^2 then the negative gradient $-\nabla f$ gives rise to a one-parameter family of diffeomorphisms φ_t on M for which the value of f decreases as t increases. For $a \in \mathbb{R}$, we let $M^a = \{p \in M : f(p) \leq a\}$. We have then the following result.

Proposition 12.9 *Let M be a compact finite dimensional manifold, and let $f : M \to \mathbb{R}$ be a function of class C^2. Let $a, b \in \mathbb{R}$ with $a < b$, and assume that $\{a \leq f \leq b\}$ contains no critical points of f. Then M^a is a deformation retract of M^b.*

Proof. Let us define the following vector field Y on the closure of $M^b \setminus M^a$:

$$Y(p) = -\frac{\nabla f(p)}{\|\nabla f(p)\|^2}.$$

Indeed, since we are assuming that f has no critical points in $\{a \le f \le b\}$, this vector field is well defined, of class C^1, and admits a C^1 extension to a neighbourhood of its domain. Let us now consider the initial value problem

$$\begin{cases} \dfrac{dp}{dt} = Y(p(t)), \\[2mm] p(0) = p_0 \in M^b \setminus M^a. \end{cases} \tag{12.8}$$

By the regularity of Y, problem (12.8) possesses a local solution $\varphi(t, p_0)$ and, as long as this is defined, by the definition of gradient we have that

$$\frac{df(\varphi(t, p_0))}{dt} = df(p(t))[Y(p(t))] = -\frac{1}{\|\nabla f(p(t))\|^2}(\nabla f(p(t)), \nabla f(p(t))) = -1.$$

If the maximal interval of definition of (12.8) contains $[0, T)$, then integrating the last equality one finds

$$f(p(0)) - f(\varphi(T, p_0)) = T, \tag{12.9}$$

and hence if $p_0 \in M^b \setminus M^a$ then $\varphi(t, p_0)$ is defined for $t \in [0, f(p) - a]$. We can consider now the homotopy

$$h(t, p) = \begin{cases} \varphi(t(f(p) - a), p) & \text{for } p \in M^b \setminus M^a \\ p & \text{for } p \in M^a. \end{cases}$$

As one can easily check, this is a continuous map and a deformation retract of M^b onto M^a. This concludes the proof. ∎

We deal next with critical points of f. We recall that a critical point p of a function f of class C^2 is said to be nondegenerate if the second differential has no kernel. Working in a local coordinate system $(x_1, \ldots, x_n)$ for which, say, $x(p) = 0$, then the nondegeneracy condition is equivalent to the invertibility of the Hessian matrix $((\partial^2 f / \partial x_i \partial x_j)(0))_{ij}$. As one can easily see, this condition is independent of the choice of coordinates, since we have vanishing of all the first partial derivatives at $x = 0$. The number of negative eigenvalues of the Hessian matrix, which is well-defined by the above comments, is called the *Morse index* of f at p.[1] In the next lemma we show that near a nondegenerate critical point, every function assumes a simple and standard form.

Lemma 12.10 *(Morse lemma) Let $f : M \to \mathbb{R}$ be of class C^2, and let p be a nondegenerate critical point of f. Then there exists a local system of coordinates*

---

[1] This definition clearly extends to the infinite-dimensional case as well, with obvious modifications.

$(y_1, \ldots, y_n)$ *near p such that, in these coordinates one has*

$$f(y) = f(p) - y_1^2 - \cdots - y_\lambda^2 + y_{\lambda+1}^2 + \cdots + y_n^2, \qquad (12.10)$$

where λ denotes the Morse index of f at p.

Proof. First of all, we show that if f has the form in (12.10), then λ must coincide with the Morse index of f. This is indeed rather easy to see, since the Hessian of f at 0 is

$$\frac{\partial^2 f}{\partial y_i \partial y_j}(0) = \begin{pmatrix} -2 & & & & & \\ & \ddots & & & 0 & \\ & & -2 & & & \\ & & & 2 & & \\ & 0 & & & \ddots & \\ & & & & & 2 \end{pmatrix},$$

where the upper-left block has dimension λ, so the claim follows immediately.

Next, we claim that given a C^1 function $\tilde{f}$ for which $\tilde{f}(0) = 0$, one can write that $\tilde{f}(x) = \sum_{j=1}^n x_j g_j(x_1, \ldots, x_n)$ for some continuous functions $(g_j)_j$ such that $g_j(0) = \partial f / \partial x_j(0)$. In fact, it is sufficient to notice that in a neighbourhood of 0 there holds

$$\tilde{f}(x) = \int_0^1 \frac{\mathrm{d}}{\mathrm{d}t} \tilde{f}(tx)\,\mathrm{d}t = \int_0^1 \sum_j \frac{\partial \tilde{f}}{\partial x_j}(tx)\,\mathrm{d}t,$$

so the claim is valid with $g_j = \int_0^1 \partial \tilde{f}/\partial x_j(tx)\,\mathrm{d}t$. We can apply this formula to both $f - f(p)$ and $\partial f/\partial x_i$, so that we deduce

$$f(x) = f(p) + \sum_{i,j=1}^n x_i x_j h_{ij}(x), \qquad h_{ij}(0) = \frac{1}{2}\frac{\partial^2 f}{\partial x_i \partial x_j}(0).$$

Let us now prove the statement by induction on n, assuming that in some coordinates $u = (u_1, \ldots, u_n)$ we have

$$f(u) = f(p) \pm u_1^2 \pm u_2^2 \cdots \pm u_{r-1}^2 + \sum_{i,j \geq r} u_i u_j H_{ij}(u_1, \ldots, u_n), \qquad (12.11)$$

for some continuous functions $H_{ij} = H_{ji}$. After a linear change of variables in $(u_r, \ldots, u_n)$, we can assume that $H_{rr}(0) = 0$. Therefore near zero we have a well defined and continuous function $g(u) = H_{rr}^{1/2}(u)$. Now consider the new coordinates

$$v_i = u_i, \quad i \neq r; \qquad v_r(u) = g(u)\left[u_r + \sum_{i>r} \frac{u H_{ir}(u)}{H_{rr}(u)}\right].$$

At the origin we have $\partial v_r / \partial u_r = g(0)$, and hence it turns out that

$$\frac{\partial v}{\partial u}(0) = \begin{pmatrix} -1 & & & & & & \\ & \ddots & & & & 0 & \\ & & -1 & & & & \\ & * & & g(0) & & * & \\ & & & & 1 & & \\ & 0 & & & & \ddots & \\ & & & & & & 1 \end{pmatrix},$$

and therefore the transformation $u \mapsto v$ is locally invertible near the origin. Moreover, collecting the indices in (12.11) which are strictly greater than r and noticing that

$$v_r^2 = H_{rr}(u) \left[u_r^2 + \left(\sum_{i>r} \frac{u H_{ir}(u)}{H_{rr}(u)} \right)^2 + 2u_r \sum_{i>r} \frac{u H_{ir}(u)}{H_{rr}(u)} \right],$$

we find immediately

$$f(v) = f(p) \pm v_1^2 \pm v_2^2 \cdots \pm v_r^2 + \sum_{i,j>r} H'_{ij}(v) v_i v_j,$$

for some smooth functions H'_{ij}. This concludes the proof. ∎

Proposition 12.11 (Handle body decomposition) *Suppose p is a nondegenerate critical point of f with Morse index λ, and suppose there exists $\varepsilon > 0$ such that f has no critical points with values between $c - \varepsilon$ and $c + \varepsilon$ except for p, where we have set $c = f(p)$. Then $M^{c+\varepsilon}$ has the homotopy type of $M^{c-\varepsilon}$ with a λ dimensional cell attached.*

Regarding the terminology, *a topological space X with a λ-cell attached* is a topological space $\tilde{X}$ with an equivalence relation $\sim$. The space $\tilde{X}$ is the union of X and the closed λ-cell $\overline{B}_1^\lambda$. We also define a continuous map $h : \partial \overline{B}_1^\lambda$ into X, which is a homeomorphism onto its image. The equivalence relation $\sim$ identifies every point $y \in \partial \overline{B}_1^\lambda$ with its image through the map h.

Proof. By the Morse lemma, we can choose a coordinate system $\phi : U \to \mathbb{R}^n$, $q \mapsto (u_1, \ldots, u_n)$, near p such that

$$f(u) = c - u_1^2 - u_2^2 - \cdots - u_\lambda^2 + u_{\lambda+1}^2 + \cdots + u_n^2$$

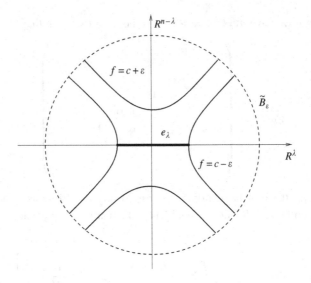

Figure 12.2

in a neighbourhood of 0. Now choose ε so small that the assumption holds for this value of ε and such that the image of ϕ contains the ball $\tilde{B}_\varepsilon = \{(u_1, \ldots, u_n) : \sum_i u_i^2 \leq 2\varepsilon\}$. Now let e_λ denote the set

$$e_\lambda = \{q \in U \ : \ u_1^2(q) + \cdots + u_\lambda^2(q) \leq \varepsilon, u_{\lambda+1}^2(q) + \cdots + u_n^2(q) = 0\}, \tag{12.12}$$

which is homeomorphic to the λ dimensional Euclidean ball. Clearly, see Figure 12.2, $e_\lambda \cap M^{c-\varepsilon}$ is the boundary of e_λ and e_λ is attached to $M^{c-\varepsilon}$ according to the definition above.

We now construct a new function $F : M \to \mathbb{R}$ defined in the following way. First we choose a smooth cutoff function $\eta : \mathbb{R}_+ \to \mathbb{R}$ which satisfies

$$\begin{cases} \mu(r) = 0 & \text{for } r \geq 2\varepsilon \\ \mu(0) > \varepsilon \\ \mu'(0) = 0 \\ -1 < \mu'(r) \leq 0 & \text{for every } r. \end{cases} \tag{12.13}$$

Let F coincide with f outside U, while

$$F = f - \mu(u_1^2 + \cdots + u_\lambda^2 + 2u_{\lambda+1}^2 + 2u_n^2) \qquad \text{in } U.$$

It is convenient to define $\xi, \eta : U \to [0, +\infty)$ by

$$\xi = u_1^2 + \cdots + u_\lambda^2, \qquad \eta = u_{\lambda+1}^2 + \cdots + u_n^2,$$

so in this way we can write that

$$f(q) = c - \xi(q) + \eta(q), \quad F(q) = c - \xi(q) + \eta(q) - \mu(\xi(q) + 2\eta(q)), \qquad q \in U.$$

We now proceed with some different steps.

Step 1. $F^{-1}((-\infty, c+\varepsilon])$ coincides with $f^{-1}((-\infty, c+\varepsilon])$. To prove this fact, we notice that when $\mu(\xi + 2\eta) = 0$, then F coincides with f, and this is true in particular if $\xi + 2\eta \geq 2\varepsilon$. On the other hand, when $\xi + 2\eta \leq 2\varepsilon$, we have

$$F \leq f = c - \xi + \eta \leq c + \frac{1}{2}\xi + \eta \leq c + \varepsilon,$$

which proves the claim.

Step 2. The critical points of F coincide with those of f. To see this, it is sufficient to restrict our attention to the set U, namely to show that in U the only critical point of F is p. In U indeed we have

$$\frac{\partial F}{\partial \xi} = -1 - \mu'(\xi + 2\eta) < 0, \qquad \frac{\partial F}{\partial \eta} = 1 - 2\mu'(\xi + 2\eta) \geq 1.$$

Since we have $dF = \partial F/\partial \xi \, d\xi + \partial F/\partial \eta \, d\eta$, $d\xi = 2\sum_{i=1}^{\lambda} u_i du_i$ and $d\eta = 2\sum_{i=\lambda+1}^{n} u_i du_i$, then $dF = 0$ implies $u = 0$, so also step 2 is proved.

Step 3. $F^{-1}((-\infty, c-\varepsilon))$ is a deformation retract of $M^{c+\varepsilon}$. By step 1 and by the fact that $F \leq f$, it follows that $F^{-1}([c-\varepsilon, c+\varepsilon]) \subseteq f^{-1}([c-\varepsilon, c+\varepsilon])$. Therefore, since F and f coincide outside U, in $F^{-1}([c-\varepsilon, c+\varepsilon])$ there is no critical point of F except for p. On the other hand we have that $F(p) = c - \mu(0) < c - \varepsilon$, so $F^{-1}([c - \varepsilon, c + \varepsilon])$ contains no critical points, and step 3 is proved. We now write $F^{-1}((-\infty, c - \varepsilon)) = M^{c-\varepsilon} \cup H$, where H is the closure of $F^{-1}((-\infty, c - \varepsilon]) \setminus M^{c-\varepsilon}$. We claim next that if e_λ is defined in (12.12), then $e_\lambda \subseteq H$. In fact, since $\partial F/\partial \xi < 0$ then $F(q) \leq F(p) < c - \varepsilon$ for any $q \in e_\lambda$, but also $f(q) \leq c - \varepsilon$. Hence the claim follows.

Step 4. $F^{-1}((-\infty, c - \varepsilon))$ is a deformation retract of $M^{c+\varepsilon}$.

We have to prove that there exists a continuous map $r : [0, 1] \times (M^{c-\varepsilon} \cup H) \to M^{c-\varepsilon} \cup e_\lambda$ such that $r(0, \cdot)$ is the identity on $M^{c-\varepsilon} \cup H$, such that $r(t, \cdot)_{M^{c-\varepsilon} \cup e_\lambda}$ is the identity for every t and such that $r(1, \cdot)$ has values in $M^{c-\varepsilon} \cup e_\lambda$. We divide the proof into three cases.

Case 1. $\xi \leq \varepsilon$. We can use the coordinates given by the Morse lemma and define r_t by

$$r_t(u_1, \ldots, u_n) = (u_1, \ldots, u_\lambda, t u_{\lambda+1}, \ldots t u_n).$$

Since $\partial F/\partial \eta \geq 0$, we deduce that r_t maps $F^{-1}((-\infty, c - \varepsilon))$ into itself.

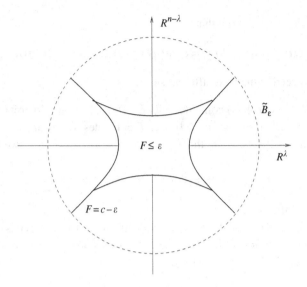

Figure 12.3

Case 2. $\tilde{B}_\varepsilon \cap \{\varepsilon \le \xi \le \eta + \varepsilon\}$. We can still use the above coordinates and we set

$$r_t(u_1, \ldots, u_n) = (u_1, \ldots, u_\lambda, s_t u_{\lambda+1}, \ldots, s_t u_n), \qquad s_t = t + (1-t) \left[\frac{\xi - \varepsilon}{\eta} \right]^{1/2}.$$

We notice that r_1 maps this region into $f^{-1}(c - \varepsilon)$ and is a continuous map.
Case 3. $\{\xi \ge \eta + \varepsilon\} \cup (M \setminus \tilde{B}_\varepsilon)$. In this case we just take r_t to be the identity for every $t \in [0, 1]$.

It is easy to check that r_t is continuous through the three different regions, and that it satisfies the required properties, so step 4 is also complete.
The conclusion of the proposition follows from the above four steps and the deformation lemma. ∎

Corollary 12.12 *Let p be a nondegenerate critical point of f with Morse index λ. let $c = f(p)$ and assume that $Z_c = \{p\}$. Then for ε sufficiently small we have that*

$$H_q(M^{c+\varepsilon}, M^{c-\varepsilon}) \simeq \begin{cases} \mathbb{Z} & \text{for } q = \lambda \\ 0 & \text{for } q \ne \lambda \end{cases}.$$

Remark 12.13 If a regular function $f : M \to \mathbb{R}$ possesses only a finite number of critical points $z_1, \ldots, z_m$ at the level c, and if all of them are nondegenerate and of Morse indices respectively $\lambda_1, \ldots, \lambda_m$, then $M^{c+\varepsilon}$ has the homotopy type of $M^{c-\varepsilon}$ with k cells attached, of dimension respectively $\lambda_1, \ldots, \lambda_m$. Moreover, for ε sufficiently small $H_q(M^{c+\varepsilon}, M^{c-\varepsilon}) \simeq Z^{i_q}$ where i_q is the number of points in $\{z_1, \ldots, z_m\}$ with Morse index equal to q. ∎

We introduce next some inequalities concerning the relative homologies of pairs, beginning with a couple of definitions.

Definition 12.14 *The rank of an Abelian group G is the maximal number k for which $\sum_{i=1}^{k} n_i g_i = 0$ with $(n_i)_i \subseteq \mathbb{Z}$ and $(g_i)_i \subseteq G$ implies $g_i = 0$ for every i.*

For example, if $G = Z^p \oplus Z_q$ for some integers p and q, then rank $G = p$.

Definition 12.15 *Given a pair of spaces (B, A) and an integer q we set*

$$\beta_q(B, A) = \text{rank } H_q(B, A), \qquad \chi(B, A) = \sum (-1)^q \beta_q(B, A).$$

Here we are assuming that all the ranks are finite, and for the second definition that β_q is nonzero except for a finite number of q. The number $\beta_q(B, A)$ is called the q-th Betti number of (B, A), while $\chi_q(B, A)$ is called the Euler-Poincaré characteristic of (B, A).

Lemma 12.16 *If $A \subseteq B \subseteq C$ and if q is an integer, then we have*

$$\beta_q(C, A) \leq \beta_q(C, B) + \beta_q(B, A).$$

Proof. Let us recall that, given a homomorphism $\varphi : G \to H$ between two Abelian groups, by the fundamental theorem of homomorphism then $G \simeq \ker(\varphi) \oplus im(\varphi)$, and therefore rank $G = \text{rank } \ker(\varphi) + \text{rank } im(\varphi)$.

Let us now consider the exact sequence

$$\cdots \xrightarrow{\partial_{q+1}} H_q(B, A) \xrightarrow{\varphi_q} H_q(C, A) \xrightarrow{\psi_q} H_q(C, B) \xrightarrow{\partial_q} \cdots,$$

where φ_q and ψ_q are induced by natural inclusions. By the above observation we have that rank $H_q(C, A) = \text{rank } \ker(\psi_q) + \text{rank } im(\psi_q)$, so by the exactness of the sequence of the triple (C, B, A) it follows that

$$\beta_q(C, A) = \text{rank } im(\varphi_q) + \text{rank } im(\psi_q), \qquad (12.14)$$

from which we immediately deduce

$$\text{rank } im(\psi_q) \leq \text{rank} H_q(C, B) = \beta_q(C, B). \qquad (12.15)$$

Moreover, applying the above reasoning to $G = H_q(B, A)$ and to $\varphi = \varphi_q$ we obtain that

$$\text{rank im}(\varphi_q) \le \text{rank } H_q(B, A) = \beta_q(B, A). \qquad (12.16)$$

Therefore, from (12.14)–(12.16) we find $\beta_c(C, A) \le \beta_q(C, B) + \beta_q(B, A)$, which concludes the proof. ∎

Lemma 12.17 *Let A, B, C be as in Lemma 12.16. Then one has*

$$\chi(C, A) = \chi(C, B) + \chi(B, A).$$

Proof. Let us consider the exact sequence

$$\cdots \xrightarrow{\partial_{q+1}} H_q(B, A) \xrightarrow{\varphi_q} H_q(C, A) \xrightarrow{\psi_q} H_q(C, B) \xrightarrow{\partial_q} H_{q-1}(B, A) \longrightarrow \cdots.$$

Then, recalling (12.14) we find

$$\beta_q(C, B) = \text{rank im}(\psi_q) + \text{rank im}(\partial_q),$$
$$\beta_q(B, A) = \text{rank im}(\partial_{q+1}) + \text{rank im}(\partial_{q+1}),$$
$$\beta_q(C, A) = \text{rank im}(\varphi_q) + \text{rank im}(\psi_q),$$

and hence

$$\beta_q(C, A) = \beta_q(C, B) + \beta_q(B, A) - \text{rank im}(\partial_q) - \text{rank im}(\partial_{q+1}).$$

Taking the sum in q with alternate signs we obtain

$$\chi(C, A) = \sum_q (-1)^q \beta_q(C, A) = \sum_q (-1)^q \beta_q(C, B) + \sum_q (-1)^q \beta_q(B, A)$$
$$- \sum_q \text{rank im}(\partial_q) - \sum_q \text{rank im}(\partial_{q+1}).$$

Summing over q, one easily checks that the last two terms cancel, and therefore the last formula becomes

$$\chi(C, A) = \sum_q (-1)^q \beta_q(C, B) + \sum_q (-1)^q \beta_q(B, A) = \chi(C, B) + \chi(B, A),$$

so the proof is concluded. ∎

We have next an extension of Lemma 12.16.

Lemma 12.18 *Given a pair of spaces (B, A), we set*

$$\mathcal{B}_q(B, A) = \beta_q(B, A) - \beta_{q-1}(B, A) + \cdots \pm \beta_0(B, A).$$

Then, if $A \subseteq B \subseteq C$, we have the inequality

$$\mathcal{B}_q(C,A) \le \mathcal{B}_q(C,B) + \mathcal{B}_q(B,A).$$

Proof. Given an exact sequence

$$\cdots \xrightarrow{\varphi_{q+1}} G_q \xrightarrow{\varphi_q} G_{q-1} \longrightarrow \cdots \xrightarrow{\varphi_1} G_0 \xrightarrow{\varphi_0} 0,$$

by the fundamental theorem of homomorphism we find

$$\text{rank im}(\varphi_{q+1}) = \text{rank } G_q - \text{rank im}(\varphi_q)$$
$$= \text{rank } G_q - [\text{rank } G_{q-1} - \text{rank im}(\varphi_{q-1})]$$
$$= \text{rank } G_q - \text{rank } G_{q-1} + \text{rank im}(\varphi_{q-1}).$$

Continuing in this way from the last formula we deduce

$$\text{rank im}(\varphi_{q+1}) = \text{rank } G_q - \text{rank } G_{q-1} + \cdots \pm \text{rank } G_0, \tag{12.17}$$

so clearly the right hand side is non-negative.

Applying this argument to the exact sequence of the triple (C, B, A)

$$\cdots \xrightarrow{\partial_{q+1}} H_q(B,A) \longrightarrow H_q(C,A) \longrightarrow H_q(C,B) \longrightarrow \cdots,$$

taking $\varphi_{q+1} = \partial_{q+1}$ we get

$$\beta_q(B,A) - \beta_q(C,A) + \beta_q(C,B) - \beta_{q-1}(B,A) + \cdots \ge 0.$$

Collecting the terms involving the pairs (B,A), (C,A) and (C,B) we obtain the conclusion. ∎

Corollary 12.19 *Suppose that we have the inclusion of spaces $X_0 \subseteq X_1 \subseteq \cdots \subseteq X_n$. Then*

$$\beta_q(X_n, X_0) \le \sum_{i=1}^n \beta_q(X_i, X_{i-1}), \qquad \mathcal{B}_q(X_n, X_0) \le \sum_{i=1}^n \mathcal{B}_q(X_i, X_{i-1}),$$

$$\chi(X_n, X_0) = \sum_{i=1}^n \chi(X_i, X_{i-1}).$$

Proof. The proof follows immediately using Lemmas 12.16, 12.17, 12.18 and an induction procedure on n. ∎

We can now state the main result of this chapter.

Theorem 12.20 *Suppose M is a compact finite dimensional manifold, and that $f : M \to \mathbb{R}$ is a function of class C^2 whose critical points are all nondegenerate. Then for any non-negative integer q we have the following relations*

$$C_q(M) \geq \beta_q(M) \qquad \text{(Weak Morse inequalities;)}$$

$$\sum_{i \geq 0}(-1)^{q-i}C_i(M) \geq \sum_{i \geq 0}(-1)^{q-i}\beta_i(M) \qquad \text{(Strong Morse inequalities;)}$$

$$\sum_{i \geq 0}(-1)^i C_i(M) = \chi(M). \tag{12.18}$$

In these formulas, $C_i(M)$ denotes the number of critical points of f with Morse index equal to i.

Proof. Let $c_1 < c_2 \cdots < c_k$ denote the critical levels of f, which are only finitely many according to our assumptions. Now choose real numbers $a_0, a_1, \ldots, a_k$ such that $a_i < c_{i+1} < a_{i+1}$ for every $i = 0, \ldots, k-1$. In particular we have that $M^{a_0} = \emptyset$ and that $M^{a_k} = M$. By Proposition 12.9 we have that, for any integers i, q and any small $\varepsilon > 0$ $H_q(M^{c_i+\varepsilon}, M^{c_i-\varepsilon}) \simeq H_q(M^{a_i}, M^{a_{i-1}})$. Then it is sufficient to apply Remark 12.13 and Corollary 12.19. ∎

Remark 12.21 (i) By subtraction the strong Morse inequalities immediately imply the weak ones.

(ii) If a, b, with $a < b$ are two regular values of f, then the Morse relations still hold if we replace $\beta_q(M)$ with $\beta_q(a,b) := \text{rank } H_q(M^b, M^a)$ and count only the critical points with values between a and b.

(iii) Under suitable assumptions on the function f, it is possible to cover also the case of manifolds with boundary, see [134]. Precisely, a function $f \in C^2(\overline{M})$, where M is a manifold with boundary, is said to satisfy the *general boundary conditions on M* if both f and its restriction to ∂M are Morse functions, and if it has no critical point on ∂M. In this case, at every critical point of $f|_{\partial M}$ the gradient of f is always nonzero and points either inward on M or outward. Then, in the Morse relations, $C_q(M)$ has to be substituted with $C_q(M) + c_q(M)$, where $c_q(M)$ represents the number of critical points of $f|_{\partial M}$ with Morse index q and for which the gradient points inward. Results of this type, but in the framework of the Lusternik–Schnirelman category, have been obtained by P. Majer [121]. ∎

Example 12.22 (1) If $\beta_0(a, b) = 1$ and $\beta_q(a, b) = 0$ for every $q > 0$, then we have $C_1 \geq C_0 - 1$. In particular if f has two nondegenerate minima then there exists at least a saddle point.

(2) If $C_{q+1} = 0$ then clearly $\beta_{q+1} = 0$. Using the strong inequalities with index $q + 1$ we find that

$$-C_q + C_{q-1} + \cdots \pm C_0 \geq -\beta_q + \beta_{q-1} + \cdots \pm \beta_0.$$

Similarly, the strong inequalities with index q imply the reversed inequality and hence we get

$$C_q - C_{q-1} + \cdots \pm C_0 = \beta_q - \beta_{q-1} + \cdots \pm \beta_0.$$

Analogously, if $C_{q-1} = 0$ we have $\beta_{q-1} = 0$ and as before we find

$$C_{q-2} - C_{q-3} + \cdots \pm C_0 = \beta_{q-2} - \beta_{q-3} + \cdots \pm \beta_0.$$

It follows that, if both C_{q+1} and C_{q-1} vanish then $C_q = \beta_q$. ∎

We mention next some extensions of Theorem 12.20, which allow us to cover both the infinite dimensional case, and the presence of degenerate (but still isolated) critical points. We refer the reader to [72], where some further extensions and several applications are given.

We consider now the case of a Banach (or Hilbert) manifold M, see Section 6.1, and of a function $f : M \to \mathbb{R}$ of class C^1. Let also $p \in M$ be an isolated critical point of f. Then, if r is so small that $B_r(p)$ contains no critical points of f except for p, one can define the *critical groups of f at p* by

$$C_q(f, p) := H_q(M^c \cap B_\varepsilon(p), (M^c \setminus \{p\} \cap B_\varepsilon(p))), \qquad q \geq 0, \quad (12.19)$$

where $c = f(p)$. By the excision property of the relative homology groups, we see that this definition is independent of r (taking this sufficiently small). If f is of class C^2 (on a Hilbert manifold) and p is a nondegenerate critical point of f with Morse index λ, the Morse lemma still holds, and in suitable local coordinates u (with $u(p) = 0$) one has $f(u) = c - \|u_-\|^2 + \|u_+\|^2$, where $u = u_- + u_+$ with $u_\pm \in E_\pm$ and $E_+ \oplus E_- \simeq T_p M$. This allows us to prove that in this case: [2]

$$C_q(f, p) \simeq \begin{cases} \mathbb{Z} & q = \lambda \\ 0 & q \neq \lambda. \end{cases} \quad (12.20)$$

In particular, if the index λ is infinite, then all the critical groups vanish.

Using pseudogradient vector fields and the Palais–Smale condition, one can employ the deformation lemma to deal with noncritical levels, and the critical groups for the critical ones. Then the following general result can be proved.

---

[2] In finite dimension this follows from the construction in the proof of Proposition 12.11.

Theorem 12.23 *Let M be a Banach manifold and let $f \in C^1(M)$ satisfy the $(PS)_c$ condition for every $c \in [a,b]$, where a, b are regular values of f with $a < b$. Assume that the critical points of f in $M^a \setminus M^b$ are only finitely many, denoted by $\{z_1, \dots, z_l\}$. Then, letting $M_q(a,b) = \sum_{i=1}^{l} \text{rank } C_q(f, z_i)$ and $\beta_q(a,b) = \text{rank } H_q(M^b, M^a)$, for every integer q one has*

$$\sum_{j=0}^{q}(-1)^{q-j}M_j(a,b) \geq \sum_{j=0}^{q}(-1)^{q-j}\beta_q(a,b),$$

$$\sum_{j\geq 0}(-1)^j M_j(a,b) = \sum_{j\geq 0}(-1)^j \beta_j(a,b).$$

Here we are assuming that all the numbers $M_q(a,b)$, $\beta_q(a,b)$ are finite and that the series converge.

12.3 An application: bifurcation for variational operators

In this section we apply Morse theory to prove a bifurcation theorem due to Krasnoselski [110] and improved by R. Böhme [59] and A. Marino [128]. We follow the proof given in [129].

We consider an equation of the form

$$\lambda u - T(u) = 0, \tag{12.21}$$

where u belongs to a Hilbert space X, λ is a real parameter and T is an operator from a neighbourhood of zero in X with values in X and such that $T(0) = 0$. Our main goal is to prove the following result.

Theorem 12.24 *Let $U \subseteq X$ be a neighbourhood of 0, and let $a : U \to X$ be a function of class C^2 with $a(0) = 0$. Letting $T = \nabla a$, assume T is compact and that $T(0) = 0$. Then every nonzero eigenvalue of $T'(0)$ is a bifurcation point for (12.21).*

Before beginning the proof, we need some preliminary notation and lemmas. We set

$$b_\lambda(u) = \frac{\lambda}{2}\|u\|^2 - a(u),$$

so that $\nabla b_\lambda = \lambda \text{Id} - T$. Therefore the solutions of (12.21) are critical points of b_λ. Without loss of generality we can assume that T has a positive eigenvalue λ_0. Let us also introduce the following subsets of $\mathbb{R} \times U$:

$$\Omega = \{(\lambda, u) \in \mathbb{R} \times (U \setminus \{0\}) : b_\lambda(u) = 0\};$$

$$\Omega_\gamma = \{(\lambda, u) \ : \ \lambda_0 - \gamma \leq \lambda \leq \lambda_0 + \gamma, u \in U \setminus \{0\}, b_\lambda(u) = 0\}.$$

Letting P denote the natural projection of $\mathbb{R} \times X$ onto X, we also set

$$P(\Omega_\gamma) = M_\gamma,$$

and one can easily check that M_γ is relatively closed in $U \setminus \{0\}$. We next define

$$U_\lambda^- = \{u \in U \ : \ b_\lambda(u) \leq 0\}, \qquad B_\delta = B_\delta(0).$$

Lemma 12.25 *In the above assumptions, suppose that λ_0 is not a bifurcation value for (12.21). Then there exist two positive numbers γ, δ such that in $M_\gamma \cap B_\delta$ the operator T satisfies*

$$\|T(u)\|^2 \|u\|^2 - (T(u)|u)^2 > 0, \tag{12.22}$$

namely there is no point $u \in B_\delta \setminus \{0\}$ with $b_\lambda(u) = 0$, with $|\lambda - \lambda_0| \leq \gamma$ and with $T(u)$ parallel to u.

Proof. Suppose by contradiction that there exist three sequences $(u_n)_n \subseteq U \setminus \{0\}, (\lambda_n)_n, (\mu_n)_n \subseteq \mathbb{R}$ such that

$$\mu_n u_n - T(u_n) = 0, \qquad \frac{\lambda_n}{2}\|u_n\|^2 - a(u_n) = 0,$$

$$\lim_{n \to \infty} u_n = 0, \qquad \lim_{n \to \infty} \lambda_n = \lambda_0.$$

Taking the scalar product of the first equation with u_n we get $\rho(u_n) = \mu_n$, where $\rho(u)$ is defined as

$$\rho(u) = \frac{(T(u)|u)}{(u|u)}.$$

Since $(\lambda_n, u_n) \in \Omega$ and since $\lim_{n \to \infty}(\lambda_n, u_n) = (\lambda_0, 0)$, from the C^2-continuity of a we find that

$$\mu_n = \lambda_n \frac{(T(u_n)|u_n)}{2a(u_n)} = \lambda_n \frac{(a''(0)u_n|u_n) + o(\|u_n\|^2)}{(a''(0)u_n|u_n) + o(\|u_n\|^2)},$$

so that $\lim_{n \to \infty} \mu_n = \lambda_0$. This is clearly in contradiction to the assumption that λ_0 is not a bifurcation value. ∎

In the next lemma we obtain a uniform positive lower bound of the left hand side of (12.22) in terms of $\|u\|$.

Lemma 12.26 *In the above assumptions (taking $\lambda_0 > 0$), there exist two positive numbers γ, δ (with $B_\beta \subseteq U$) and a real function $\sigma : (0, \delta] \to \mathbb{R}$ which* *is positive, continuous monotone, with $\sigma(t) \to 0$ as $t \to 0$ and such that in $M_\gamma \cap B_\delta$ we have*

$$\|T(u)\|^2 \|u\|^2 - (T(u), u)^2 \geq \sigma(\|u\|).$$

Proof. Let us take two small positive numbers γ, δ such that we have the condition (12.22) in $M_\gamma \cap B_\delta$, by Lemma 12.25. By the C^2-continuity of a we can also assume that $\rho(u) \geq \lambda_0/2$ in this set.

At this point, we first prove that $u \mapsto T(u) - \rho(u)u$ maps the closed sets of $M_\gamma \cap B_\delta$ into closed sets of X. Indeed, let $(u_n)_n$ be a sequence of points in a closed set K of $M_\gamma \cap B_\delta$ such that the image sequence $y_n = T(u_n) - \rho(u_n)u_n$ converges to some point y_*. By compactness, we can extract a sequence, still denoted by $(u_n)_n$ such that both $T(u_n)$ and $\rho(u_n)$ converge. Then we have

$$u_n = \frac{1}{\rho(u_n)}(T(u_n) - y_n).$$

Since $\rho(u_n) \geq \lambda_0/2$, we deduce that u_n converges to some $u_* \in K$ such that $T(u_*) - \rho(u_*)u_* = y_*$. This proves our claim.

Therefore, the set $M_\gamma \cap B_\delta$ is mapped onto a closed set which, by (12.22), does not contain the origin and therefore has a positive distance $\tilde{\sigma}(\gamma)$ from it. The function $\tilde{\sigma}$ satisfies all the required properties in the statement except for the continuity. But it is easy to construct a continuous function $\sigma \leq \tilde{\sigma}$ which also satisfies the last requirement. ∎

Let us now fix the numbers γ and δ from Lemma 12.26. Let $0 < \varepsilon < \gamma/2$ and let $\theta : \mathbb{R} \to \mathbb{R}$ be a smooth function supported in $(-\gamma/2, \gamma/2)$ and which is identically equal to 1 in the interval $[-\varepsilon, \varepsilon]$. Let us define

$$\omega(u) = \theta\left(\frac{\lambda_0}{2} - \frac{a(u)}{\|u\|^2}\right), \qquad u \in U \setminus \{0\}.$$

Let us observe that if $0 < \|u\| \leq \delta$ and $\omega(u) \neq 0$, then $-\gamma/2 < \lambda_0/2 - a(u)/\|u\|^2 < \gamma/2$, and hence there exists $\lambda \in (\lambda_0 - \gamma, \lambda_0 + \gamma)$ such that $\lambda/2 - a(u)/\|u\|^2 = 0$, namely $u \in M_\gamma \cap B_\delta$.

Let us now consider the operator $D : U \to X$ defined as

$D(u)$
$$= \begin{cases} \omega(u)\dfrac{\|u\|^2 T(u) - (T(u)|u)u}{\|u\|^2 \|T(u)\|^2 - (T(u)|u)^2}\dfrac{\|u\|^2}{2} & \text{for } \|u\|^2\|T(u)\|^2 - (T(u)|u)^2 \neq 0 \\ 0 & \text{otherwise.} \end{cases}$$

One easily checks that D is of class C^1 in an open set $\tilde{U}$ containing $B_\delta \setminus \{0\}$. Moreover one immediately verifies that

$$(u|D(u)) = 0 \quad \forall u, \qquad (T(u)|D(u)) = 1 \quad \forall u \neq 0. \qquad (12.23)$$

Lemma 12.27 *In the above notation, consider the following ordinary differential equation*

$$\frac{\mathrm{d}\eta}{\mathrm{d}\lambda} = D(\eta), \qquad in \ \tilde{U}. \tag{12.24}$$

Then, for any choice of $\overline{\lambda}$ and of $\overline{u} \in B_\delta \setminus \{0\}$ problem (12.24) has a unique solution $\eta(\lambda, \overline{u})$ with initial condition $\eta(\overline{\lambda}, \overline{u}) = \overline{u}$. This solution is defined on the whole real line and verifies

$$\|\eta(\lambda, \overline{u})\| = \|\overline{u}\| \qquad for \ any \ \lambda.$$

If in addition one has $b_{\overline{\lambda}}(\overline{u}) = 0$ for some $\overline{\lambda} \in [\lambda_0 - \varepsilon, \lambda_0 + \varepsilon]$, then also $b_\lambda(\eta(\lambda, \overline{u})) = 0$ in the whole interval $[\lambda_0 - \varepsilon, \lambda_0 + \varepsilon]$.

Proof. Since D is of class C^1 in $\tilde{U}$, we have local existence and uniqueness of solutions. On the other hand, using the first of (12.23), we have that

$$\frac{\mathrm{d}}{\mathrm{d}\lambda} \|\eta(\lambda, \overline{u})\|^2 = 2\left(\eta(\lambda, \overline{u}) \mid \frac{\mathrm{d}\eta(\lambda, \overline{u})}{\mathrm{d}\lambda}\right) = 2(\eta(\lambda, \overline{u}) \mid D(\eta(\lambda, \overline{u}))) = 0.$$

It follows that every solution has constant norm and hence the set $\{\eta(\lambda, \overline{u})\}$, when λ varies in the interval of definition, has closure contained in $\tilde{U}$. Moreover, since $\omega(\eta(\lambda, \overline{u})) \neq 0$ only when $\eta(\lambda, \overline{u}) \in M_\gamma \cap B_\delta$, we deduce from Lemma 12.26 and from the expression of D that $D(\eta(\lambda, \overline{u}))$ stays bounded, so the solution is indeed globally defined.

Let us now prove the second statement. Let $b_{\overline{\lambda}}(\overline{u}) = 0$, namely $\overline{\lambda}/2 - a(\overline{u})/\|\overline{u}\|^2 = 0$. From the definition of ω we have that $\omega(\overline{u}) = \theta((\lambda_0 - \overline{\lambda})/2) = 1$ since we are assuming $|\lambda_0 - \overline{\lambda}| \leq \varepsilon$. Moreover, since the argument of θ does not exceed $\varepsilon/2$ in absolute value, there is a neighbourhood of $\overline{u}$ in which ω is identically equal to 1. Therefore, in a neighbourhood of $\overline{\lambda}$, the first definition of D can be applied, with $\omega(u) = 1$. It follows that

$$\frac{\mathrm{d}}{\mathrm{d}\lambda} b_\lambda(\eta(\lambda, \overline{u})) = (\nabla b_\lambda(\eta) \mid D(\eta)) = \lambda(\eta \mid D(\eta)) - (T(\eta) \mid D(\eta)).$$

Using (12.23) we get $\mathrm{d}/\mathrm{d}\lambda b_\lambda(\eta(\lambda, \overline{u})) = 0$, namely that $\lambda \mapsto b_\lambda(\eta(\lambda, \overline{u}))$ is constant in $[\lambda_0 - \varepsilon, \lambda_0 + \varepsilon]$. This proves the lemma. ∎

Lemma 12.28 *Under the above assumptions, suppose that λ_0 is not a bifurcation point for (12.21). Then, for sufficiently small ε and δ, the spaces $U^-_{\lambda_0-\varepsilon} \cap B_\delta$ and $U^-_{\lambda_0+\varepsilon} \cap B_\delta$ are homeomorphic, and the homeomorphism maps the origin into itself.*

Proof. We let $\psi : \mathbb{R}^2 \times B_\delta \to B_\delta$ be such that if $\|\overline{u}\| > 0$, then $\psi(\overline{\lambda}, \lambda, \overline{u}) = \eta(\lambda, \overline{u})$ (where $\eta(\lambda, \overline{u})$, as above, denotes the solution of (12.24) with initial condition $u(\overline{\lambda}) = \overline{u}$), while $\psi(\overline{\lambda}, \lambda, 0) = 0$ for every $\overline{\lambda}, \lambda$. Next, one can easily show that $\psi(\overline{\lambda}, \lambda, \cdot)$ is continuous for $\overline{u} \neq 0$, but since $\|\psi(\overline{\lambda}, \lambda, u)\| = \|u\|$ we have continuity also at the origin. By Lemma 12.27, when $\overline{\lambda} \in [\lambda_0 - \varepsilon, \lambda_0 + \varepsilon]$, ψ preserves the sign of b_λ in the whole interval $[\lambda_0 - \varepsilon, \lambda_0 + \varepsilon]$: then the map $\overline{u} \mapsto \psi(\lambda_0 - \varepsilon, \lambda_0 + \varepsilon, \overline{u})$ admits the function $u \mapsto \psi(\lambda_0 + \varepsilon, \lambda_0 - \varepsilon, \overline{u})$ as inverse, and it gives the desired homeomorphism from $U^-_{\lambda_0 - \varepsilon} \cap B_\delta$ onto $U^-_{\lambda_0 + \varepsilon} \cap B_\delta$.   ∎

We are now in a position to prove the main theorem.

Proof of Theorem 12.24. Let $\lambda_0 > 0$ be an eigenvalue of $T'(0)$, and suppose by contradiction that λ_0 is not a bifurcation value for (12.21). Then by Lemma 12.28 if ε and δ are chosen small enough, then we have a sequence of isomorphisms

$$H_q(U^-_{\lambda_0 - \varepsilon} \cap B_\delta, U^-_{\lambda_0 - \varepsilon} \cap B_\delta \setminus \{0\}) \simeq H_q(U^-_{\lambda_0 + \varepsilon} \cap B_\delta, U^-_{\lambda_0 + \varepsilon} \cap B_\delta \setminus \{0\}).$$

On the other hand, if λ_0 is an eigenvalue of $T'(0)$, the Morse index of b_λ at $u = 0$ changes when λ crosses λ_0. Thus, by (12.19) and (12.20), the preceding isomorphism between homology groups cannot hold for every q.   ∎

Remark 12.29 Under some further nondegeneracy conditions, it is possible to show that the bifurcation set is (locally) a branch, see [11]. However, in general, this might not be the case, see [59].   ∎

Remark 12.30 Following [145], we can reformulate the preceding arguments in the following suggestive way. The stationary point $u = 0$ of b_λ can be associated, for any fixed λ, to a multi-index $\ell_\lambda := [m^0, m^1, \ldots, m^q, \ldots]$, where m^q denote the Betti numbers of the homology groups $H_q(U^-_\lambda \cap B_\delta, U^-_\lambda \cap B_\delta \setminus \{0\})$. According to the previous calculation, if 0 is nondegenerate, one finds that $m^q = \delta^s_q$, where δ is the Kronecker symbol and s is the number of eigenvalues of $T'(0)$ greater that λ, counted with their multiplicity. Now, letting λ_0 be an isolated eigenvalue of $T'(0)$, the multi-index ℓ_λ changes as λ crosses λ_0. On the other hand, if λ_0 is not a bifurcation point, then Lemma 12.28 implies that $\ell_{\lambda_0 - \varepsilon} = \ell_{\lambda_0 + \varepsilon}$ provided $\varepsilon > 0$ is small enough, getting a contradiction. It is worth pointing out the difference between the multi-index ℓ_λ and the Leray–Schauder index. The former changes when crossing any eigenvalue of $T'(0)$, including those with even multiplicity. The latter, which equals $(-1)^s$ (see Theorem 3.20), changes only when crossing an odd eigenvalue.   ∎

12.4 Morse index of mountain pass critical points

In this section we want to evaluate the Morse index of critical points found using the MP theorem.

12.4.1 An abstract result

Let us consider a functional $J \in C^1(E, \mathbb{R})$ which has the mountain pass (MP, in short) geometry, namely that satisfies the assumptions (MP-1) and (MP-2) introduced in Section 8.1. With the same notation used there, let us set

$$\Gamma = \{\gamma \in C([0,1], E) : \gamma(0) = 0, \ \gamma(1) = e\},$$

and

$$c = \inf_{\gamma \in \Gamma} \max_{t \in [0,1]} J(\gamma(t)).$$

According to the MP theorem 8.2, we know that if (MP-1) and (MP-2) and $(PS)_c$ hold, then $c > 0$ is a critical level of J, namely there exists $z \in Z_c = \{u \in E : J(u) = c, \ J'(u) = 0\}$.

The min-max characterization of the MP critical level c allows us to establish some further properties of the MP critical points. Let us introduce some notation. If $J \in C^2(E, \mathbb{R})$ and z is a critical point of J, we set

- $E^0(z) = \mathrm{Ker} J''(z)$,
- $E^-(z) = \{v \in E : (J''(z)v \mid v) < 0\}$,
- $E^+(z) = \{v \in E : (J''(z)v \mid v) > 0\}$.

We say that z is *nondegenerate if* $E^0(z) = \{0\}$ and we denote the Morse index by $m(z)$.

Theorem 12.31 *In addition to the assumptions of the MP theorem, let $J \in C^2(E, \mathbb{R})$ and suppose that Z_c is discrete. Then there exists $z \in Z_c$ such that $m(z) \leq 1$. Furthermore, if $Z_c = \{z\}$ and z is nondegenerate, then $m(z) = 1$.*

Let us give an outline of the proof in the case in which $Z_c = \{z\}$ and z is nondegenerate. We follow [10]. For simplicity of notation, we take $z = 0$ and write $E^\pm$ instead of $E^\pm(z)$.

Arguing by contradiction, let $E = E^- \oplus E^+$ with $\dim(E^-) \geq 2$. By the Morse lemma, up to a smooth change of coordinates, we have that

$$J(u) = c - \|u^-\|^2 + \|u^+\|^2,$$

where $u = u^- + u^+$, $u^\pm \in E^\pm$. Given $\beta > \alpha > 0$, let $U_{\alpha,\beta}$ denote the neighbourhood of $z = 0$

$$U_{\alpha,\beta} = \{u = u^- + u^+ \; : \; \|u^-\| < \alpha, \; \|u^+\| < \beta\}.$$

For any $u \in \overline{U}_{\alpha,\beta}$ with $\|u^+\| = \beta$, one finds that

$$J(u) \geq c - \alpha^2 + \beta^2.$$

Thus, given $d > c$, there exists $\varepsilon > 0$ such that if $\alpha < \beta \leq \varepsilon$ one has

$$\inf\{J(u) : u \in \overline{U}_{\alpha,\beta}, \quad \|u^+\| = \beta\} \geq d.$$

Taking α, β possibly smaller, we can also suppose that

$$\inf\{J(u) : u \in \overline{U}_{\alpha,\beta}\} > 0.$$

Let us point out that, in particular, this and the fact that $J(0) = 0$ and $J(e) \leq 0$ imply $0, e \notin U_{\alpha,\beta}$. Let $\delta \in \,]0, d - c[$. By the definition of the MP critical level c we can find $\gamma \in \Gamma$ such that

$$\sup\{J(\gamma(t)) : t \in [0, 1]\} \leq c + \delta.$$

According to the deformation Lemma 9.12 and Remark 9.13, we can find a deformation η such that

(a) $\eta \circ \gamma \in \Gamma$,
(b) $\eta(E^{c+\delta} \setminus U_{\alpha,\beta}) \subset E^{c-\delta}$.

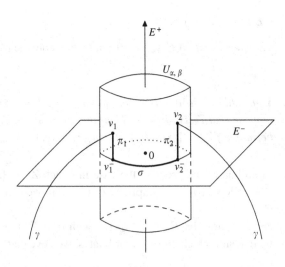

Figure 12.4 The modified part of γ, namely the path $\pi_1 \cup \sigma \cup \pi_2$, is in bold.

If $\gamma(t) \cap U_{\alpha,\beta} = \emptyset \ \forall \ t \in [0, 1]$, then we get a contradiction directly from (a) and (b) above. Then, let us suppose that γ intersects $U_{\alpha,\beta}$. Since $\gamma(0) = 0$ and $\gamma(1) = e$ and $0, e \notin U_{\alpha,\beta}$ (see before) we infer that there exist $t_1 < t_2 \notin]0, 1[$ such that $v_i := \gamma(t_i), i = 1, 2$, belong to $\partial U_{\alpha,\beta}$, while $\gamma(t) \notin \overline{U}_{\alpha,\beta}$ for all $t < t_1$ and $t > t_2$. Since

$$J(v_i) \leq c + \delta < d \leq \inf\{J(u) : u \in \overline{U}_{\alpha,\beta}, \quad \|u^+\| = \beta\},$$

it follows that $\|v_i^-\| = \alpha$ and $\|v_i^+\| < \beta, i = 1, 2$. Let $\pi_i \subset \partial U_{\alpha,\beta}$ denote the segments joining v_i and v_i^-. Since $\dim(E^-) \geq 2$ we can connect v_1^- and v_2^- with an arc σ contained in $\partial U_{\alpha,\beta} \cap E^-$. Finally, consider the path $\widetilde{\gamma}$ defined as follows

$$\widetilde{\gamma}(t) = \begin{cases} \gamma(t) & \text{if } t \in [0, t_1] \cup [t_2, 1] \\ \pi_1 \cup \sigma \cup \pi_2 & \text{if } t \in]t_1, t_2[. \end{cases}$$

Remark that $\widetilde{\gamma} \in G$. Since for $v \in \partial U_{\alpha,\beta} \cap E^-$ one has that $\|v^-\| = \alpha$ and $\|v^+\| = 0$, then

$$J(v) = c - \alpha^2, \qquad \forall \, v \in \partial U_{\alpha,\beta} \cap E^-.$$

Therefore, one has that

$$\sup_{\sigma} J(u) < c.$$

Moreover, one readily checks that $\sup_{\pi_i} J(u) \leq c + \delta$. In conclusion, the path $\widetilde{\gamma} \in \Gamma$ is such that

$$\sup_{\widetilde{\gamma}} J(u) \leq c + \delta, \qquad \widetilde{\gamma}(t) \cap U = \emptyset, \qquad \forall t \in [0, 1].$$

Using (a) and (b) we infer that $\eta \circ \widetilde{\gamma} \in \Gamma$ and $\sup\{J(\eta \circ \widetilde{\gamma}(t)) : t \in [0, 1]\} \leq c - \delta$. This is in contradiction to the definition of c, and proves that $m(z) \leq 1$. To rule out that $m(u) = 0$ it suffices to remark that in such a case u is a local strict minimum for J and $E = E^+$. Now one can take $U_{\alpha,\beta}$ such that $\inf\{J(v) : v \in \partial U_{\alpha,\beta}\} \geq d > c$. Repeating the previous arguments one readily reaches again a contradiction. ∎

Remark 12.32 The case of a degenerate MP critical point is investigated in [97, 106]. The Morse index of a linking critical point has been studied in [114]. In this framework, we also recall a result by H. Amann [6] which asserts that the Leray–Schauder index of a local minimum is 1. ∎

12.4.2 Some applications

Theorem 12.31 can be used to prove multiplicity results. We will explicitly discuss a couple of examples. The first one deals with the coercive Dirichlet boundary value problem

$$\begin{cases} -\Delta u = \lambda u - h(u) & x \in \Omega \\ u = 0 & x \in \partial\Omega, \end{cases} \qquad (\boldsymbol{D_{\lambda^-}})$$

with $h(u) \sim |u|^{p-1}$, $p > 1$, see also Theorem 10.22. Below λ_k, $k = 1, 2, \ldots$, denote, as usual, the kth eigenvalue of $-\Delta$ on $E = H_0^1(\Omega)$.

Theorem 12.33 *Suppose that h is Hölder continuous, such that $h(0) = 0$ and $h(u) \to +\infty$ as $|u| \to \infty$. Then for every $\lambda > \lambda_2$, (D_λ^-) has at least three nontrivial solutions.*

Proof. (Sketch) We will prove this claim under the additional assumption that

$$\lambda - h(u) > f_\lambda'(u), \qquad \forall\, u \neq 0. \qquad (12.25)$$

The general case requires some more work and a Lyapunov–Schmidt reduction (see [17]). By this assumption, there exist two positive numbers M_1, M_2 such that $\lambda u - h(u)$ is negative (respectively positive) for $u > M_1$ (respectively for $u < -M_2$) and positive in $(0, M_1)$ (respectively negative in $(-M_2, 0)$). Therefore, if we define the function $\tilde{f}_\lambda : \mathbb{R} \to \mathbb{R}$ as

$$\tilde{f}_\lambda(u) = \begin{cases} \lambda u - h(u) & \text{for } u \in [-M_2, M_1] \\ 0 & \text{for } u \in (-\infty, -M_2] \cup [M_1, +\infty), \end{cases}$$

then it is easy to check that the solutions of (D_λ^-) are the critical points of

$$J_\lambda(u) = \frac{1}{2}\|u\|^2 - \int_\Omega F_\lambda(u)\,dx, \qquad F_\lambda(u) = \int_0^u \tilde{f}_\lambda(s)\,ds.$$

Hereafter, it is understood that we have carried out the same truncation made in Section 10.4 dealing with (10.12), which is indeed a specific case of (D_λ^-).

Let us first prove that, for $\lambda > \lambda_1$, J_λ has two minima which give rise to a positive and to a negative solution of (D_λ^-). As in Example 5.11, let us consider the positive part $\tilde{f}_\lambda^+$ of $\tilde{f}_\lambda$. Since $h(u) \to +\infty$ as $|u| \to \infty$ then $\tilde{f}_\lambda^+$ is bounded. Then the functional

$$J_\lambda^+(u) = \frac{1}{2}\|u\|^2 - \int_\Omega F_\lambda^+(u)\,dx, \qquad F_\lambda^+(u) = \int_0^u \tilde{f}_\lambda^+(s)\,ds,$$

has a global minimum u_λ^+. Since

$$\lim_{u \to 0} \frac{f(u)}{u} = \lambda > \lambda_1,$$

it follows that $u_\lambda^+ \neq 0$ and by the maximum principle one infers that $u_\lambda^+ > 0$. Let us now prove that u_λ^+ is indeed a local minimum for J_λ, too. Since u_λ^+ satisfies (D_λ^-) and $u_\lambda^+ \neq 0$, it follows that 1 is an eigenvalue of the linear problem

$$-\Delta u_\lambda^+ = a_\lambda(x)u_\lambda^+, \qquad \text{where} \quad a_\lambda(x) = \lambda - h(u_\lambda^+(x)).$$

In other words, using the notation introduced in Section 1.4.1, there exists $k \geq 1$ such that $\lambda_k[a_\lambda] = 1$. Furthermore, since $u_\lambda^+ > 0$ we have that $\lambda_1[a_\lambda] = 1$, see Theorem 1.13(i). Using the monotonicity property of the eigenvalues (see (EP-1) in Section 1.4.1), it follows that

$$\lambda_1[f_\lambda'(u_\lambda^+)] > \lambda_1[a_\lambda] = 1.$$

This implies that u_λ^+ is a strict minimum of J_λ and that it is nondegenerate with Morse index $m(u_\lambda^+) = 0$. Similarly, substituting $\tilde{f}_\lambda^+$ with its negative part, one finds that for $\lambda > \lambda_1$, (D_λ) possesses a negative solution $u_\lambda^- < 0$ which is also a nondegenerate minimum of J_λ. Applying the MP theorem, we find a third critical point u_λ^*, different from $u_\lambda^\pm$. If the MP critical point u_λ^* coincides with $0 \in E$ and J_λ has no other critical point than $u_\lambda^\pm$ and 0, then Theorem 12.31 applies yielding that $m(0) \leq 1$. On the other hand, if $\lambda > \lambda_2$ then the Morse index of 0 is greater than or equal to 2, a contradiction. This completes the proof. ∎

Remark 12.34 The last result (assuming in particular no upper bound on p) can be extended to some cases in which the function h also depends on x. For example, if there exists $\bar{u} > 0$ such that for every $x \in \Omega$ one has $\lambda u - h(x, u) < 0$, $\forall u \geq \bar{u}$ (and a similar condition for negative u), then the solutions of the problem with truncated right hand side also solve the original problem, and the multiplicity result will hold unchanged. ∎

As a second application, we consider a Dirichlet problem with a jumping nonlinearity. By this we mean a boundary value problem like

$$\begin{cases} -\Delta u = \beta u^+ - \alpha u^- + g(u) + t\varphi_1 & x \in \Omega \\ u = 0 & x \in \partial\Omega, \end{cases} \qquad (D_t)$$

where $t \in \mathbb{R}$, $\alpha < \lambda_1 < \beta$, $g : \mathbb{R} \longrightarrow \mathbb{R}$ is smooth and such that

$$\lim_{|s|\to\infty} \frac{g(s)}{s} = 0. \qquad (12.26)$$

Above $u^+ = \max\{u, 0\}$, $u^- = u^+ - u$ and $\varphi_1 > 0$ is such that $\Delta\varphi_1 = \lambda_1\varphi_1$ in Ω, $\varphi_1 = 0$ on $\partial\Omega$. The name 'jumping nonlinearity' is due to the fact that the limits as $u \to \pm\infty$ of the nonlinearity in (D_t) are different and the interval (α, β) contains an eigenvalue of the Laplacian. For some results on these problems we refer to [20, Chapter 4, Section 2]. In particular, it is proved that there exists $T \in \mathbb{R}$ such that

(i) if $t < T$, (D_t) has at least two solutions,
(ii) if $t < T$, (D_t) has at least one solution,
(iii) if $t > T$, (D_t) has no solution.

We shall prove the following.

Theorem 12.35 *Suppose that $\alpha < \lambda_1 < \lambda_k < \beta < \lambda_{k+1}$ for some $k \geq 2$. Then, there exists $T^* \in \mathbb{R}$ such that if $t < T^*$, (D_t) has at least three solutions.*

The proof is based on the following lemma. Below we set $f(x, u) = \beta u^+ - \alpha u^- + g(u)$.

Lemma 12.36 *There exists $T^* \in \mathbb{R}$ such that:*

(i) *if $t < T_1$, (D_t) has a positive solution $u_t > 0$ and a negative solution $v_t < 0$;*
(ii) *$\lambda_1[f_u(v_t)] > 1$, while $\lambda_k[f_u(u_t)] < 1 < \lambda_{k+1}[f_u(u_t)]$.*

Proof. Consider the problem

$$\begin{cases} -\Delta u = \beta u + g(u) + t\varphi_1 & x \in \Omega \\ u = 0 & x \in \partial\Omega. \end{cases} \tag{12.27}$$

Since $\beta \neq \lambda_j$ for all $j \geq 1$, and g is sublinear, it follows from Theorem 3.23 and Remark 3.24 that (12.27) has a solution u_t. Setting

$$\psi_t = u_t - \frac{t}{\lambda_1 - \beta}\varphi_1,$$

one checks that ψ_t solves $-\Delta\psi_t = \beta\psi_t + g(u_t)$. By elliptic regularity, one has that $\|\psi_t\|_{C^1} \leq c$, for some constant $c > 0$ independent of t. It follows that there exists $t_1 < 0$ such that $u_t > 0$ provided $t < t_1$ and therefore u_t is a solution of (D_t). In a similar way, considering the problem (12.27) with β substituted by α, one shows that (D_t) possesses a negative solution v_t, provided t is the smaller of some $t_2 < 0$. This proves that (i) holds, provided $t < \min\{t_1, t_2\}$.

To prove (ii) we argue as follows. Since $u_t = \psi_t + t/(\lambda_1 - \beta)\varphi_1$ and $\|\psi_t\|_{C^1} \leq c$, we infer that $f_u(u_t) \to \beta$ in $L^\infty(\Omega)$, as $t \to -\infty$. We can now use the continuity property of eigenvalues, see (EP-2) in Section 1.4.1, to find that

$$\begin{cases} \lambda_k[f_u(u_t)] \to \lambda_k[\beta] = \lambda_k\beta^{-1} < 1 & (t \to -\infty), \\ \lambda_{k+1}[f_u(u_t)] \to \lambda_{k+1}[\beta] = \lambda_{k+1}\beta^{-1} > 1 & (t \to -\infty). \end{cases}$$

Similarly, $f_u(v_t) \to \alpha$ in $L^\infty(\Omega)$, as $t \to -\infty$ and thus $\lambda_1[f_u(v_t)] \to \lambda_1[\alpha] = \lambda_1\alpha^{-1} > 1$. Thus there exists $t_3 < 0$ such that (ii) holds provided $t < t_3$. Taking $T^* = \min\{t_1, t_2, t_3\}$, the lemma follows.   ■

Proof of Theorem 12.35. Let J_t denote the Euler functional of (D_t) in $E = H_0^1(\Omega)$. From Lemma 12.36(i) it follows that if $t < T^*$, J_t has two critical points u_t and v_t. Furthermore, statement (ii) of the same lemma implies that u_t and v_t are nondegenerate and the Morse index of u_t, v_t, is k, respectively 0. In particular, v_t is a strict local minimum of J_t. Evaluating J_t on the half-line $r\varphi_1$ one finds

$$J_t(r\varphi_1) = \frac{1}{2}r^2 \int_\Omega |\nabla\varphi_1|^2 \, dx - \int_\Omega F(r\varphi_1) \, dx - rt$$
$$\leq \frac{1}{2}r^2\lambda_1 - \frac{1}{2}r^2\beta - rt.$$

Since $\beta > \lambda_1$ it follows that $J_t(r\varphi_1) \to -\infty$ as $r \to +\infty$. Hence J_t satisfies the geometric conditions of the MP theorem. Moreover, it is straight forward to check that the (PS) condition holds. Thus J_t has a MP critical point $u_t^* \neq v_t$. If the only critical points of J_t are u_t and v_t, then $u_t^* = u_t$. Therefore, it is nondegenerate and its Morse index is k. If $k > 2$, this is in contradiction with Theorem 12.31.   ■

12.5 Exercises

(i) (a) Similarly to the proof of (12.2), use the exactness of the Mayer-Vietoris sequence to show that $H_q(\mathbb{T}^n) \simeq \mathbb{Z}^{d_{n,q}}$, for $n \geq 0$ and $q = 0, \ldots, n$, where

$$d_{n,q} := \binom{n}{q}$$

stand for the binomial coefficients.

(b) Using Theorem 12.20, find the Euler characteristic of the two-dimensional torus $T = S^1 \times S^1$, the sphere $S^{n-1} \subset \mathbb{R}^n$ and the projective space $\mathbb{RP}^2 = S^2/\{-Id, Id\}$. Using the result in (i), show that every Morse function on $\mathbb{T}^n$ possesses at least 2^n critical points.

(ii) The Morse inequalities can also be found when $f \in C^2(\mathbb{R}^n, \mathbb{R})$ has critical manifolds. See e.g. R. BOTT, Annals of Math. **60** (1954), 248-261. We say that $N \subset \mathbb{R}^n$ is a *non-degenerate critical manifold* for f if every $x \in N$ is a non-degenerate critical point for the restriction of f to the subspace orthogonal to $T_x N$. The Morse index of x as a critical point of f (which is independent of $x \in N$) is, by definition, the Morse index of N. If U_ε denotes an ε-neighborhood of N, and N has Morse index k, it is possible to prove that the critical homology groups verify, for $\varepsilon \ll 1$, $H_q(M^c \cap U_\varepsilon, M^c \cap U_\varepsilon \setminus N) \simeq H_{q-k}(N)$. Taking into account these facts, prove the following statements, under the assumption that f has only non-degenerate critical manifolds and $N = S^1$.

   (a) $H_q(M^c \cap U_\varepsilon, M^c \cap U_\varepsilon \setminus N) = \mathbb{Z}$ if $q = k$, $k+1$, otherwise is 0.
   (b) $\beta_q(M) = C_q(M) + C_{q-1}(M)$.
   (c) $C_q(M) \geq \sum_{j=0}^{q}(-1)^{q-j}\beta_j(M)$.

(iii) (the VON KARMAN equations for a clamped plate) Let $\Omega \subset \mathbb{R}^2$ be bounded, let E denote the closure of $C_0^\infty(\Omega)$ in $W^{2,2}(\Omega)$ with respect to

$$(u \mid v)_E = \int_\Omega \left[uv + \sum DuDv + \sum_{\|\alpha\|=2} D^\alpha u D^\alpha v \right] dx$$

and define $[f, g] = f_{xx}g_{yy} + f_{yy}g_{xx} - 2f_{xy}g_{xy}$. Consider the non-linear eigenvalue problem

$$\begin{cases} \Delta^2 u &= -\tfrac{1}{2}[v, v], & x \in \Omega, \\ \Delta^2 v &= \lambda[h, v] + [u, v], & x \in \Omega, \\ u, v &\in E, \end{cases} \quad (VK)$$

where Δ^2 denotes the bi-harmonic operator.

   (a) Define the operator $C : E \mapsto E$ by setting
       $(C(u,v) \mid w)_E = -\int_\Omega [u, v]wdx$. Prove that C is compact, such that $C(tu, tv) = t^2 C(u,v)$ and that the form $(C(u,v) \mid w)_E$ is symmetric.
   (b) Deduce that $Lu := C(h, u)$ is selfadjoint and compact.
   (c) Show that the weak solutions of (VK) are solutions of
       $S_\lambda(u) := u - \lambda Lu + \tfrac{1}{2}C(u, C(u,u)) = 0$, $u \in E$.
   (d) Show that S_λ is the gradient of the functional

       $$J_\lambda(u) = \frac{1}{2}\|u\|_E^2 - \frac{1}{2}\lambda(Lu \mid u)_E + \frac{1}{8}\|C(u,u)\|_E^2.$$

   (e) Apply Theorem 12.24 to deduce that any characteristic value of L is a bifurcation point for (VK).
   (f) Carry out a global study of the critical point of J_λ, in dependence of the parameter λ.

(iv) We consider the example introduced by Böhme in [59]. Let $\phi : \mathbb{R}^2 \to \mathbb{R}$ be a smooth function which is 2π-periodic in both its variables, with

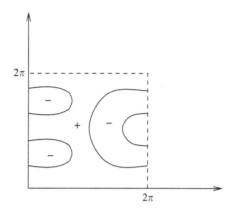

Figure 12.5

zero average on $[0, 2\pi] \times [0, 2\pi]$ and for which the positivity and negativity are depicted as in Figure 12.5.

Consider the map $\Phi : \mathbb{R}^2 \to \mathbb{R}$ defined by

$$\Phi(t, s) = \int_0^s \phi(t, z)dz, \qquad t, s \in \mathbb{R}.$$

Then Φ is also smooth and 2π-periodic in both the variables. In polar coordinates (r, θ), define the function $f : \mathbb{R}^2 \to \mathbb{R}$ by

$$f(x, y) = \begin{cases} \Phi(r^{-1}, \theta)e^{-r^{-2}} & \text{if } r > 0, \\ 0 & \text{if } r = 0. \end{cases}$$

Show that $\lambda = 0$ is a bifurcation value for the equation

$$\nabla f(x, y) + \lambda(x, y) = 0, \tag{12.28}$$

but that there is no continuous curve of solutions to (12.28) branching from $\lambda = 0$.

PART IV

Appendices

Appendix 1

Qualitative results

In this appendix we discuss some results concerning symmetry, classification and a priori estimates for solutions of some elliptic equations.

A1.1 The Gidas–Ni–Nirenberg symmetry result

We present here a result by Gidas, Ni and Nirenberg [99], concerning symmetry of solutions to some elliptic equations on balls of $\mathbb{R}^n$. The arguments rely on a procedure called the *moving plane method* which goes back to Alexandrov [3] and Serrin [161]. For simplicity we will treat only a simple example, omitting further extensions in order to avoid technicalities. The result we want to discuss is the following.

Theorem A1.1 *Let $\Omega = B_R(0) \subseteq \mathbb{R}^n$, and let $u \in C^2(\overline{\Omega})$ be a solution of*

$$\begin{cases} -\Delta u = f(u) & in\ \Omega \\ u = 0 & on\ \partial\Omega \\ u > 0 & in\ \Omega, \end{cases} \tag{A1.1}$$

where $f : [0, +\infty) \to \mathbb{R}$ is of class C^1. Then u is radially symmetric in Ω and moreover, letting $r = |x|$, one has $(\partial u/\partial r)(r) < 0$ for $r \in (0, R)$.

To prove this result, we need some preliminary lemmas. The first is a variant of the classical Hopf lemma, where no restriction on the sign of the coefficient $c(x)$ is assumed.

Lemma A1.2 *Let Ω be an open set of $\mathbb{R}^n$, and let $u \in C^2(\Omega)$, $u \geq 0$ be a solution of the differential inequality*

$$L(u) \equiv \sum_{i,j=1}^{n} a_{ij}(x) \frac{\partial^2 u}{\partial x_i \partial x_j} + \sum_{i=1}^{n} b_i(x) \frac{\partial u}{\partial x_i} + c(x)u \leq 0 \qquad in\ \Omega,$$

241

where the coefficients of L are uniformly bounded, and L is uniformly elliptic, in the sense that there exists $c_0 > 0$ such that $\sum_{i,j=1}^{n} a_{ij}\xi_i\xi_j \geq c_0|\xi|^2$ for all $\xi \in \mathbb{R}^n$. Suppose there exists a ball $B_r(Q)$ contained in Ω, and let $P \in \partial B_r(Q) \cap \partial\Omega$. Suppose u is continuous in $\Omega \cup P$, and that $u(P) = 0$. Then, if $u \not\equiv 0$ in B_r and if v is a unit outward normal to Ω at P, then one has $(\partial u/\partial v)(P) < 0$.

Proof. Without loss of generality, we can assume that $v = (1, 0, \ldots, 0)$. Consider the function $v(x) = e^{-\alpha x_1}u(x)$, for some $\alpha > 0$. Then one has

$$0 \geq L(u) = e^{\alpha x_1}L'(v) + vL(e^{\alpha x_1}),$$

where L' is a suitable elliptic operator with no zero-order term. Hence, computing $L(e^{\alpha x_1})$ we find that

$$0 \geq L'(v) + (a_{11}\alpha^2 + b_1\alpha + c)v.$$

For α sufficiently large, the coefficient $a_{11}\alpha^2 + b_1\alpha + c$ is positive and hence, since $v \geq 0$, we have $L'(v) \leq 0$. Hence, by the classical Hopf lemma we have that $(\partial v/\partial x_1)(P) < 0$, so the conclusion follows from the fact that $\partial u/\partial x_1 = e^{\alpha x_1}(\partial v/\partial x_1)$. ∎

Lemma A1.3 *Let Ω, f, u be as in Theorem A1.1, let $P \in \partial\Omega$, and assume that $v_1(P) > 0$, where $v_1(P)$ is the first component of the outward unit normal v to $\partial\Omega$ at P. Then there exists $\delta > 0$ such that $\partial u/\partial x_1 < 0$ in $\Omega \cap \{|x - P| < \delta\}$.*

Proof. Since $u \geq 0$ in Ω, we have clearly $\partial u/\partial v \leq 0$ at the whole $\partial\Omega$, and hence $\partial u/\partial x_1 \leq 0$ in a neighbourhood of P in $\partial\Omega$. Assuming by contradiction that the conclusion of the lemma is false, there would be a sequence $(P_n)_n \subseteq \Omega$, with $P_n \to P$ such that $(\partial u/\partial x_1)(P_n) \geq 0$. We write $P_n = (P_{n,1}, P'_n) \in \mathbb{R} \times \mathbb{R}^{n-1}$. For n sufficiently large there exists a unique $\tilde{P}_n \in \partial\Omega$ such that $P'_n = \tilde{P}'_n$ with $\tilde{P}_n \to P$ as $n \to +\infty$. Since $(\partial u/\partial x_1)(\tilde{P}_n) \leq 0$, by Lagrange's theorem there exists $\hat{P}_n$ belonging to the segment $[P_n, \tilde{P}_n]$ such that $(\partial u/\partial x_1)(\hat{P}_n) = 0$, see Figure A1.1, and it also must be $(\partial^2 u/\partial x_1^2)(\hat{P}_n) \leq 0$.

Then, since $(\hat{P}_n)_n$ also converges to P as $n \to +\infty$, we have that $(\partial u/\partial x_1)(P) = (\partial^2 u/\partial x_1^2)(P) = 0$.

Suppose now that $f(0) \geq 0$. Then u satisfies $\Delta u + f(u) - f(0) \leq 0$, so using Lagrange's theorem with f we find that

$$\Delta u + c_1(x)u \leq 0 \tag{A1.2}$$

for some bounded function $c_1(x)$. Then, applying Lemma A1.2 we find that $(\partial u/\partial x_1)(P) < 0$, which is a contradiction to our assumptions.

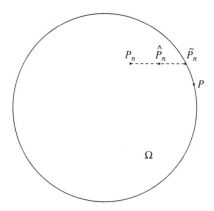

Figure A1.1

Suppose now that $f(0) < 0$. Still by our contradiction assumption we have that $(\partial u/\partial x_1)(P) = 0$, and since u is constant on $\partial \Omega$ it follows that $\nabla u(P) = 0$. Let $\phi(x) = \text{dist}(x, \partial \Omega)$: notice that, defining $v(x) = u(x)/\phi(x)$, one easily finds that $v \in C^1(\overline{\Omega})$ and that v is of class C^2 inside Ω, at the points where the distance from the boundary is smooth (and in particular near the boundary). By the assumptions on u, with some elementary computations one finds that $v(P) = 0$ and that $(\partial^2 u/\partial x_i \partial x_j)(P) = v_i v_j \delta$ for some constant δ. It is indeed possible to determine this constant by noticing that $-f(0) = \Delta u(P) = \delta |v|^2 = \delta$. Therefore we also find that $(\partial^2 u/\partial x_1^2)(P) = -f(0)v_1^2 > 0$, which is again a contradiction. This concludes the proof of the lemma. ∎

Before stating the next result, we introduce some notation. For $\lambda > 0$ we let

$$T_\lambda = \{x_1 = \lambda\} \qquad \Sigma_\lambda = \{x \in \Omega \; : \; x_1 > \lambda\},$$

and we also define

$$x_\lambda = (2\lambda - x_1, x_2, \ldots, x_n), \quad x \in \Sigma_\lambda \qquad \Sigma'_\lambda = \{x_\lambda \; : \; x \in \Sigma_\lambda\},$$

see Figure A1.2. Notice that x_λ and Σ'_λ are nothing but the reflections of the point x and the set Σ_λ through the plane T_λ.

Lemma A1.4 *Let* Ω, f, u *be as in Theorem A1.1, and suppose that for some* $\lambda \in (0, R)$ *there holds*

$$\frac{\partial u}{\partial x_1}(x) \leq 0, \qquad u(x_\lambda) \leq u(x), \quad x \text{ in } \Sigma_\lambda \qquad u(\cdot) \not\equiv u(\cdot_\lambda) \text{ in } \Sigma_\lambda.$$

Then $u(\cdot) < u(\cdot_\lambda)$ *in* Σ_λ *and* $\partial u/\partial x_1 < 0$ *in* $\Omega \cap T_\lambda$.

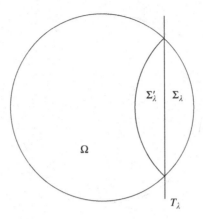

Figure A1.2

Proof. In Σ'_λ define the function $v(x) = u(x_\lambda)$ (notice that $x_\lambda \in \Sigma_\lambda$). Clearly v satisfies $\Delta v + f(v) = 0$ and $\partial v/\partial x_1 \geq 0$. Subtracting the equation satisfied by u we find that

$$\Delta(v - u) + f(v) - f(u) \geq 0 \qquad \text{in } \Sigma'_\lambda.$$

Using Lagrange's theorem we find that $w = v - u$ satisfies $\Delta w(x) + c(x)w(x) \geq 0$ for some bounded function $c(x)$. Moreover we have that $w \leq 0$, $w \not\equiv 0$.

Since $w = 0$ on $T_\lambda \cap \Omega$, from the maximum principle and Lemma A1.2 it follows that $w < 0$ in Σ'_λ and that $\partial w/\partial x_1 > 0$ on $T_\lambda \cap \Omega$. But on T_λ we have $\partial w/\partial x_1 = (\partial v/\partial x_1) - (\partial u/\partial x_1) = -2(\partial u/\partial x_1)$, so the conclusion follows. ∎

Proof of Theorem A1.1. We consider the following conditions

$$\frac{\partial u}{\partial x_1} < 0, \quad u(x) < u(x_\lambda) \qquad \text{in } \Sigma_\lambda. \tag{A_λ}$$

From Lemma A1.3 we know that (A_λ) holds for $\lambda \in (0, R)$ sufficiently close to R. Let us now define

$$\mu = \inf\{\tilde\lambda \in (0, R) \ : \ (A_\lambda) \text{ holds for } \lambda \in (\tilde\lambda, R)\}.$$

Our aim is to show that $\mu = 0$. Suppose by contradiction that $\mu > 0$. By continuity we have that

$$\frac{\partial u}{\partial x_1} \leq 0, \quad u(x) \leq u(x_\lambda) \qquad \text{in } \Sigma_\mu.$$

For any point $x_0 \in \partial\Sigma_\mu \setminus T_\mu \ (\subseteq \partial\Omega)$ we have that $(x_0)_\mu \in \Omega$, because we are assuming $\mu > 0$. Since $u(x_0) = 0$ and $u((x_0)_\mu) > 0$, it follows that $u \not\equiv u(\cdot_\lambda)$

in Σ_μ. Then we can apply Lemma A1.4 to obtain

$$u(x) < u(x_\mu) \quad \text{in } \Sigma_\mu \quad \text{and} \quad \frac{\partial u}{\partial x_1} < 0 \quad \text{in } T_\mu \cap \Omega.$$

Therefore (A_μ) also holds. Since $\partial u / \partial x_1 < 0$ on $T_\mu \cap \Omega$, from Lemma A1.3 we see that there exists $\varepsilon > 0$ such that

$$\frac{\partial u}{\partial x_1} < 0 \quad \text{in } \Omega \cap \{x_1 > \mu - \varepsilon\}. \tag{A1.3}$$

On the other hand, from our definition of μ then there exists a sequence $\lambda_j \nearrow \mu$ and a sequence $(x_j)_j$ with $x_j \in \Sigma(\lambda_j)$ and $u(x_j) \geq u((x_j)_{\lambda_j})$. We can assume that, passing to a subsequence, $x_j \to x \in \overline{\Sigma(\mu)}$. Then also $(x_j)_{\lambda_j} \to x_\mu$, and $u(x) \geq u(x_\mu)$. Since (A_μ) holds, it must be $x \in \partial \Sigma_\mu$. If x does not belong to the plane T_μ, then $x_\mu \in \Omega$, and hence $0 = u(x) < u(x_\mu)$, which is impossible. Therefore $x \in T_\mu$ and $x = x_\mu$. On the other hand, for j sufficiently large the segment $[(x_j)_{\lambda_j}, x_j]$ is contained in Ω and by Lagrange's theorem there exists y_j in this segment such that $\partial u / \partial x_1 \geq 0$. Since $y_j \to x$, we obtain a contradiction to (A1.3). Therefore we have that $\mu = 0$, so (A_λ) holds for every $\lambda \in (0, R)$.

We claim next that if $\partial u / \partial x_1 = 0$ at some point of $T_0 \cap \Omega$, then u must be symmetric with respect to the plane T_0. In fact, from Lemma A1.4 it follows that $u(x) = u(x_0)$ for any $x \in \Sigma(0)$, which is equivalent to our claim.

The conclusion of the theorem follows from the last result, since the above arguments can be repeated reflecting through planes perpendicular to any direction ν. ∎

Remark A1.5

(i) The above result can be extended to some nonautonomous case as well, when $f = f(r, u)$, assuming f, $\partial f / \partial u$ continuous and f nonincreasing in r. A dependence on ∇u is also allowed, provided f stays monotone when we reflect u with respect to a plane.

(ii) There is a version of Theorem A1.1 concerning solutions defined on the whole $\mathbb{R}^n$, see Theorem 2 in [100]. For example, this result applies to positive solutions of $-\Delta u + u = u^p$ in $\mathbb{R}^n$ for any $p > 1$, assuming that u tends to zero at infinity.

(iii) The moving plane method can also be applied to more general domains, assuming that the reflection Σ'_λ of Σ_λ stays inside the domain. In particular, for convex domains, it is possible to consider reflections near the boundary.

(iv) For other symmetry results using different methods, see for example [140] and references therein. ∎

A1.2 A Liouville type theorem by Gidas and Spruck

We next present a classification result due to Gidas and Spruck [98], following the arguments of Chen and Li [78], which use crucially the moving plane method.

Theorem A1.6 *Consider the equation*

$$-\Delta u = u^p \qquad in \ \mathbb{R}^n. \tag{A1.4}$$

Suppose $n \geq 3$, $p < (n+2)/(n-2)$ and that $u \in C^2(\mathbb{R}^n)$ is a non-negative solution of (A1.4). Then $u \equiv 0$ on $\mathbb{R}^n$.

For proving the theorem, we need some preliminaries. We let $\tilde{u}$ denote the *Kelvin transform* of u, namely the function defined by

$$\tilde{u}(x) = \frac{1}{|x|^{n-2}} u\left(\frac{x}{|x|}\right), \qquad x \in \mathbb{R}^n \setminus \{0\}.$$

Then $\tilde{u}$ satisfies

$$\Delta \tilde{u} + \frac{1}{|x|^{n+2-p(n-2)}} \tilde{u}^p = 0 \qquad in \ \mathbb{R}^n \setminus \{0\}. \tag{A1.5}$$

As in the previous section, for $\lambda > 0$ we define

$$\Sigma_\lambda = \{x \in \mathbb{R}^n \ : \ x_1 > \lambda\}, \qquad T_\lambda = \partial \Sigma_\lambda,$$

and again we set

$$x_\lambda = (2\lambda - x_1, x_2, \ldots, x_n) \qquad \text{if } x = (x_1, x_2, \ldots, x_n).$$

Define the functions $\tilde{u}_\lambda(x) = \tilde{u}(x_\lambda)$, $w_\lambda(x) = \tilde{u}_\lambda(x) - \tilde{u}(x)$, and $\overline{w}_\lambda(x) = w_\lambda(x)/g(x)$, where $g(x) = \log(-x_1 + 3)$. Since $\tilde{u}$ has a singularity at the origin, w_λ is singular at the point $x^\lambda = (2\lambda, 0, \ldots, 0)$. Therefore, we consider w_λ defined on the set $\tilde{\Sigma}_\lambda = \Sigma_\lambda \setminus \{x^\lambda\}$.

Lemma A1.7 *For λ sufficiently large, if $\inf_{\tilde{\Sigma}_\lambda} \overline{w}_\lambda < 0$, then the infimum is achieved. Moreover, there exists $R_0 > 0$ (independent of λ) such that, if x_0 is a minimum point of $\overline{w}_\lambda$ on $\tilde{\Sigma}_\lambda$ with $\overline{w}_\lambda(x_0) < 0$ and if $\lambda > 0$, then $|x_0| < R_0$.*

Proof. Let us prove the first assertion. Since $\tilde{u}(x) > 0$ for $x \neq 0$ and since $\Delta \tilde{u} \leq 0$, in $B_1(0)$ we have (compare for example u to $\min_{\partial B_1(0)} -\gamma/|x|^{n-2}$ for $\gamma > 0$ arbitrarily small)

$$x \geq \min_{\partial B_1(0)} \tilde{u} := \varepsilon_0 > 0.$$

Since $\tilde{u}$ tends to zero at infinity, we can find λ sufficiently large such that $\tilde{u}(x) \leq \varepsilon_0$ for every $x \in B_1(x^\lambda)$. For such values of λ, we have clearly $\overline{w}_\lambda \geq 0$ on $B_1(x^\lambda) \setminus x^\lambda$. Hence, if $\inf_{\tilde{\Sigma}} \overline{w}_\lambda < 0$, then the infimum is attained in $\Sigma_\lambda \setminus B_1(x_\lambda)$ since $\overline{w}_\lambda$ tends to zero at infinity and since $\overline{w}_\lambda = 0$ on T_λ.

To prove the second statement we notice that $\tilde{u}$ satisfies

$$\Delta\tilde{u}(x) + \frac{1}{|x|^{n+2-p(n-2)}}\tilde{u}^p(x) \leq 0 \qquad \text{for } x \in \Sigma_\lambda \quad \text{and} \quad \lambda \geq 0.$$

It follows from Lagrange's theorem and some elementary computations that $\Delta w_\lambda(x) + c(x)w_\lambda(x) \leq 0$, and

$$\Delta\overline{w}_\lambda + \frac{2}{g}\nabla g \cdot \nabla\overline{w}_\lambda + \left(c(x) + \frac{\Delta g}{g}(x)\right)\overline{w}_\lambda \leq 0, \qquad \text{(A1.6)}$$

where $c(x) = (p/|x|^{n+2-p(n-2)})\psi(x)^{p-1}$, for some function $\psi(x)$ between $\tilde{u}(x)$ and $\tilde{u}(x_\lambda)$. Since $\tilde{u}(x)$ is asymptotic to $|x|^{2-n}$ for $|x| \to +\infty$ (and the same holds for $\tilde{u}_\lambda$), it follows also that ψ has the same asymptotics, and hence $c(x)$ is asymptotic to $|x|^{-4}$. This implies that $c(x) + (\Delta g/g)(x) < 0$ for x large. Hence the conclusion follows from (A1.6). ■

Proof of Theorem A1.6. By Lemma A1.7 we know that for λ sufficiently large we have $\overline{w}_\lambda \geq 0$ in $\tilde{\Sigma}_\lambda$. Let μ denote the infimum of the non-negative λ such that this property holds.

If $\mu > 0$, we show that $\overline{w}_\mu \equiv 0$ in $\tilde{\Sigma}_\mu$. Indeed, applying the maximum principle to $\Delta w_\lambda(x) + c(x)w_\lambda(x) \leq 0$, we obtain $\overline{w}_\mu > 0$ in $\tilde{\Sigma}_\mu$. Let now $\lambda_k \searrow \mu$ be such that $\overline{w}_{\lambda_k} < 0$ somewhere in $\tilde{\Sigma}_{\lambda_k}$. To reach a contradiction, by the second part of Lemma A1.7 it is sufficient to show that $\inf_{\tilde{\Sigma}_{\lambda_k}} \overline{w}_{\lambda_k}$ can be achieved at some point x_k, and that the x_k stay bounded from the singularities x^{λ_k} of w_{λ_k}. But we notice that there exist $\varepsilon > 0$ and $\delta > 0$ such that

(a) $\overline{w}_\mu \geq \varepsilon$ in $B_\delta(x^\mu) \setminus \{x^\mu\}$
(b) $\lim_{\lambda \to \mu} \inf_{B_\delta(x^\lambda)} \overline{w}_\lambda \geq \inf_{B_\delta(x^\mu)} \overline{w}_\mu \geq \varepsilon$.

The second claim is trivial while, reasoning as before, we obtain (a) from the fact that $\overline{w}_\mu > 0$ in $\tilde{\Sigma}_\mu$ and $\Delta w_\mu \leq 0$. Therefore $\overline{w}_\mu \equiv 0$.

If we have instead $\mu = 0$, then we can repeat the above argument for λ near $-\infty$. If the procedure stops before reaching zero, we get symmetry of $\tilde{u}$ in x_1 by the last argument. If we again reach zero, we have both $\tilde{u} \geq \tilde{u}_0$ and $\tilde{u} \leq \tilde{u}_0$, so the symmetry in x_1 follows again.

Since we can repeat the procedure for any direction, it follows that $\tilde{u}$ must be symmetric around some point of $\mathbb{R}^n$. From (A1.5), we see that either $\tilde{u} \equiv 0$, or $\tilde{u}$ must be symmetric around the origin, which means that also u is symmetric with respect to the origin. But since we can choose the origin arbitrarily, it must be $u \equiv 0$. ∎

Remark A1.8 The above proof can also be applied to the case $p = n + 2/n - 2$, and yields a classification of the solutions of (A1.4) as functions of the form

$$u(x) = \frac{(n(n-2)\lambda^2)^{(n-2)/4}}{(\lambda^2 + |x - x_0|^2)^{(n-2)/2}}$$

for some $\lambda > 0$ and some $x_0 \in \mathbb{R}^n$. ∎

A1.3 An application

We apply the previous results to obtain existence and a priori estimates for some nonlinear elliptic equation which is not variational.

Theorem A1.9 *Let* $\Omega \subseteq \mathbb{R}^n$ *be a convex, smooth bounded domain, and let* $p \in (1, (n+2)/(n-2))$. *Consider the following problem*

$$\begin{cases} -\Delta u = u^p + f(u, |\nabla u|) & in \ \Omega \\ u = 0 & on \ \partial\Omega \\ u > 0 & in \ \Omega. \end{cases} \tag{A1.7}$$

Assume f *is smooth in* $\overline{\Omega} \times [0, \infty] \times \mathbb{R}^n$, *bounded with bounded derivatives, and that* $f(0, 0) > 0$ *for every* $x \in \Omega$. *Then problem* (A1.7) *admits a solution.*

Remark A1.10 We notice that, since the problem is not variational and since the right hand side is unbounded (and also superlinear), neither the above min-max methods nor a direct fixed point argument can be applied.

We also refer the interested reader to the paper [94] where some techniques different from those we discuss are presented. ∎

Proof of Theorem A1.9. First of all, we claim that there exists $\delta > 0$ depending only on Ω such that for every solution of (A1.7)

$$u(x) = \max_{\Omega} u \quad \Rightarrow \quad \text{dist}(x, \partial\Omega) \geq \delta. \tag{A1.8}$$

To see this, we notice that Lemma A1.3 and the reflection method apply also to our case (near $\partial\Omega$), see Remark A1.5 (i) and (ii). In particular, we obtain that any solution of (A1.7) is strictly decreasing (while approaching the boundary) in a suitable neighbourhood of $\partial\Omega$. It follows that the maximum of any solution to (A1.7) is attained at a distance greater than δ from the boundary.

We want to show now that the solutions of (A1.7) satisfy an a priori bound which depends only on Ω, p and f. In fact, assume by contradiction that there exists a sequence of solutions $(u_n)_n$ for which $M_n := \sup_\Omega u_n \to +\infty$. For any integer n, let x_n denote a point where the maximum of u_n is attained. Let $\gamma = (1/2)(p-1)$, let $\Omega_n := M_n^\gamma (\Omega - x_n)$, and let $v_n : \Omega_n \to \mathbb{R}$ be defined by

$$u_n(x) = M_n v_n(M_n^\gamma(x + x_n)).$$

We notice that $v_n(0) = 1$ for every n, and that $v_n(x) \in [0,1]$ for every x. Moreover, since $\text{dist}(x_n, \partial\Omega) \geq \delta$, we have that Ω_n invades the whole $\mathbb{R}^n$. Also, by the definition of γ, the functions v_n satisfy the equation

$$-\Delta v_n = v_n^p + \frac{1}{M_n^p} f(v_n, M_n^{1+\gamma} |\nabla v_n|) \qquad \text{in } \Omega_n.$$

Since f is bounded, by elliptic regularity results, see [101], one can check that v_n converges locally in $\mathbb{R}^n$ to an entire non-negative solution of $\Delta v + v^p = 0$, so by Theorem A1.6 we obtain a contradiction.

We want to prove next also a bound from below on the solutions to (A1.7). We claim that there exists a small constant $\sigma > 0$, depending only on Ω, p and f such that every solution of (A1.7) is greater or equal to $\sigma \varphi_1$, where $\varphi_1 > 0$ stands for the first eigenfunction of $-\Delta$ in Ω (with Dirichlet boundary conditions). In fact, assuming the contrary, suppose that there exists a sequence $(u_n)_n$ of solutions to (A1.7) such that $\sup_\Omega(u_n/\varphi_1) \to 0$ as $n \to +\infty$. Let $\sigma_n = \max_{\overline{\Omega}}(u_n/\varphi_1) \to 0$. Then at a maximum point x_n of this ratio (which might also lie on $\partial\Omega$), we have $u_n = \sigma_n \varphi_1$, $\nabla u_n = \sigma_n \nabla \varphi_1$, and $\Delta u_n \geq \sigma_n \Delta \varphi_1$, which implies

$$f(\sigma_n \varphi_1(x_n), \sigma_n |\nabla \varphi_1|) = -\Delta u_n(x_n) - u_n(x_n)^p \leq -\sigma_n \Delta \varphi_1 - \sigma_n^p \varphi_1(x_n)^p$$

$$= \sigma_n \lambda_1 \varphi_1(x_n) - \sigma_n^p \varphi_1(x_n)^p.$$

Since we are assuming $\sigma_n \to 0$ as $n \to +\infty$, we get a contradiction since $f(0,0) > 0$. Therefore we also have an a priori bound from below on the solutions to (A1.7).

We consider a smooth cutoff function $\chi : \mathbb{R} \to [0,\infty)$ for which

$$\begin{cases} \chi(t) = 0 & \text{for } t \leq 0 \\ \chi(t) = 1 & \text{for } t \geq 1 \\ \chi(t) \in [0,1] & \text{for every } t. \end{cases}$$

Fix a large $M \in \mathbb{R}$, and for $s \in [0,1]$, consider the family of problems

$$\begin{cases} -\Delta u = \chi(u - M)M_+^p + (1 - \chi(u-M))u_+^p + f_s(x, u, \nabla u) & \text{in } \Omega \\ u = 0 & \text{on } \partial\Omega, \end{cases}$$

$$(P_M, s)$$

where

$$f_s(u, \nabla u) = sf(u, \nabla u) + (1 - s), \qquad s \in [0, 1].$$

From elliptic regularity estimates, we have that for any fixed value of $M_0 > 0$, the positive solutions to $(P_{M,1})$ with $M \leq M_0$ are uniformly bounded from above by a constant depending only on M_0, p and Ω. Moreover, by the above a priori estimates, if we choose M_0 sufficiently large, all the positive solutions to $(P_M, 1)$ with $M \geq M_0$ coincide with those of (A1.7).

Setting $X = C^{2,\alpha}(\overline{\Omega})$, problem (P_M, s) can be written in the following abstract form

$$u = K_{M,s}u, \qquad u \in X,$$

where $K_{M,s}$ is the solution operator of the Laplace equation with Dirichlet boundary data and with right hand side equal to that in (P_M, s).

We fix now the set $A_{C,\sigma}$ defined as

$$A_{C,\sigma} = \{u \in X \ : \ \|u\|_{L^\infty(\Omega)} < C, u > \sigma\varphi_1\}.$$

From the above estimates, we know that $\mathrm{Id} - K_{M_0,1} \neq 0$ on the boundary of $A_{C,\sigma}$ if we choose C sufficiently large and σ sufficiently small.

We now consider the following homotopy, for $t \in [0, 1]$

$$u = K_{t(M_0+1)+(t-2),1}u, \qquad u \in X.$$

From the comments after the definition of f_s, and with a similar reasoning concerning the a priori lower bounds on positive solutions, we have that also $\mathrm{Id} - K_{t(M_0+1)+(t-2),1} \neq 0$ on the boundary of $A_{C,\sigma}$ for all $t \in [0, 1]$, and in particular for $t = 0$. Similarly, we can obtain a priori bounds, both from above and from below, for positive solutions of $u = K_{-2,s}u$ when s varies from 1 to 0, so that also $\mathrm{Id} - K_{-2,s} \neq 0$ on the boundary of $A_{C,\sigma}$ for all $s \in [0, 1]$. From the homotopy invariance of the Leray–Schauder degree it then follows that

$$\deg(\mathrm{Id} - K_{M_0,1}, A_{C,\sigma}, 0) = \deg(\mathrm{Id} - K_{-2,0}, A_{C,\sigma}, 0).$$

Notice that the operator $\mathrm{Id} - K_{-2,0}$ is affine (with $K_{-2,0}$ compact in X, indeed constant) and corresponds to the problem

$$\begin{cases} -\Delta u = 1 & \text{in } \Omega, \\ u = 0 & \text{on } \partial\Omega, \end{cases}$$

so if C is sufficiently large and σ sufficiently small it follows that the last equation has a (unique) solution in $A_{C,\sigma}$. Therefore we can apply Lemma 3.19 to find that $\deg(\mathrm{Id} - K_{M_0,1}, A_{C,\sigma}, 0) = 1$, and hence problem (A1.7) is solvable. ∎

Remark A1.11 L^∞ a priori bounds similar to the preceding ones, jointly with degree theoretic arguments or global bifurcation results, can be used to find positive solutions of superlinear equations. Consider a boundary value problem like

$$\begin{cases} -\Delta u = \lambda u + f(u) & \text{in } \Omega \subset \mathbb{R}^n \\ u = 0 & \text{on } \partial\Omega, \end{cases} \tag{A1.9}$$

where Ω is a bounded domain and, roughly, f is smooth, $f(u) \sim |u|^{p-1}u$, $1 < p < (n+2)/(n-2)$. Using the arguments of Section 4 it is possible to show that from λ_1 bifurcates an unbounded branch $\mathcal{C}_0$ of positive solutions of (A1.9) such that $\mathcal{C}_0 \subset (-\infty, \lambda_1) \times C^{0,\alpha}(\bar{\Omega})$. Moreover, arguing as before, one can find L^∞ a priori bounds for these positive solutions, for all λ in any compact interval of $\mathbb{R}$. This implies that $\mathcal{C}_0$ has a projection on $\mathbb{R}$ which covers all $(-\infty, \lambda_1)$, namely that (A1.9) has a positive solution for all $\lambda < \lambda_1$. In particular, $\mathcal{C}_0$ has to intersect the axis $\lambda = 0$, yielding a positive solution to the problem $-\Delta u = f(u)$ in Ω. For results of such a type, we refer for example to [66] and [94]. ∎

Appendix 2

The concentration compactness principle

In this appendix we discuss a celebrated result by P.-L. Lions, see [118] which is useful in several contexts where some lack of compactness occurs. Below, we recall the abstract result and then we discuss applications to semilinear elliptic problems in $\mathbb{R}^n$.

A2.1 The abstract result

We state now the main result reported in this appendix.

Theorem A2.1 *Suppose $(\mu_l)_l$ is a sequence of non-negative probability (Radon) measures on $\mathbb{R}^n$. Then, passing to a subsequence, one of the following three alternatives holds.*

(i) Compactness: for any $\varepsilon > 0$ there exists $R > 0$ and $(x_l)_l \subseteq \mathbb{R}^n$ such that

$$\int_{\mathbb{R}^n \setminus B_R(x_l)} \mathrm{d}\mu_l < \varepsilon.$$

(ii) Vanishing: for any $R > 0$ there holds

$$\lim_{l \to +\infty} \left(\sup_{x \in \mathbb{R}^n} \int_{B_R(x)} \mathrm{d}\mu_l \right) = 0.$$

(iii) Dichotomy: there exists $\lambda \in (0, 1)$ such that the following property holds: for any $\varepsilon > 0$ there exists $R > 0$ and $(x_l)_l \subseteq \mathbb{R}^n$ such that, given any $\overline{R} > R$, there exist two non-negative measures $\mu_{l,1}, \mu_{l,2}$ such that $\mu_{l,1}$ is supported in $B_R(x_l)$, $\mu_{l,2}$ is supported in $\mathbb{R}^n \setminus B_{\overline{R}}(x_l)$, $\mu_{l,1} + \mu_{l,2} \leq \mu_l$ and such that

$$\lim_{l \to \infty} \left(\left| \int_{\mathbb{R}^n} \mathrm{d}\mu_{l,1} - \lambda \right| + \left| \int_{\mathbb{R}^n} \mathrm{d}\mu_{l,2} - (1 - \lambda) \right| \right) < \varepsilon.$$

Roughly speaking, the theorem asserts that either the measures $(\mu_l)_l$ stay concentrated near the points $(x_l)_l$, or that they do not concentrate near any point of $\mathbb{R}^n$, or that some fraction $\lambda \in (0, 1)$ concentrates near some $(x_l)_l$ and that the remaining part spreads away from these points.

Proof. Consider the functions $M_l : [0, +\infty) \to [0, 1]$ defined by

$$M_l(r) = \sup_{x \in \mathbb{R}^n} \int_{B_r(x)} d\mu_l, \qquad l = 1, 2, \ldots.$$

We notice that the functions M_l are nondecreasing in r and that, since every μ_l is a probability measure, there holds $\lim_{r \to +\infty} M_l(r) = 1$ for every l. It follows that the M_l are locally bounded in $BV([0, +\infty))$ and hence, passing to a subsequence, they converge almost everywhere on $[0, +\infty)$ to a function $M(r)$ which is bounded (with values in $[0, 1]$), non-negative and nondecreasing. Since M is nondecreasing, we can assume that it is continuous from the left, so we have

$$M(r) \leq \liminf_{l \to +\infty} M_l(r) \qquad \text{for every } r \in [0, +\infty). \tag{A2.1}$$

We then set

$$\lambda = \lim_{r \to +\infty} M(r) \in [0, 1].$$

We consider now three different cases.

Case 1: $\lambda = 1$. We prove that in this case we have concentration. In fact, let $\varepsilon > 0$ be given, and let R be such that $M(R) \geq 1 - \varepsilon/4$. This and (A2.1) imply that for l sufficiently large one has $M_l(R) > 1 - \varepsilon/2$. By the definition of M_l, then there exists a point x_l of $\mathbb{R}^n$ such that $\int_{B_R(x_l)} d\mu_l \geq 1 - \varepsilon$. Since the integral of μ_l over $\mathbb{R}^n$ is equal to 1 we obtain compactness.

Case 2: $\lambda = 0$. By (A2.1) we have $\liminf_{l \to +\infty} \int_{B_R(x)} d\mu_l = 0$ for any $R > 0$, which immediately implies the vanishing alternative.

Case 3: $\lambda \in (0, 1)$. Given any $\varepsilon > 0$, we choose $R > 0$ such that $M(R) > \lambda - \varepsilon/4$, so for l large enough we have that there exists $x_l \in \mathbb{R}^n$ for which

$$\int_{B_R(x_l)} d\mu_l > \lambda - \frac{\varepsilon}{4}. \tag{A2.2}$$

Let now $\overline{R} > R$. Then, since M_l converges to M almost everywhere and since $M(r) \nearrow_{r \to +\infty} \lambda$, we can find $R_l \to +\infty$ such that

$$M_l(R) \leq M_l(R_l) \leq \lambda + \frac{\varepsilon}{4} \qquad \text{for } l \text{ sufficiently large.} \tag{A2.3}$$

Now we define $\mu_{l,1} = \mu_l \llcorner B_R(x_l)$ and $\mu_{l,2} = \mu_l \llcorner (\mathbb{R}^n \setminus B_{R_l}(x_l))$. Clearly $\mu_{l,1}$ and $\mu_{l,2}$ are non-negative measures supported respectively in $B_R(x_l), \mathbb{R}^n \setminus B_{\overline{R}}(x_l)$

for l large, and satisfy $\mu_{l,1} + \mu_{l,2} \leq \mu_l$. Moreover by (A2.2) and (A2.3) we have

$$\left| \int_{\mathbb{R}^n} d\mu_{l,1} - \lambda \right| + \left| \int_{\mathbb{R}^n} d\mu_{l,2} - (1 - \lambda) \right| = \left| \int_{B_R(x_l)} d\mu_l - \lambda \right| + \left| \int_{B_{R_l}(x_l)} d\mu_l - \lambda \right| < \varepsilon.$$

This concludes the proof.   ∎

A2.2 Semilinear elliptic equations on $\mathbb{R}^n$

We provide here some applications of Theorem A2.1 to classes of semilinear elliptic equations on $\mathbb{R}^n$. Consider the following problem

$$\begin{cases} -\Delta u + u = a(x)u^p & \text{in } \mathbb{R}^n \\ u(x) \to 0 & \text{as } |x| \to +\infty. \end{cases} \quad (A2.4)$$

We assume that $p \in (1, (n + 2)/(n - 2))$, and that a is positive, smooth and such that

$$\lim_{|x| \to +\infty} a(x) = 1. \quad (a0)$$

We analyse two model cases, namely situations in which a satisfies one of the following two properties

$$a(x) > 1 \qquad \text{for every } x \in \mathbb{R}^n, \quad (a1)$$

$$a(x) < 1 \qquad \text{for every } x \in \mathbb{R}^n. \quad (a2)$$

Solutions of (A2.4) can be found as critical points of the functional $J_a : E \to \mathbb{R}$, $E = H^1(\mathbb{R}^n)$, defined as

$$J_a(u) = \frac{1}{2} \int_{\mathbb{R}^n} (|\nabla u|^2 + u^2) - \frac{1}{p+1} \int_{\mathbb{R}^n} a(x)|u|^{p+1}.$$

Since the embedding $H^1(\mathbb{R}^n) \hookrightarrow L^{p+1}(\mathbb{R}^n)$ is not compact, the Palais–Smale condition fails in general. In fact, if u_0 denotes a positive (radial) ground state constructed in Theorem 11.3 (corresponding to $a(x) \equiv 1$), and if $|x_n| \to +\infty$, then the sequence $u_n = u_0(\cdot - x_n)$ is a Palais–Smale sequence which does not admit any (strongly) convergent subsequence. To see this we notice that, by (a0) there holds clearly

$$J_a(u_n) = J(u_0) + \frac{1}{p+1} \int_{\mathbb{R}^n} (1 - a(x))\|u_n\|^{p+1} \to J(u_0),$$

where $\| \cdot \|$ stands for the standard norm of $H^1(\mathbb{R}^n)$, and where $J = J_{a \equiv 1}$, as in Chapter 11. Moreover, given any $v \in E$, using some computations we have that

$$|J'_a(u_n)[v]| = \left| \int_{\mathbb{R}^n} (a(x) - 1)u_n v \right| \leq \|v\| \, \|(a(x + x_n) - 1)u_0\|_{L^2(\mathbb{R}^n)} = o(1)\|v\|.$$

However, one can hope that still some *specific* Palais–Smale sequence might admit converging subsequences. We are going to consider the two cases (**a1**) and (**a2**) separately.

A2.2.1 An existence result under assumption (*a1*)

Throughout this subsection we assume that the coefficient $a(x)$ satisfies (**a1**), in addition to (**a0**). We are going to prove the following result.

Theorem A2.2 *Suppose $p \in (1, (n + 2)/(n - 2))$, and let $a : \mathbb{R}^n \to \mathbb{R}$ be smooth and satisfy* (**a0**), (**a1**). *Then problem* (A2.4) *admits a positive solution.*

Proof. We will look for solutions as constrained maxima of some functional over the manifold $S = \{u \in E : \|u\| = 1\}$, the unit sphere of E. We let $\widetilde{J}_a : S \to \mathbb{R}$ denote

$$\widetilde{J}_a(u) = \int_{\mathbb{R}^n} a(x)|u|^{p+1}.$$

Then, as in Example 6.5(iii), one finds that a constrained critical point of $\widetilde{J}_a$ on S, after a suitable rescaling, gives rise to a solution of (A2.4).

By the Sobolev embedding $E \hookrightarrow L^{p+1}(\mathbb{R}^n)$ and by the boundedness of a, the functional $\widetilde{J}_a$ is bounded from above and from below on S. Indeed, letting $S_{p,n}$ denote the best Sobolev constant for which

$$\|u\|_{L^{p+1}(\mathbb{R}^n)} \leq S_{p,n}\|u\|,$$

by (**a0**) (which implies $\|a\|_\infty < \infty$) we have that

$$\widetilde{J}_a(u) \leq \left(\max_{\mathbb{R}^n} a \right) S_{p,n}^{p+1} \qquad \text{for every } u \in S. \tag{A2.5}$$

Furthermore, using as test in $\widetilde{J}_a$ a function which realizes the optimal Sobolev inequality we also find

$$\sup_S \widetilde{J}_a > S_{p,n}^{p+1}. \tag{A2.6}$$

We let u_n be a maximizing sequence, and we introduce the following sequence of probability measures on $\mathbb{R}^n$

$$\mu_n = \frac{a(x)|u_n|^{p+1}}{\int_{\mathbb{R}^n} a(x)|u_n|^{p+1}} dx.$$

We can apply Theorem A2.1, and we will show that the concentration alternative occurs.

Suppose by contradiction that $\lambda \neq 1$, and let us begin by considering the case $\lambda = 0$, namely assuming that vanishing holds. We let δ be an arbitrary small number, and we let R be such that $|a(x) - 1| < \delta$ for $|x| > R$. By the vanishing assumption we have that $\int_{B_R} a(x)|u_n|^{p+1} \to 0$ as $n \to +\infty$, and hence by our choice of R we deduce that

$$\tilde{J}_a(u_n) = \int_{B_R} a(x)|u_n|^{p+1} + \int_{\mathbb{R}^n \setminus B_R} a(x)|u_n|^{p+1}$$

$$\geq (1 - \delta)S_{p,n}^{p+1} + o(1) \quad \text{as } n \to +\infty.$$

Since $(u_n)_n$ is a maximizing sequence, by the arbitrarity of δ, we reach a contradiction to (A2.5).

We now consider the case $\lambda \in (0, 1)$. We choose a small $\varepsilon > 0$, and we let R, $(x_l)_l$ be given by the dichotomy condition. We also choose $\overline{R}$ such that $|a(x) - 1| < \varepsilon$ for $|x| > \overline{R}/4$. Then we have that either $B_R(x_l) \subseteq \mathbb{R}^n \setminus B_{\overline{R}/4}$, or $\mathbb{R}^n \setminus B_{\overline{R}}(x_l) \subseteq \mathbb{R}^n \setminus B_{\overline{R}/4}$. Assume the former holds (the other alternative requires only obvious changes): then we have that

$$\int_{B_R(x_l)} a(x)|u_n|^{p+1} < (1 + \varepsilon) \int_{B_R(x_l)} |u_n|^{p+1}. \qquad (A2.7)$$

We consider now a new sequence of functions $\overline{u}_n$ defined as

$$\overline{u}_n(x) = \chi_n(x)u_n(x),$$

where $\chi_n(x)$ is given by

$$\chi_n(x) = \begin{cases} 0 & \text{for } |x - x_n| \leq R \\ \dfrac{|x - x_n| - R}{\overline{R} - R} & \text{for } R \leq |x - x_n| \leq \overline{R} \\ 1 & \text{for } |x - x_n| \geq \overline{R}. \end{cases}$$

Hence, since $\nabla \chi_n$ tends to zero as $\overline{R} - R$ tends to infinity, it is easy to see that

$$\|u_n\|^2 = \|\chi_n u_n\|^2 + \|(1 - \chi_n)u_n\|^2 + o_{\overline{R}-R}(1)\|u_n\|^2$$

$$= \|\overline{u}_n\|^2 + \|u_n - \overline{u}_n\|^2 + o_{\overline{R}-R}(1),$$

since $u_n \in S$. Furthermore, by the dichotomy assumption we have that

$$\int_{B_{\overline{R}}(x_n) \setminus B_R(x_n)} a(x)|u_n|^{p+1} < 2\varepsilon,$$

and also

$$\tilde{J}(\overline{u}_n) = \lambda \sup_S \tilde{J} + O(\varepsilon) + o(1), \qquad \tilde{J}(u_n - \overline{u}_n) = (1 - \lambda) \sup_S \tilde{J} + O(\varepsilon) + o(1),$$

$$\tilde{J}(u_n) = \tilde{J}(\overline{u}_n) + \tilde{J}(u_n - \overline{u}_n) + O(\varepsilon),$$

where $o(1) \to 0$ as $n \to +\infty$. We now define $\tilde{u}_n = (u_n - \overline{u}_n)/\|u_n - \overline{u}_n\| \in S$. We notice that, since $\lambda \in (0, 1)$, for ε sufficiently small we have $J(u_n - \overline{u}_n) \neq 0$, so by the Sobolev embedding also $u_n - \overline{u}_n \neq 0$, so $\tilde{u}_n$ is well defined.

Then, by the homogeneity of $\tilde{J}$ and the above formulas we have that

$$\tilde{J}(\tilde{u}_n) = \frac{\tilde{J}(u_n - \overline{u}_n)}{\|u_n - \overline{u}_n\|} = \frac{\tilde{J}(u_n) - \tilde{J}(\overline{u}_n) + O(\varepsilon)}{\left(\|u_n\|^2 - \|\overline{u}_n\|^2\right)^{(p+1)/2}}$$

$$= \frac{(1 - \lambda)\sup_S \tilde{J} + O(\varepsilon) + o(1)}{\left(1 - \|\overline{u}_n\|^2 + o_{\overline{R}-R}(1)\right)^{(p+1)/2}}.$$

By (A2.7), the Sobolev embedding and the dichotomy assumption it follows that

$$\tilde{J}(\tilde{u}_n) \geq \frac{(1 - \lambda)\sup_S \tilde{J} + O(\varepsilon) + o(1)}{\left(1 - \lambda^{2/(p+1)} + o_{\overline{R}-R}(1) + O(\varepsilon)\right)^{(p+1)/2}}$$

$$= \frac{(1 - \lambda)}{\left(1 - \lambda^{2/(p+1)}\right)^{(p+1)/2}} \sup_S \tilde{J} + o_{\overline{R}-R}(1) + O(\varepsilon) + o(1).$$

If we choose ε small, $\overline{R} - R$ and n large, we reach a contradiction because the coefficient in front of $\sup_S \tilde{J}$ is greater than 1. Therefore we have the compactness alternative.

We prove next that, given any $\varepsilon > 0$ sufficiently small, if R and $(x_l)_l$ are given by Theorem A2.1 then $(x_l)_l$ stays bounded. In fact, by (**a0**), if it were $|x_l| \to +\infty$ up to a subsequence, then by the Sobolev embedding theorem we would have

$$\tilde{J}(u_n) = \int_{B_R(x_n)} a(x)|u_n|^{p+1} + O(\varepsilon) \leq S_{p,n} + o(1) + O(\varepsilon),$$

and this contradicts the fact that $\sup_S \tilde{J} > S_{p,n}^{p+1}$.

Finally, letting $\bar{u}$ denote the weak limit of the u_n, since concentration holds we have that

$$\int_{\mathbb{R}^n} a(x)|u_n|^{p+1} = \int_{B_R(x_l)} a(x)|u_n|^{p+1} + O(\varepsilon)$$

$$= \int_{B_R(x_l)} a(x)|\bar{u}|^{p+1} + O(\varepsilon) + o(1),$$

by the compactness of the Sobolev embedding on bounded sets. Since ε is arbitrary, by the Brezis–Lieb lemma we obtain that $u_n \to \bar{u}$ strongly in E. Since all the u_n can be taken to be non-negative (replacing u_n by $|u_n|$ if necessary), $\bar{u}$ also is non-negative, and hence positive by the maximum principle. This concludes the proof.   ∎

A2.2.2 An existence result under assumption (*a2*)

We give now another application of the concentration compactness theorem, considering the case in which $a(x)$ satisfies the assumption (*a2*). We discuss a particular case of a result by D. Cao [69].

Theorem A2.3 *Suppose $p \in (1, (n+2)/(n-2))$, and let $a : \mathbb{R}^n \to \mathbb{R}$ be smooth, satisfy (*a0*), (*a2*), and also the following condition*

$$a(x) > 2^{-(p-1)/2} \qquad \text{for every } x \in \mathbb{R}^n. \tag{A2.8}$$

Then problem (A2.4) admits a positive solution.

Proof. To prove this result we apply a reversed linking argument. We work again on the constraint $S = \{u \in E : \|u\| = 1\}$. First of all, we characterize the maximizing sequences for $\tilde{J}$, proving that compactness holds, but that their weak limit is zero. We notice first that, by (*a0*) and by (*a2*), $\sup_S \tilde{J} = S_{p,n}^{p+1}$.

Let $(u_n)_n$ be a maximizing sequence for $\tilde{J}$, and let us apply Theorem A2.1 to the sequence of measures

$$\mu_n = \frac{a(x)|u_n|^{p+1}}{\int_{\mathbb{R}^n} a(x)|u_n|^{p+1}} dx.$$

Let us suppose that dichotomy holds, let $\varepsilon > 0$ be a small number, let R, $(x_n)_n$ be given by Theorem A2.1, and let $\bar{R} > R$. Then, if χ_n, $\bar{u}_n$ and $\tilde{u}_n$ are as in the proof of Theorem A2.2, we can repeat the above arguments to find

$$\widetilde{J}(\tilde{u}_n) \geq \frac{(1-\lambda)}{\left(1-\lambda^{2/(p+1)}\right)^{(p+1)/2}} \sup_S \widetilde{J} + o_{\overline{R}-R}(1) + O(\varepsilon) + o(1) > \sup_S \widetilde{J}$$

if we choose ε small, and $\overline{R} - R, n$ sufficiently large.

Assume now that vanishing holds. Then we can decompose $\mathbb{R}^n$ into a sequence of disjoint unit cubes $(Q_i)_i$ centred at the points with (relative) integer coordinates. Then, by the Sobolev embedding (which holds with the same constant $C_{p,n}$ for every Q_i) and by the normalization on the u_n we get

$$1 = \sum_i \int_{Q_i} \left(|\nabla u_n|^2 + u_n^2\right) \geq \frac{1}{C_{p,n}} \sum_i \left(\int_{Q_i} |u_n|^{p+1}\right)^{2/(p+1)}.$$

By the vanishing assumption, and since $a(x)$ is bounded from above and from below by positive constants, we have that each integral $\int_{Q_i} |u_n|^{p+1}$ tends to zero as n tends to infinity uniformly with respect to the index i. Therefore from this fact and the last formula we find

$$\int_{\mathbb{R}^n} |u_n|^{p+1} = \sum_i \int_{Q_i} |u_n|^{p+1} = o(1) \left(\sum_i \left(\int_{Q_i} |u_n|^{p+1}\right)^{2/(p+1)}\right) = o(1),$$

contradicting the fact that $(u_n)_n$ is a maximizing sequence for $\widetilde{J}$.

It follows that $(\mu_n)_n$ satisfies the compactness alternative. Letting $(x_n)_n$ denote the sequence given by Theorem A2.1, we want to show that $|x_n| \to +\infty$. In fact, if x_n stays bounded, we can reason as in the proof of Theorem A2.2 to show that u_n converges strongly in E to some function $\bar{u}$ (with unit norm). On the other hand, we have that $a(x) < 1$ for any $x \in \mathbb{R}^n$, which implies

$$\widetilde{J}(u_n) \to_{n \to +\infty} \widetilde{J}(\bar{u}) < S_{p,n}^{p+1},$$

contradicting the facts that $(u_n)_n$ is maximizing and that $\sup_S \widetilde{J} = S_{p,n}^{p+1}$.

Applying the Brezis–Lieb lemma, as for the end of the proof of Theorem A2.2, one can show that, from the concentration alternative, $u_n(\cdot - x_n)$ converges strongly in E to some function $\tilde{u}$, which must be a maximizer of the Sobolev quotient $\|u\|_{L^{p+1}}/\|u\|_E$, and hence a ground state of (11.1) (which must be positive and radial up to a translation). By a result of Kwong [111] we know that $\tilde{u}$ is unique, and we call this solution u_0.

Consider now the following function $\beta : E \to \mathbb{R}^n$ defined as

$$\beta(u) = \int_{\mathbb{R}^n} |u|^{p+1} \frac{x}{|x|} \arctan(|x|) \, dx.$$

Since the coefficient of $|u|^{p+1}$ in the last integral is bounded, the function β is well defined on E. Furthermore, by the above discussion and from the exponential decay of u_0, we deduce that

$$(u_n) \text{ maximizing for } \tilde{J} \Rightarrow u_n(\cdot - x_n) \to u_0 \text{ in } E \text{ and } \beta(u_n) \neq 0 \text{ for } n \text{ large.}$$
$$\text{(A2.9)}$$

We now fix $L > 0$ and define

$$N_L = \{u_0(\cdot - y) : y \in B_L(0)\}, \qquad C = \{u \in S : \beta(u) = 0\}.$$

As $L \to +\infty$ one has that

$$\beta(u_0(\cdot - y)) = \frac{\pi}{2} S_{p,n}^{p+1} \frac{y}{|y|} + o_L(1) \qquad \text{for every } y \in \partial B_L(0),$$

where $o_L(1)$ tends to zero as L tends to infinity uniformly in $y \in \partial B_L(0)$. Then, letting $\mathcal{H}_L = \{h \in C(N_L, S) : h(u) = u, \forall u \in \partial N_L\}$, using the solution property of the degree it follows that ∂N_L and C link.

Then we define the max-min value

$$\rho_L = \sup_{h \in \mathcal{H}_L} \inf_{y \in N_L} \tilde{J}(h(y)).$$

Since $a(x)$ tends to 1 at infinity, with easy computations one finds that

$$\inf_{u \in \partial N_L} \tilde{J}(u) \to S_{p,n}^{p+1} \qquad \text{as } L \to +\infty.$$

Furthermore, from $\sup_S \tilde{J} = S_{p,n}^{p+1}$ and (A2.9), since C and ∂N_L link we have that $\rho_L < \inf_{u \in \partial N_L} \tilde{J}(u) < S_{p,n}^{p+1}$ for L sufficiently large. Moreover, using the map $\bar{h} : y \mapsto u_0(\cdot - y)$ and the assumption (A2.8) we find that $\rho_L > 2^{-(p-1)/2} S_{p,n}^{p+1}$. Then the arguments of Chapter 8 yield the existence of a Palais–Smale sequence $(v_n)_n$ for $\tilde{J}$ at level ρ_L.

It is a standard fact, see for example [42] and references therein, that v_n can be written as $v_n = w_n / \|w_n\|$, with

$$w_n = w_0 + \sum_{i=1}^{k} u_0(\cdot - x_{i,n}) + o(1), \qquad \text{(A2.10)}$$

where $o(1) \to 0$ in E as $n \to +\infty$, where k is some non-negative integer, where $|x_{i,n}|, |x_{i,n} - x_{j,n}| \to +\infty$ as n for $i, j = 1, \ldots, k$, $i \neq j$, and where w_0 is the weak limit of the w_n and solves (A2.4). Furthermore, by the Brezis–Lieb lemma one has that

$$\tilde{J}(v_n) = \frac{1}{\|w_n\|^{p+1}} \left(\tilde{J}(w_0) + k S_{p,n}^{p+1} + o(1) \right), \qquad \text{(A2.11)}$$

with $o(1) \to 0$ as $n \to +\infty$, and also

$$\|w_n\|^2 = \|w_0\|^2 + k\|u_0\|^2 + o(1). \tag{A2.12}$$

Multiplying (11.1) by u_0 and integrating by parts, we have that u_0 satisfies $\|u_0\|^2 = S_{p,n}^{p+1} \|u_0\|^{p+1}$, which implies $\|u_0\| = S_{p,n}^{-(p+1)/(p-1)}$. Furthermore, multiplying (A2.4) by w_0 and integrating by parts, we also find that $\|w_0\|^2 = \tilde{J}(w_0)$.

Then, from the fact that $\rho_L > 2^{-(p-1)/2} S_{p,n}^{p+1}$ and from (A2.11), (A2.12) we obtain

$$\alpha S_{n,p}^{p+1} \left(\|w_0\|^2 + k S_{n,p}^{-2(p+1)/(p-1)} \right)^{(p+1)/2} < \|w_0\|^2 + k S_{n,p}^{-2(p+1)/(p-1)}.$$

Using again (A2.4), the Sobolev inequality and the fact that $a(x) \leq 1$, we find also that $\|w_0\|^2 \leq S_{n,p}^{p+1} \|w_0\|^{p+1}$. From the last formula we then deduce that either $w_0 = 0$ and

$$k^{(p+1)/2} < 2^{(p-1)/2} k,$$

or that $w_0 \neq 0$ and

$$(1 + k)^{(p-1)/2} < 2^{(p-1)/2}.$$

In the former case, we would have $k = 1$ and by (A2.10) also $\tilde{J}(v_n) \to S_{n,p}^{p+1}$, contradicting the fact that $\rho_L < S_{p,n}^{p+1}$. On the other hand in the second case, by the last formula it has to be $k = 0$, so by (A2.10) we have strong convergence of w_n to $w_0 \neq 0$, a solution of (A2.4). Since in all this construction we can work in the subset of non-negative functions without affecting the min-max value, w_0 turns out to be non-negative, and hence strictly positive by the strong maximum principle. This concludes the proof. ∎

Bibliographical remarks We mention here other results related to problem (A2.4). In [42], the authors prove the existence of solutions to (A2.4) under assumption (*a0*) and requiring that $a(x) > 1 - Ce^{-(2+\delta)|x|}$ for $|x|$ large, where C and δ are two positive constants. The method is still based on a linking argument, as in Theorem A2.3, but using a different scheme.

In [44], an improvement of this result is given, replacing the last condition with $a(x) > 1 - Ce^{-\delta|x|}$ near infinity. The proof is based on the method of *critical points at infinity* developed by Bahri and Coron, see for example [38] and [41].

For results concerning problems with the critical exponent, see for example [49].

Appendix 3

Bifurcation for problems on $\mathbb{R}^n$

In this appendix we will discuss some bifurcation problems on $\mathbb{R}^n$ that cannot be handled by the theory developed in Chapters 2 and 4, because the linear part has an essential spectrum. In Section A3.1 we will deal with a problem that still possesses a principal eigenvalue; in Section A3.2 we will be concerned with a case in which there are no eigenvalues and the bifurcation occurs from the bottom of the essential spectrum. We anticipate that we will focus on some specific problems which, however, highlight the main features of the theory. The interested reader can find a broad discussion on these topics in the survey paper [167] which also contains a wide list of references.

A3.1 Bifurcation for problems on $\mathbb{R}^n$ in the presence of eigenvalues

Let us consider the following elliptic problem on $\mathbb{R}^n$

$$-\Delta u + q(x)u = \lambda u - u^p, \qquad u \in W^{1,2}(\mathbb{R}^n), \qquad \textbf{(P)}$$

where $p > 1$, $q \in L^2(\mathbb{R}^n)$ and

$$\liminf_{|x|\to\infty} q(x) = 0.$$

It is well known [54] that the spectrum of the linear problem

$$-\Delta u + q(x)u = \lambda u, \qquad u \in W^{1,2}(\mathbb{R}^n), \qquad \textbf{(L)}$$

contains eigenvalues if and only if

$$\Lambda := \inf\left\{ \int_{\mathbb{R}^n} [|\nabla u|^2 + qu^2]\,dx : u \in W^{1,2}(\mathbb{R}^n),\ \|u\|_{L^2} = 1 \right\} < 0. \quad \text{(A3.1)}$$

Moreover, if $\Lambda < 0$ then Λ is the lowest eigenvalue of $\textbf{(L)}$.

Problem $\textbf{(P)}$ will be approximated by problems on balls $B_{R_k} = \{x \in \mathbb{R}^n : |x| < R_k\}$,

$$-\Delta u + q(x)u = \lambda u - u^p, \qquad u \in W_0^{1,2}(B_{R_k}), \qquad R_k \to \infty. \qquad \textbf{(P}_k\textbf{)}$$

In the sequel it is understood that the solutions u_k of (P_k) are extended to all of $\mathbb{R}^n$ by setting $u_k(x) \equiv 0$ for $|x| > R_k$. Let λ_{R_k} denote the first (lowest) eigenvalues of

$$-\Delta u + q(x)u = \lambda u, \qquad u \in W_0^{1,2}(B_{R_k}),$$

given by

$$\lambda_{R_k} = \inf \left\{ \int_{B_{R_k}} [|\nabla u|^2 + qu^2]\, dx : u \in W_0^{1,2}(B_{R_k}),\ \|u\|_{L^2} = 1 \right\}.$$

Let us remark that $\lambda_{R_k} \downarrow \Lambda$ as $R_k \to \infty$. In particular, if (A3.1) holds, one has that $\lambda_{R_k} < 0$ provided $R_k \gg 1$.

Equation (P_k) can be written in the form $u = \lambda T_k(u)$, $u \in X = L^2(B_{R_k})$, with T_k compact. Let Σ^k denote the set of (λ, u) with $\lambda > 0$ and $u > 0$ such that $u = \lambda T_k(u)$.

Equation (P_k) can be faced by means of the Rabinowitz global bifurcation theorem, see Theorem 4.8. Here, since we are dealing with positive solutions, the arguments used in Section 4.4 allow us to say that the branch emanating from $(\lambda_{R_k}, 0)$ is unbounded. In other words, there exists an unbounded connected component $\Sigma_0^k \subset \Sigma^k$ such that $(\lambda_{R_k}, 0) \in \overline{\Sigma}_0^k$. Furthermore, the fact that the nonlinearity has the specific form $\lambda u - u^p$ allows us to say that the bifurcation is supercritical, see Example 2.12, and that $(\lambda, u) \in V$ implies that $\lambda > \lambda_{R_k}$, namely that the branch emanating from $(\lambda_{R_k}, 0)$ lies on the right of λ_{R_k}, see also Fig. A3.1.

In order to consider the limit of Σ^k as $k \to \infty$, we will use a topological result by G. T. Whyburn [170]. First, some definitions are in order. Let Y be a metric space and let Y_k be a sequence of subsets of Y. We define $\liminf Y_k$ as the set of $y \in Y$ such that every neighbourhood of y has nonempty intersection with all but a finite number of Y_k. On the other hand, $\limsup Y_k$ is the set of $y \in Y$ such that every neighbourhood of y has nonempty intersection with infinitely many of the Y_k.

Lemma A3.1 *Suppose that Y_k are connected and such that*

(i) $\bigcup Y_k$ *is precompact,*
(ii) $\liminf Y_k \neq \emptyset$.

Then $\limsup Y_k$ *is precompact and connected.*

In order to use this lemma, we take $E = W^{1,2}(\mathbb{R}^n)$, endowed with the standard norm

$$\|u\|^2 = \int_{\mathbb{R}^n} [|\nabla u|^2 + u^2]\, dx.$$

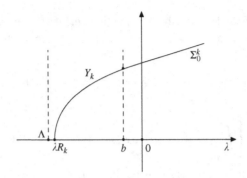

Figure A3.1

Fix $b < 0$, let $Y = [\Lambda, b] \times E$ and let Y_k be the connected component of $\{(\lambda, u) \in \overline{\Sigma}_0^k : \lambda \in [\Lambda, b]\}$ such that $(\lambda_{R_k}, 0) \in \overline{\Sigma}_0^k$.

We also let $\Pi : \mathbb{R} \times E$ be defined by setting $\Pi(\lambda, u) = \lambda$. It is not difficult to check that $\Pi(\overline{\Sigma}_0^k) = [\lambda_{R_k}, +\infty)$. Since $(\lambda_{R_k}, 0) \in \overline{\Sigma}_0^k$ and $\lambda_{R_k} \to \Lambda$, then $(\Lambda, 0) \in \liminf Y_k$ and thus Lemma A3.1(ii) holds. Moreover, one has that $b \in \Pi(\overline{\Sigma}_0^k)$ for all $k \gg 1$. In order to prove that $\bigcup Y_k$ is precompact, we need a preliminary lemma.

Lemma A3.2 *Let* (A3.1) *hold. For all* $b < 0$ *there exists* $\Psi = \Psi_b \in L^2(\mathbb{R}^n) \cap L^\infty(\mathbb{R}^n)$, $\Psi > 0$, *such that* $u < \Psi$ *for all* $(\lambda, u) \in Y_k$, *for all* $k \gg 1$.

Proof. We will indicate the main ideas of the argument.

Step 1. Fix a with $b < a < 0$. Since $\liminf_{|x| \to \infty} q(x) = 0$ and $a < 0$, the support of $(q(x) - a)^-$ (the negative part of $q - a$) is compact and is contained in the ball B_ρ, for some $\rho > 0$. We define a piecewise linear continuous function $\gamma_\alpha(t), t \in \mathbb{R}$, such that

$$\gamma_\alpha(t) = \begin{cases} -\alpha & t \le \rho \\ 0 & t \ge \rho + 1. \end{cases}$$

Let

$$\mu_\alpha = \inf \left\{ \int_{\mathbb{R}^n} [|\nabla u|^2 + \gamma_\alpha(|x|)u^2] \, dx : u \in E, \ \|u\|_{L^2} = 1 \right\}.$$

Since $\gamma_\alpha \le 0$, it follows that $\mu_\alpha \le \inf\{\int_{\mathbb{R}^n} |\nabla u|^2 \, dx : u \in E, \ \|u\|_{L^2} = 1\} = 0$. Furthermore, consider a smooth function $\phi \in E$ with support contained in B_ρ and such that $\|\phi\|_{L^2} = 1$. Then

$$\mu_\alpha \le \int_{B_\rho} [|\nabla \phi|^2 + \gamma_\alpha \phi^2] \, dx = \int_{B_\rho} |\nabla \phi|^2 \, dx - \alpha,$$

and thus there exists $\alpha^* > 0$ such that $\mu_\alpha < 0$ for all $\alpha > \alpha^*$. This implies that for $\alpha > \alpha^*$, μ_α is the principal eigenvalue of

$$-\Delta u + \gamma_\alpha(|x|)u = \mu u, \qquad u \in E.$$

We denote by $\varphi_\alpha > 0$ the (normalized) eigenfunction corresponding to $\mu_\alpha < 0$. In addition, we notice that μ_α depends continuously upon α.

Step 2. From the preceding step it follows that we can find $\alpha_0 > 0$ such that $\mu_0 := \mu_{\alpha_0}$ verifies $b - a < \mu_0 < 0$. We define a function $\psi \in C^2(\mathbb{R}^n) \cap E$ by setting $\psi(x) = \varphi_{\alpha_0}(x)$ for all $|x| \geq \rho + 1$; in the ball $B_{\rho+1}$ ψ is arbitrary, but positive. One shows that there exists $C > 0$ such that $C\psi$ is a super-solution of (P_k) for all $k \geq 1$ and all $\lambda \geq b$. Roughly, it is easy to check that for $C > 0$ sufficiently large one has that $-\Delta(C\psi) + q(C\psi) \geq \lambda(C\psi) - (C\psi)^p$ for all $|x| \leq \rho + 1$. For $|x| > \rho + 1$, one remarks that $\gamma_\alpha \equiv 0$, so $-\Delta\psi = \mu_0\psi$ and one finds $-\Delta\psi + q\psi = (\mu_0 + q)\psi$. The definition of ρ implies that $q > a$ for all $|x| > \rho$ and thus $-\Delta\psi + q\psi \geq (\mu_0 + a)\psi \geq b\psi$. Then for $\lambda \leq b$ we get $-\Delta\psi + q\psi \geq \lambda\psi - \psi^p$ for all $|x| > \rho + 1$, and the claim follows.

Step 3. One proves that $\Psi = C\psi$ is such that $u \leq \Psi$ for all $(\lambda, u) \in Y_k$ with k large. For $\lambda \leq b$, set $\tilde{f}_\lambda(u) := \lambda u - qu - u^p$ and take $M > 0$ such that $\tilde{f}_\lambda$ is strictly increasing for $u \in [0, \max \Psi]$. Let v_k be the solution of

$$\begin{cases} -\Delta v_k + Mv_k = \tilde{f}_b(\Psi) + M\Psi & |x| < R_k \\ v_k = 0 & |x| = R_k. \end{cases}$$

We want to show that for all $\lambda \leq b$, v_k is a super-solution of (P_k) but not a solution. Since $\tilde{f}_b(\Psi) + M\Psi \geq 0$ then $v_k \in \mathcal{P}_k$, where $\mathcal{P}_k$ denotes the *interior* of the positive cone in $C_0^1(B_{R_k})$. From the preceding step we know that $-\Delta\Psi \geq b\Psi - q\Psi - \Psi^p = \tilde{f}_b(\Psi)$. From this one easily infers

$$\begin{cases} -\Delta(\Psi - v_k) + M(\Psi - v_k) \geq 0 & |x| < R_k \\ \Psi - v_k > 0 & |x| = R_k, \end{cases}$$

and the maximum principle yields

$$\Psi(x) > v_k(x), \qquad \forall \, |x| < R_k. \tag{A3.2}$$

Since $\tilde{f}_\lambda + M$ is strictly increasing, it follows that $\tilde{f}_\lambda(\Psi) + M\Psi > \tilde{f}_\lambda(v_k) + Mv_k$. This and the fact that $\tilde{f}_b \geq \tilde{f}_l$ provided $\lambda \leq b$, imply

$$-\Delta v_k = \tilde{f}_b(\Psi) + M\Psi - Mv_k \geq \tilde{f}_\lambda(\Psi) + M\Psi - Mv_k > \tilde{f}_\lambda(v_k), \quad |x| < R_k.$$

This proves our claim.

Finally, let us prove that $u < v_k$ for all $(\lambda, u) \in Y_k$. Consider the set $\tilde{Y}_k = \{(\lambda, v_k - u) : (\lambda, u) \in Y_k\}$. Since $(\lambda_{R_k}, 0) \in Y_k$ then $(\lambda_{R_k}, v_k) \in \tilde{Y}_k$, and thus $\tilde{Y}_k \cap ([\Lambda, b] \times \mathcal{P}_k) \neq \emptyset$. Let us check that $\tilde{Y}_k \subset [\Lambda, b] \times \mathcal{P}_k$. Otherwise, there exists $(\lambda^*, u^*) \in Y_k$ such that $v_k - u^* \in \partial\mathcal{P}_k$. Since v_k is not a solution of (P_k^*) it follows that $v_k \geq u^*$ but $v_k \not\equiv u^*$ in B_{R_k}. This implies $-\Delta(v_k - u^*) + M(v_k - u^*) \geq \widetilde{f}_\lambda(v_k) + Mv_k - \widetilde{f}_\lambda(u^*) + Mu^* \geq 0$. By the maximum principle we infer that $v_k > u^*$, namely $v_k - u^* \in \mathcal{P}_k$, while $v_k - u^* \in \partial\mathcal{P}_k$. This proves that $u < v_k$ and thus, using (A3.2) we get $u < v_k < \Psi$ for all $|x| < R_k$, and the proof is completed. ∎

Let us point out that we do not know whether $u < \Psi$ for all $(\lambda, u) \in \Sigma_0^k$, with $\lambda \in [\Lambda, b]$. The proof only works for $(\lambda, u) \in Y_k$.

The preceding lemma allows us to show the following.

Lemma A3.3 $\bigcup Y_k$ *is precompact.*

Proof. Let $(\lambda_j, u_j) \in \bigcup Y_k$. We can assume that $\lambda_j \to \lambda$, for some $\lambda \in [\Lambda, b]$. From Lemma A3.2 it follows there is $c_1 > 0$ such that

$$\|u_j\|_{L^2} \leq c_1, \qquad \forall j.$$

From (P_k) we also get

$$\int_{\mathbb{R}^n} |\nabla u_j|^2 \, dx + \int_{\mathbb{R}^n} q u_j^2 \, dx = \lambda_j \int_{\mathbb{R}^n} u_j^2 \, dx - \int_{\mathbb{R}^n} u_j^{p+1} \, dx. \qquad \text{(A3.3)}$$

Let us remark that, without loss of generality, we can suppose that $p + 1 < 2^*$, otherwise we can use (A3.2) jointly with a truncation argument. From (A3.3) it follows that $\exists c_2 > 0$ such that $\|u_j\| \leq c_2$ and hence, up to a subsequence, $u_j \rightharpoonup u$ in E. One immediately verifies that u satisfies

$$\int_{\mathbb{R}^n} \nabla u \cdot \nabla \phi \, dx + \int_{\mathbb{R}^n} q u \phi \, dx = \lambda \int_{\mathbb{R}^n} u\phi \, dx - \int_{\mathbb{R}^n} u^p \phi \, dx, \quad \forall \phi \in C_0^\infty(\mathbb{R}^n).$$
$$\text{(A3.4)}$$

Set $G_\lambda(u) = \lambda u - qu - u^p$. From (A3.3) we get

$$\|u_j\|^2 = \int_{\mathbb{R}^n} u_j^2 \, dx + \int_{\mathbb{R}^n} G_{\lambda_j}(u_j) u_j \, dx. \qquad \text{(A3.5)}$$

Moreover, by density, we can set $\phi = u_j$ in (A3.4) yielding

$$\int_{\mathbb{R}^n} \nabla u_j \cdot \nabla u \, dx = \int_{\mathbb{R}^n} G_\lambda(u) u_j \, dx. \qquad \text{(A3.6)}$$

Similarly, letting $\phi = u$, we get $\int_{\mathbb{R}^n} |\nabla u|^2 u \, dx = \int_{\mathbb{R}^n} G_\lambda(u) u \, dx$ and hence

$$\|u\|^2 = \int_{\mathbb{R}^n} G_\lambda(u) u \, dx + \int_{\mathbb{R}^n} u^2 \, dx. \qquad \text{(A3.7)}$$

Using (A3.5), (A3.6) and (A3.7), we infer

$$\|u_j - u\|^2 = \|u_j\|^2 + \|u\|^2 - 2\int_{\mathbb{R}^n} \nabla u_j \cdot \nabla u \, dx - 2\int_{\mathbb{R}^n} u_j u \, dx$$

$$= \int_{\mathbb{R}^n} u_j^2 \, dx + \int_{\mathbb{R}^n} G_{\lambda_j}(u_j) u_j \, dx + \int_{\mathbb{R}^n} G_\lambda(u) u \, dx + \int_{\mathbb{R}^n} u^2 \, dx$$

$$\quad - 2\int_{\mathbb{R}^n} G_\lambda(u) u_j \, dx - 2\int_{\mathbb{R}^n} u_j u \, dx$$

$$= \int_{\mathbb{R}^n} [G_{\lambda_j}(u_j) - G_\lambda(u)] u_j \, dx + \int_{\mathbb{R}^n} G_\lambda(u)[u - u_j] \, dx$$

$$\quad + \int_{\mathbb{R}^n} u[u - u_j] \, dx + \int_{\mathbb{R}^n} u_j[u_j - u] \, dx.$$

Since $u_j < \Psi \in L^2(\mathbb{R}^n)$ we find

$$\|u_j - u\|^2 \le \int_{\mathbb{R}^n} |G_{\lambda_j}(u_j) - G_\lambda(u)| \Psi \, dx + \int_{\mathbb{R}^n} |G_\lambda(u)||u - u_j| \, dx$$

$$\quad + \int_{\mathbb{R}^n} |u||u - u_j| \, dx + \int_{\mathbb{R}^n} \Psi |u_j - u| \, dx.$$

Since

$$|G_{\lambda_j}(u_j) - G_\lambda(u)| \le |\lambda_j - \lambda| \, |u_j - u| + |q| \, |u_j - u| + |u_j^p - u^p|,$$

also taking into account that $u_j \rightharpoonup u$ in E, it readily follows that all the integrals in the right hand side of the preceding equation tend to zero. Thus $\|u_j - u\|^2 \to 0$, proving that $u_j \to u$ strongly in E. ∎

We are now ready to prove the main result of this section.

Theorem A3.4 *If* (A3.1) *holds, then there exists a connected set* $\Sigma_0 = \{(\lambda, u) \in \mathbb{R} \times E\}$ *such that*

(a) u *is a positive solution of* (**P**),
(b) $(\Lambda, 0) \in \overline{\Sigma}_0$ *and* $\Pi\overline{\Sigma}_0 \supset [\Lambda, 0)$.

Proof. We set $\Sigma_0 = \limsup Y_k \setminus \{(\Lambda, 0)\}$. According to Lemma A3.1, Σ_0 is connected and it is easy to check that any $(\lambda, u) \in \Sigma_0$ is a non-negative solution of (P). To prove (a) we need to show that $u > 0$. We have already remarked that for each $k \ge 1$, $(\lambda, u) \in \Sigma_k$ implies that $\lambda > \lambda_{R_k}$, and this yields that $(\lambda, u) \in \Sigma_0 \Rightarrow \lambda \ge \Lambda$. Suppose that there exist $(\lambda_j, u_j) \in Y_{k_j}$ such that $(\lambda_j, u_j) \to (\lambda, 0)$ as $k_j \to \infty$. Since $\lambda_R \downarrow \Lambda$ as $R \to \infty$, given $\delta > 0$

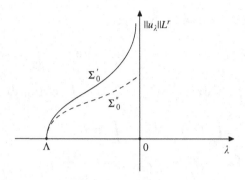

Figure A3.2 Notation: Σ_0' refers to the case $r \leq n/(n-2)$; Σ_0'' refers to the case $r > n/(n-2)$.

there exists $\ell \in \mathbb{N}$ such that $\lambda_{R_{k_j}} < \Lambda + \delta < \lambda$, for all $k_j \geq \ell$. Then u_j is a super-solution of

$$-\Delta u + qu = (\Lambda + \delta)u - u^p, \qquad u \in W_0^{1,2}(B_{R_\ell}). \tag{A3.8}$$

One can also find $\varepsilon_j \ll 1$ such that $\varepsilon_j \varphi_1$ is a sub-solution of (A3.8) such that $\varepsilon_j \varphi_1 \leq u_j$ in B_{R_ℓ} and thus there exists a positive solution $\widetilde{u}_j$ of (A3.8). Since $u_j \to 0$, then also $\widetilde{u}_j \to 0$ and therefore $\Lambda + \delta$ is a bifurcation point of positive solutions of (A3.8). This is not possible, since the unique bifurcation point of positive solutions of (A3.8) is $\lambda_{R_\ell} < \Lambda + \delta$. This contradiction proves that $u > 0$.

Since $(\Lambda, 0) \in \limsup Y_k$ it follows immediately that $(\Lambda, 0) \in \overline{\Sigma}_0$. As already remarked before, $b \in \Pi(\overline{\Sigma}_0^k)$ for all $k \gg 1$ and all $b \in (\Lambda, 0)$. Repeating the arguments carried out in Lemma A3.3, it follows that $b \in \Pi(\overline{\Sigma}_0)$. Finally, from the fact that Σ_0 is connected one deduces that $[\Lambda, 0) \subset \Pi(\overline{\Sigma}_0)$. ∎

Remark A3.5 It is possible to complete the statement of Theorem A3.4 by showing that as $\lambda \uparrow 0$ the solutions u_λ such that $(\lambda, u_\lambda) \in \Sigma_0$ satisfy:

(i) $\|u_\lambda\|_{L^r} \leq$ constant if $r > n/(n-2)$,

(ii) $\|u_\lambda\|_{L^r} \to \infty$ if $r \leq n/(n-2)$. ∎

By similar arguments one can handle sublinear problems on $\mathbb{R}^n$.

Theorem A3.6 *Let $\rho \in L^\infty$, and suppose that $\exists\, U \in L^\infty \cap L^2$ such that $-\Delta U = \rho$ in $\mathbb{R}^n$. Then, for all $0 < q < 1$ the problem*

$$-\Delta u = \lambda \rho(x)u^q, \qquad u \in W^{1,2}(\mathbb{R}^n),$$

possesses a branch Σ of positive solutions bifurcating from $(0,0)$ and such that $\Pi(\overline{\Sigma}) = [0, +\infty)$.

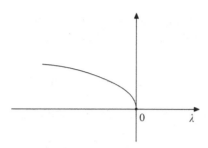

Figure A3.3 Bifurcation diagram of $-u'' = \lambda u + |u|^{p-1}u$.

Theorem A3.4 is a particular case of results dealing with a more general class of equations, see [15], where we refer to for more details and further results. See also [91]. Theorem A3.6 is a particular case of a result by H. Brezis and S. Kamin [61].

A3.2 Bifurcation from the essential spectrum

Consider the problem

$$-\Delta u + V(x)u = \lambda u + h(x)|u|^{p-1}u, \qquad u \in W^{1,2}(\mathbb{R}^n), \qquad (A3.9)$$

where $p > 1$, V is bounded and $h > 0$. If we assume

$$V \in L^\infty, \qquad V(x) \geq 0, \qquad \lim_{|x| \to \infty} V(x) = 0, \qquad (A3.10)$$

then the spectrum of the linearized problem

$$-\Delta v + V(x)v = \lambda v, \qquad v \in W^{1,2}(\mathbb{R}^n)$$

is the whole half-line $[0, \infty)$ and coincides with its essential spectrum, which is the set of all points of the spectrum that are not isolated, jointly with the eigenvalues of infinite multiplicity. Clearly, none of the bifurcation results proved so far apply to (A3.9). In order to have an idea of the results we can expect, let us consider the elementary case in one dimension when $V \equiv 0$ and $h \equiv 1$:

$$-u'' = \lambda u + |u|^{p-1}u, \quad u \in W^{1,2}(\mathbb{R}),$$

which can be studied in a straightforward way by a phase-plane analysis.
It follows that from $\lambda = 0$, the bottom of the essential spectrum of $-v'' = \lambda v$, $v \in W^{1,2}(\mathbb{R})$, bifurcates a family of solutions (λ, u_λ), $\lambda < 0$, of $-u'' = \lambda u + |u|^{p-1}u$, with $(\lambda, u_\lambda) \to (0, 0)$ as $\lambda \uparrow 0$.

In order to prove a similar result for (A3.9) with V and h possibly depending on x, we will follow [167] and use variational tools. Let V satisfy (A3.10) and suppose that h verifies

$$h \in L^\infty, \qquad h(x) > 0, \qquad \lim_{|x|\to\infty} h(x) = 0. \qquad (\text{A3.11})$$

Let $1 < p < (n+2)/(n-2)$ and set $E = W^{1,2}(\mathbb{R}^n)$,

$$\|u\|_\lambda^2 = \int_{\mathbb{R}^n} [|\nabla u|^2 + V(x)u^2 - \lambda u^2] \, dx, \qquad u \in E.$$

Consider the functional $J_\lambda : E \mapsto \mathbb{R}$,

$$J_\lambda(u) = \frac{1}{2}\|u\|_\lambda^2 - \frac{1}{p+1}\int_{\mathbb{R}^n} h|u|^{p+1}\, dx.$$

Clearly, J_λ is of class C^2 and its critical points give rise to (weak and, by regularity results, strong) solutions of (A3.9) such that $\lim_{|x|\to\infty} u(x) = 0$. Moreover, for each fixed $\lambda < 0$, $\|\cdot\|_\lambda$ is a norm equivalent to the usual one in $W^{1,2}(\mathbb{R}^n)$. To find critical points of J_λ we can use, for example, the method of the 'natural constraint', discussed in section 6.4 and used in section 7.6 to prove the existence of solutions of a class of superlinear BVPs on a bounded domain, see Theorem 7.14. Using the notation introduced there, we set

$$\Phi(u) = \frac{1}{p+1}\int_{\mathbb{R}^n} h|u|^{p+1}\, dx, \qquad \Psi(u) = \int_{\mathbb{R}^n} h|u|^{p+1}\, dx,$$

$$M_\lambda = \{u \in W^{1,2}(\mathbb{R}^n) \setminus \{0\}; \|u\|_\lambda^2 = \Psi(u)\},$$

and

$$\widetilde{J}_\lambda(u) = \frac{1}{2}\Psi(u) - \Phi(u) = \frac{p-1}{p+1}\int_{\mathbb{R}^n} h|u|^{p+1}\, dx.$$

It is easy to check that, as for Theorem 7.14, $M_\lambda \neq \emptyset$ is a smooth manifold of codimension one in E and the critical points of $\widetilde{J}_\lambda$ constrained on M_λ also satisfy $J_\lambda' = 0$. One has the following result.

Lemma A3.7 *If (A3.11) holds, then Φ is weakly continuous and Φ' is compact.*

Proof. Let $u_k \rightharpoonup u$ in E. Given $\varepsilon > 0$, from (A3.11) it follows that there exists $R > 0$ such that

$$\int_{|x|\geq R} h(|u_k|^{p+1} - |u|^{p+1}) \, dx \leq \varepsilon.$$

Since $W^{1,2}(B_R)$ is compactly embedded in $L^{p+1}(B_R)$, we get

$$\int_{|x|<R} h(|u_k|^{p+1} - |u|^{p+1}) \, dx \leq \varepsilon,$$

provided $k \gg 1$. Putting together the two preceding inequalities, it follows that Φ is weakly continuous. The proof that Φ' is compact is similar and will be omitted. ∎

The preceding lemma allows us to repeat the arguments carried out in the proof of Theorem 7.14 yielding that

$$m(\lambda) = \min\{\widetilde{J}_\lambda(u) : u \in M_\lambda\}$$

is achieved.

In order to estimate $m(\lambda)$ we strengthen assumptions (A3.10) by requiring

$$|x|^2 V(x) \in L^\infty, \qquad V(x) \geq 0. \tag{A3.12}$$

Moreover, we suppose that h verifies (A3.11) and $\exists\, K > 0,\ y \in \mathbb{R}^n,\ \tau \in [0, 2[$ such that

$$h(x) \geq K|x|^{-\tau}, \quad \forall x \in C = \{tx : t \geq 1,\ |x - y| \leq 1\}. \tag{A3.13}$$

Fix the function $\phi(x) = |x|e^{-|x|}$ and set $u_\alpha(x) = \phi(\alpha x)$. There holds

$$\|\nabla u_\alpha\|_{L^2}^2 = \alpha^{2-n}A_1, \quad A_1 = \int_{\mathbb{R}^n} |\nabla\phi|^2\, dx,$$

$$\|u_\alpha\|_{L^2}^2 = \alpha^{-n}A_2, \quad A_2 = \int_{\mathbb{R}^n} \phi^2\, dx,$$

$$\int_{\mathbb{R}^n} Vu_\alpha^2\, dx \leq \alpha^{2-n}A_3, \quad A_3 = c_1 \int_{\mathbb{R}^n} |x|^2\phi^2\, dx,$$

where c_1 is such that $|x|^2\,|V(x)| \leq c_1$. Then one finds

$$\|u_\alpha\|_\lambda^2 \leq A_1\alpha^{2-n} + A_3\alpha^{2-n} - \lambda A_2\alpha^{-n}.$$

Putting $\lambda = -\alpha^2$ we get

$$\|u_\alpha\|_\lambda^2 \leq A_4\alpha^{2-n}, \qquad -1 \leq \lambda < 0,\ \lambda = -\alpha$$

for some $A_4 > 0$. Moreover, using (A3.13) we deduce

$$\Psi(u_\alpha) \geq K \int_C |x|^{-\tau}|u_\alpha|^{p+1}\, dx \geq K\alpha^{\tau-n} \int_{C_\alpha} |\xi|^{-\tau}|\phi(\xi)|^{p+1}\, d\xi,$$

where $C_\alpha = \{\xi : \xi/\alpha \in C\}$. Since $C \subset C_\alpha$ if $0 < \alpha \leq 1$, it follows that there exists $A_5 > 0$ such that

$$\Psi(u_\alpha) \geq \alpha^{\tau-n}A_5, \qquad 0 < \alpha \leq 1.$$

There holds

$$
\begin{aligned}
m(\lambda) &= \min \left\{ \frac{1}{2}\Phi(u) - \Psi(u) \; : \; u \in M_\lambda \right\} \\
&= \min \left\{ \frac{p-1}{p+1}\Psi(u) \; : \; u \in M_\lambda \right\} \\
&= \min \left\{ \frac{p-1}{p+1}\Psi(u) \left[\frac{\|u\|_\lambda^2}{\Psi(u)} \right]^{(p+1)/(p-1)} \; : \; u \in E \setminus \{0\} \right\} \\
&\leq \frac{p-1}{p+1}\Psi(u_\alpha) \left[\frac{\|u_\alpha\|_\lambda^2}{\Psi(u_\alpha)} \right]^{(p+1)/(p-1)} \\
&= \frac{p-1}{p+1}\|u_\alpha\|_\lambda^{(p+1)/(p-1)}(\Psi(u_\alpha))^{-2/(p-1)}.
\end{aligned}
$$

Then, using the preceding calculation, we infer that, for some $A_6 > 0$,

$$
\begin{aligned}
0 < m(\lambda) &\leq A_6 \, \alpha^{(2-n)(p+1)/(p-1)} \cdot \alpha^{-2(\tau-n)/(p-1)} \\
&= A_6 \, |\lambda|^{((2-n)(p+1)-2(\tau-n))/2(p-1)}.
\end{aligned}
$$

Thus, if $1 < p < 1 + 2(2 - \tau)/n$, we find that $m(\lambda)/|\lambda| \to 0$ as $\lambda \to 0^-$. Let $u_\lambda \in M_\lambda$ be such that $\tilde{J}_\lambda(u_\lambda) = m(\lambda)$. Since $0 < \tilde{J}_\lambda(u_\lambda) = (p-1)/(p+1)\|u\|_\lambda^2 \leq m(\lambda)$, it follows that $\|u_\lambda\|_\lambda \to 0$ as $\lambda \to 0^-$. In conclusion, we can state the following theorem.

Theorem A3.8 *Suppose that* (A3.10), (A3.11) *and* (A3.12) *hold. If* $1 < p < 1+2(2-\tau)/n$, *then the bottom of the essential spectrum,* $\lambda = 0$, *is a bifurcation point for* (A3.9). *Precisely, for all* $\lambda < 0$ *there is a family of nontrivial solutions* u_λ *of* (A3.9) *such that* $\|u_\lambda\|_\lambda \to 0$ *as* $\lambda \to 0^-$.

The next result we are going to survey, deals with the problem

$$
-u'' = \lambda u + h(x)|u|^{p-1}u, \qquad u \in W^{1,2}(\mathbb{R}). \tag{A3.14}
$$

We will assume that $p > 1$ and h satisfies

(h.1) $\exists b > 0 \; : \; h - b \in L^1(\mathbb{R})$, and $\int_{\mathbb{R}} (h - b)\,\mathrm{d}x \neq 0$.

In order to prove that $\lambda = 0$ is a bifurcation point for (A3.14), we will transform this problem into a new one, which can be solved with the perturbation methods discussed in Appendix 5.

Let us set

$$
\begin{cases}
v(x) = \varepsilon^{-2/(p-1)}u\left(\dfrac{x}{\varepsilon}\right) \\[2mm]
\lambda = -\varepsilon^2.
\end{cases}
$$

Then equation (A3.14) becomes

$$-v'' + bv = h\left(\frac{x}{\varepsilon}\right)|v|^{p-1}v, \quad v \in W^{1,2}(\mathbb{R}). \tag{A3.15}$$

If (A3.15) has for all $|\varepsilon|$ small a family of solutions $v_\varepsilon \neq 0$, then the corresponding u_λ is a family of nontrivial solutions of (A3.14) branching off from $\lambda = 0$. Moreover, since $\lambda = -\varepsilon^2$, the bifurcation is on the left of $\lambda = 0$, the infimum of the essential spectrum.

In view of the assumption (h.1), it is convenient to rewrite (A3.15) in the following form

$$-v'' + bv = |v|^{p-1}v + \left[h\left(\frac{x}{\varepsilon}\right) - b\right]|v|^{p-1}v, \quad v \in W^{1,2}(\mathbb{R}).$$

It is possible to show that (h.1) implies that $h(x/\varepsilon) - b$ tends to zero in a suitable sense as $\varepsilon \to 0$, and hence (A3.15) can be viewed as a perturbation problem, the unperturbed problem being

$$-v'' + bv = |v|^{p-1}v, \quad u \in W^{1,2}(\mathbb{R}).$$

Using a modified version of Proposition A5.3 in Appendix 5, one can prove the following.

Theorem A3.9 *Let* (h.1) *hold. Then* (A3.14) *has a family of solutions* (λ, u_λ) *such that* $\lambda \to 0^-$ *and* $u_\lambda \to 0$ *as* $\lambda \to 0^-$ *in the* $C(\mathbb{R})$ *topology. Moreover,* $\lim_{\lambda \to 0^-} \|u_\lambda\|_{L^2(\mathbb{R})}^2 = 0$ *iff* $1 < p < 5$.

It is worth pointing out that condition (h.1) can be weakened. Moreover the partial differential equation analogue of (A3.14) can be studied. For these and further bifurcation results concerning problem (A3.14), we refer to [13, 36] or to Chapter 3 of [19] and references therein.

Appendix 4
Vortex rings in an ideal fluid

A4.1 Formulation of the problem

We consider a fluid filling all of $\mathbb{R}^3$ and we suppose it is cylindrically symmetric. If the fluid is perfect, namely inviscid and with uniform density, denoting by $\mathbf{q}$ its velocity, the continuity equation div $\mathbf{q} = 0$ leads to the existence of a *stream function* $\Psi(r, z)$ defined in the half-plane $\Pi = \{(r, z) : r > 0, \ z \in \mathbb{R}\}$ such that, in cylindrical coordinates,

$$\mathbf{q} = \left(-\frac{\Psi_z}{r}, 0, \frac{\Psi_r}{r} \right).$$

Moreover, letting

$$L\Psi = r \left(\frac{1}{r} \Psi_r \right)_r + \Psi_{zz},$$

one has that curl $\mathbf{q} = (0, \omega, 0)$ where $\omega(r, z)$, the *vorticity function*, is given by

$$\omega = -\frac{1}{r} L\Psi.$$

The momentum equations require that ω/r is constant on any surface $\Psi = $ constant, and hence there is a *vorticity function* f such that

$$\omega(r, z) = r f(\Psi(r, z)).$$

A *vortex* is a toroidal region $\mathcal{R}$, such that $\omega \neq 0$ if and only if $(r, z) \in \mathcal{R}$. Thus if A, the *vortex core*, denotes the cross section of the vortex $\mathcal{R}$, we are led to the equation

$$-L\Psi = \begin{cases} r^2 f(\Psi) & \text{in } A \\ 0 & \text{in } \Pi \setminus \overline{A}. \end{cases}$$

The preceding equation is completed by suitable boundary conditions. The first one prescribes the amount of fluid flowing between the stream surfaces $r = 0$ and ∂A. This leads to the requirement that

$$\Psi(0, z) = -k, \ \forall z \in \mathbb{R}, \quad \text{and} \quad \Psi(r, z) = 0, \ \forall (r, z) \in \partial A,$$

for some flux parameter $k \geq 0$. The second boundary condition demands that Ψ approaches at infinity the uniform stream $-\frac{1}{2}Wr^2 - k$ for some given *propagation speed* $W > 0$, and so

$$\mathbf{q} \to (0, 0, -W), \qquad \text{as } r^2 + z^2 \to +\infty.$$

Putting together the preceding equations we get the following elliptic problem on Π

$$\begin{cases} -L\,\Psi = r^2 f(\Psi) & (r, z) \in A, \quad L\Psi = 0 \quad (r, z) \in \Pi \setminus \overline{A}, \\ \Psi = -k & \text{on } r = 0, \quad \Psi = 0 \quad \text{on } \partial A, \end{cases}$$

$$\begin{cases} \Psi \to -\frac{1}{2}Wr^2 - k & \text{as } r^2 + z^2 \to +\infty \\ \Psi_r/r \to 0 & \text{as } r^2 + z^2 \to +\infty \\ \Psi_z/r \to 0 & \text{as } r^2 + z^2 \to +\infty. \end{cases}$$

It is worth pointing out that this is actually a *free boundary problem*, in the sense that the vortex core A is unknown and must be determined together with the function Ψ. To circumvent this difficulty, it is convenient to introduce, as in Section 11.3.2, the Heaviside function $h(t)$,

$$h(t) = \begin{cases} 0 & t \leq 0 \\ 1 & t > 0. \end{cases}$$

Setting $g(t) = h(t)f(t)$ and

$$\psi(r, z) = \Psi(r, z) + \frac{1}{2}Wr^2 + k,$$

the preceding problem becomes

$$\begin{cases} -L\psi = r^2 g(\psi - \frac{1}{2}Wr^2 - k) & (r, z) \in \Pi \\ \psi = 0 & \text{on } r = 0 \\ \psi \to 0 & \text{as } r^2 + z^2 \to +\infty \\ |\nabla\psi|/r \to 0 & \text{as } r^2 + z^2 \to +\infty. \end{cases} \tag{A4.1}$$

By a solution of (A4.1) we mean a function ψ of class $C^1(\Pi) \cap C^2(\Pi \setminus \partial A)$ which solves (A4.1) almost everywhere. The requirement that ψ is C^1 across the boundary of the vortex core is a consequence of the fact that $\mathbf{q}$ has to be continuous. Problem (A4.1) has the trivial solution $\psi \equiv 0$, corresponding to the uniform stream $\Psi = -\frac{1}{2}Wr^2 - k$ and to an empty vortex core, $A = \emptyset$, and the aim is to find nontrivial solutions of (A4.1). On the other hand, if ψ is a nontrivial solution of (A4.1), the maximum principle implies that $\psi(r, z) > 0$, with nonempty vortex core:

$$A = \{(r, z) \in \Pi : \psi(r, z) > \tfrac{1}{2}Wr^2 + k\} = \{(r, z) \in \Pi : \Psi(r, z) > 0\} \neq \emptyset.$$

When $g = h$, the Heaviside function, and $k = 0$, Hill discovered an explicit solution to (A4.1). Letting $\rho^2 = r^2 + z^2$, and

$$2a^2 = 15W,$$

the Hill solution is given by

$$\psi_H(r, z) = \begin{cases} \dfrac{1}{2} W r^2 \left(\dfrac{5}{2} - \dfrac{3}{2} \dfrac{\rho^2}{a^2} \right) & 0 \le \rho \le a, \\[2mm] \dfrac{1}{2} W r^2 \dfrac{a^3}{\rho^3} & \rho \ge a, \end{cases}$$

whose corresponding vortex is the solid ball $\rho < a$. Actually, it has been proved that ψ_H is the unique solution of (A4.1) for $k = 0$ and $g = h$.

To cast the problem in a suitable functional setting, it is convenient to perform a further change of variable, introduced by W.-M. Ni [136]. Consider in $\mathbb{R}^5$ cylindrical coordinates (r, z), with

$$r^2 = \sqrt{x_1^2 + \cdots + x_4^2}, \qquad z = x_5,$$

and set $\psi(r, z) = r^2 u(r, z)$. Then we have

$$L\psi = r \left(\frac{1}{r} \psi_r \right)_r + \psi_{zz} = r(2u + ru_r)_r + r^2 u_{zz}$$

$$= r^2 u_{rr} + 3r u_r + r^2 u_{zz}.$$

Thus

$$L\psi = r^2 \left(u_{rr} + \frac{3}{r} u_r + u_{zz} \right) = r^2 \Delta u,$$

where Δ denotes the Laplacian in cylindrical coordinates in $\mathbb{R}^5$. Therefore, the new unknown u satisfies $-\Delta u = g(u - \frac{1}{2} W r^2 - k)$ in $\mathbb{R}^5$. Moreover, we will find solutions u which are bounded and have a decay at infinity of the order of $|x|^{-3}$. Then $\psi(r, z) = r^2 u(r, z)$ is such that $\psi(0, z) = 0$ and satisfies the conditions at infinity in (A4.1) provided $u \to 0$ as $|x| \to \infty$. In conclusion, the problem (A4.1) is equivalent to

$$\begin{cases} -\Delta u = g(r^2 u - \frac{1}{2} W r^2 - k) & \text{in } \mathbb{R}^5 \\ u \to 0 & \text{as } |x| \to +\infty. \end{cases} \qquad (A4.2)$$

A4.2 Global existence results

In this section we will outline the main results of [22], where we refer for more details.

Assuming that g is bounded and nondecreasing, we will prove that, for given parameters W and k, (A4.2) has a solution $u \in W^{1,2}(\mathbb{R}^5)$ which is cylindrically symmetric (namely it depends only on r and z) and such that the corresponding vortex core is bounded and not empty, see Theorem A4.2 later on. Roughly, the proof of this existence result is divided into three steps.

(1) (A4.2) is approximated by problems (P_R) on balls $B_R \subset \mathbb{R}^5$ which have a pair of symmetric solutions v_R and u_R. The former is a local minimum of the corresponding Euler functional, the latter is a mountain pass critical point. This result is similar in nature to that discussed in Section 11.3.2.
(2) Uniform bounds for u_R, and for the corresponding vortex core, are provided.
(3) Passing to the limit as $R \to \infty$, one shows that u_R converges to a solution u of (A4.2) with the properties listed before.

It is worth mentioning that v_R 'blows up' as $R \to \infty$: the approximating solutions which are stable for the convergence are the MP solutions, not the minima.

To carry out this program, we begin by considering the problem on the ball $B_R = \{x \in \mathbb{R}^5 : |x| < R\}$ (to simplify notation, we take hereafter $W = 2$)

$$\begin{cases} -\Delta u = g(r^2 u - r^2 - k) & x \in B_R \\ u \to 0 & x \in \partial B_R. \end{cases} \qquad (P_R)$$

We assume that

(g) $g(t) = h(t)f(t)$, where $f \in C(\mathbb{R}), f > 0$ and is not decreasing and bounded,

and set

$$G(r, u) = \int_0^u g(r^2 s - r^2 - k)\, ds.$$

We will work in the Sobolev space $E_R = H_0^1(B_R)$ with scalar product and norm given by

$$(u \mid v)_R = \int_{B_R} \nabla u \cdot \nabla v\, dx, \qquad \|u\|_R^2 = (u \mid u)_R.$$

Define the functional $J_R : E \mapsto \mathbb{R}$ by setting

$$J_R(u) = \tfrac{1}{2}\|u\|_R^2 - \int_{B_R} G(r, u)\, dx.$$

Remark that the functional J_R is Lipschitz continuous, but not C^1. This difficulty is bypassed by using the dual variational principle, like in Section 11.3, or else by using critical point theory for Lipschitz functionals, see [72]. However, as anticipated before, (P_R) shares the same properties as Equation (11.33). Actually, assumption (g) implies that (a′) and (b′) in Section 11.3.2 hold for all $R > 0$. As for the latter, one has to remark that here $\lambda_1(R)$, the first eigenvalue of $-\Delta$ on E_R, tends to zero as $R \to \infty$; nevertheless, since g is bounded, we can take $\alpha = 0$ in (b′) and hence $\alpha < \lambda_1(R)$ holds true for every $R > 0$, as well.

Essentially the same arguments used to prove Theorem 11.23 lead to show that the following facts hold:

(i) for all $R > 0$, J_R is bounded from below, coercive and satisfies the (PS) condition;
(ii) there exists $R_0 > 0$ such that for all $R > R_0$, J_R has the mountain pass geometry.

This last property follows from (i) of Lemma 11.22 by taking R_0 such that $b/a > 2\lambda_1(R_0)\theta$, see the notation in Section 11.3.2. Or else, directly, by taking a fixed $\phi \in E_R$ with $R = 1$ such that $\int_{B_1} G(r, \phi)\,\mathrm{d}x > 0$ and evaluating J_R on the rescaled $\phi_R(x) = \phi(x/R) \in E_R$:

$$J_R(\phi_R) = \tfrac{1}{2}\|\phi_R\|_R^2 - \int_{B_R} G(r, \phi_R(x))\,\mathrm{d}x$$

$$= \tfrac{1}{2}R^3\|\phi\|_1^2 - R^5\int_{B_1} G(Rr, \phi).$$

For $R > 1$ the monotonicity of g implies that $G(Rr, \phi) \geq G(r, \phi)$ and hence

$$J_R(\phi_R) \leq \tfrac{1}{2}R^3\|\phi\|_1^2 - R^5\int_{B_1} G(r, \phi) \to -\infty, \qquad R \to \infty. \qquad (A4.3)$$

According to Theorem 11.23, it follows that for all $R \gg 1$, J_R has a local minimum $v_R > 0$ and a MP critical point $u_R > 0$.

Furthermore, as in Remark 11.24, one can show that v_R, u_R are cylindrically symmetric and

$$\frac{\partial u_R}{\partial z} < 0, \qquad \frac{\partial v_R}{\partial z} < 0, \qquad \forall z > 0. \qquad (A4.4)$$

From (A4.4) we deduce that the boundary of the approximated vortex cores $\{u_R = 1 + k/r^2\}$ and $\{v_R = 1 + k/r^2\}$ have zero measure and thus u_R, v_R solve (P_R) almost everywhere. This concludes the first step.

In view of (A4.3), one has that $J_R(v_R) \to -\infty$ and thus we focus on the MP solutions u_R. We shall show the following result.

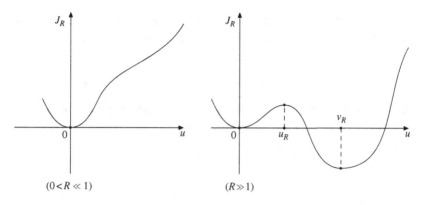

Figure A4.1 Behaviour of J_R.

Lemma A4.1

(i) *There exist $c^* > 0$, a sequence $R_m \to \infty$ and a sequence of symmetric solutions u_m of (P_{R_m}) such that $\|u_m\|_{R_m} \le c^*$.*

(ii) *Let $A_m = \{(r, z) \in B_{R_m} : u_m(r, z) = 1 + k/r^2\}$. There exists $R^* > 0$ such that $A_m \subset B_{R^*}$ for all $m \in \mathbb{N}$.*

Proof. (Sketch) To avoid technicalities, we will suppose that J_R is C^1. The case of nonsmooth functional requires some changes, but the idea of the proof is the same. For $R \ge R_0$, let $c(R)$ denote the MP critical level of J_R:

$$c(R) = \inf_{\gamma \in \Gamma_R} \max_{t \in [0,1]} J_R(\gamma(t)),$$

$$\Gamma_R = \{\gamma \in C([0, 1], E_R) : \gamma(0) = 0, J_R(\gamma(1)) < 0\}.$$

For $R' < R$ we can extend any $u \in E_{R'}$ by setting $u \equiv 0$ for $R' < |x| \le R$. As a consequence, we have $E_{R'} \subset E_R$ as well as $\Gamma(R') \subset \Gamma(R)$. It follows that $c(R)$ is a nonincreasing function of R and hence $c(R)$ is a.e. differentiable. Since

$$\int_{R_0}^{\infty} |c'(R)| \, dR \le c(R_0) - \liminf_{R \to \infty} c(R) < +\infty,$$

then there is a sequence $R_m \to \infty$ such that

$$\lim_{m \to \infty} R_m c'(R_m) = 0. \tag{A4.5}$$

We will use the min-max characterization of $c(R_m)$ to find a sequence satisfying (i). Let us begin by showing that, if $c(R)$ is differentiable at a given R, there exists a $(PS)_{c(R)}$ sequence $u_k \in E_R$ such that

$$\|u_k\|_R^2 \le c_1, \tag{A4.6}$$

with $c_1 > 0$ independent of R. This implies that there exists a MP solution $\tilde{u}_R$ such that $\tilde{u}_R \leq c_2$, with c_2 independent of R, and (i) follows by taking $u_m = \tilde{u}_{R_m}$.

One proves (A4.6) by contradiction. Set, for $\delta > 0$,

$$U_\delta = \{u \in E_R : \|u\|_R^2 \leq c_1 + \delta, \ |J_R(u) - c(R)| \leq \delta\},$$

and suppose that for some $\delta^* > 0$ and any $u \in U_{\delta^*}$ there holds $\|J_R'(u)\|_R \geq \delta^*$. Then there exists $\varepsilon \ll 1$ and a deformation $\eta : E_R \mapsto E_R$ such that $\eta(u) = u$ if $|J_R(u) - c(R)| \geq c(R)$ and

$$J_R(\eta(u)) \leq c(R) - \varepsilon, \qquad \forall u \in U_{\delta^*}, J_R(u) < c(R) + \varepsilon. \tag{A4.7}$$

For $u \in E_R$, set $u_s(x) = u(x/s) \in E_{sR}$. It is possible to show that for $s < 1$, $s \sim 1$, there holds $c(sR) = \inf_{\gamma \in \Gamma_R} \max_{u \in \gamma} E_{sR}(u_s)$. Therefore, we can find $\gamma \in \Gamma_R$ such that

$$\max_{u \in \gamma} J_{sR}(u_s) \leq c(sR) + (1 - s^5). \tag{A4.8}$$

Without loss of generality we can assume that

$$J_R(u) \geq c(R) - (1 - s^5), \tag{A4.9}$$

for all $u \in \gamma$. We claim that $u \in U_{\delta^*}$, for any such $u \in \gamma$ and $s \sim 1$. It is clear that if the claim is true, then applying the deformation η we find a contradiction, yielding (A4.6) and (i). To prove the claim, we have to show that $\forall u \in \gamma$ satisfying (A4.8) and (A4.9) there holds

$$\|u\|_R \leq c_1, \tag{A4.10}$$

and

$$J_R(u) \leq c(R) + \varepsilon. \tag{A4.11}$$

We will be sketchy, referring for the precise estimates to Proposition 3.2 in the aforementioned paper. First, (A4.8) and (A4.9) imply

$$J_{sR}(u_s) - J_R(u) \leq c(sR) - c(R) + 2\varepsilon(1 - s^5). \tag{A4.12}$$

Since $\|u_s\|_{sR}^2 = s^3 \|u\|_R^2$ and, roughly, $\int_{B_{sR}} G(u_s) \sim s^5 \int_{B_R} G(u)$, we get

$$J_{sR}(u_s) - J_R(u) \sim \tfrac{1}{2}(s^3 - 1)\|u\|_R^2 + (1 - s^5)\int_{B_R} G(u).$$

Substituting into (A4.7), dividing by $(1 - s^5)$ and letting $s \uparrow 1$ we deduce

$$-\tfrac{3}{10}\|u\|_R^2 + \int_{B_R} G(u) \leq R|c'(R)| + 2\varepsilon.$$

Then

$$\tfrac{1}{5}\|u\|_R^2 = J_R(u) - \tfrac{3}{10}|u\|_R^2 + \int_{B_R} G(u) \le J_R(u) + R|c'(R)| + 2\varepsilon.$$

Since, roughly, $J_R(u) \sim J_{sR}(u_s) \sim c(sR) \sim c(R) + Rc'(R)$ as $s \sim 1$, we finally find

$$\|u\|_R^2 \le 5 \left[c(R) + 2R|c'(R)| + 3\varepsilon \right].$$

It is clear that this, the monotonicity of $c(R)$ and (A4.5) yield a constant $c_1 > 0$, independent of R, such that (A4.10) holds, and (i) follows.

As for (ii), we first evaluate the measure of the set $A_0 = \{z : u_m(r_0, z) \ge 1/2\}$, where $r_0 > 0$ is fixed and u_m is extended to all $\mathbb{R}^5$ by setting $u_m = 0$ on $|x| > R_m$. One has (C stands for possibly different constants, independent of m and r_0)

$$|A_0| \le C \int_{A_0} u_m^{8/3}(r_0, z) \, dz \le C \int_{-\infty}^{+\infty} u_m^{8/3}(r_0, z) \, dz$$

$$\le C \int_{r_0}^{+\infty} dr \left| \frac{\partial}{\partial r} \int_{-\infty}^{+\infty} u_m^{8/3}(r, z) \, dz \right|$$

$$\le C \int_{r_0}^{+\infty} \int_{-\infty}^{+\infty} |\nabla u_m| \, u_m^{5/3} \, dr \, dz$$

$$\le C r_0^{-3} \int_{r_0}^{+\infty} \int_{-\infty}^{+\infty} |\nabla u_m| \, u_m^{5/3} r^3 \, dr \, dz$$

$$\le C r_0^{-3} \|u_m\|_{R_m} \|u_m\|_{L^{10/3}(B_{R_m})}^{5/3} \le C r_0^{-3} \|u_m\|_{R_m}^{8/3}.$$

Using (i) we find a constant $C > 0$, independent of r_0, such that

$$\text{meas}\{z : u_m(r_0, z) \ge 1/2\} \le C r_0^{-3}, \qquad \forall m. \tag{A4.13}$$

Let r_m be such that for all the points (r, z) in the vortex core A_m there holds $r \le r_m$. Then $u_m(r_m, 0) > 1$. Since $|\Delta u_m| \le \sup g \le \text{constant}$, uniformly with respect to m, it follows from elliptic regularity that $u_m(\cdot + x_m)$ is equibounded in $C_{\text{loc}}^1(\mathbb{R}^5)$ for any x_m. In particular, this implies that $u_m(r_m, z_0) \ge \tfrac{1}{2}$, for some z_0 independent of m. Then we use (A4.13) with $r_0 = r_m$ to infer that $z_0 \le C r_m^{-3}$ which implies the uniform bound for r_m, $r_m \le C z_0^{1/3}$. Similarly, if z_m is the maximum of z such that $(r, z) \in A_m$ for some r, there exists $r_0 > 0$ such that $u_m(r, z) \ge 1/2$ for all $|r - r_m| < r_0$ and (A4.13) yields $z_m \le C r_0^{-3}$ for all m. This proves that A_m are uniformly bounded and completes the proof of the lemma. ∎

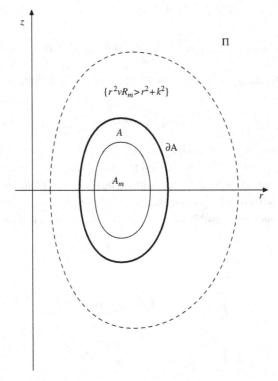

Figure A4.2 The bold line is the boundary of the vortex core A of the solution ψ.
A_m is the core of the approximating mountain-pass solutions u_m. The dashed line
is the boundary of the core of the approximating minimal solutions v_{R_m}.

We are now in a position to carry out step 3, passing to the limit on u_m. From
the uniform bound $|\Delta u_m| \leq \sup g \leq$ constant it follows that there exists u such
that $u_m \to u$ in $C^{1+\alpha}_{\mathrm{loc}}(\mathbb{R}^5)$ and u solves (A4.2). Such a u is strictly positive,
otherwise $u_m < 1$ for $m \gg 1$. This implies that $r^2 u_m < r^2 + k^2$ and thus
from (P_R) it follows that $u_m \equiv 0$, a contradiction. Since $u > 0$, the vortex core
$A = \{(r,z) \in \mathbb{R}^5 : u_m > 1 + k^2/r^2\}$ is not empty. By Lemma A4.1 $A \subset B_{R^*}$.
Moreover, one can show that u is symmetric and $\partial u/\partial z < 0$. From all these
remarks it follows that $\psi = r^2 u$ is a solution of (A4.2) in the sense indicated
before. In conclusion, we can state the following existence result.

Theorem A4.2 *Let* (g) *hold. Then there exists a symmetric* $u \in W^{1,2}(\mathbb{R}^5)$
such that

(i) $\psi = r^2 u \in C^2(\Pi) \cap C^1(\Pi \setminus \partial A)$ *solves* (A4.1) *a.e.,*
(ii) $\psi(r,z) = \psi(r,-z),\ \partial\psi/\partial z < 0\ for\ z > 0,$

(iii) *the corresponding vortex core $A = \{(r,z) \in \Pi : \psi(r,z) > r^2 + k^2\}$ is nonempty and bounded.*

Remark A4.3

(i) The solution u is found as limit of the MP solutions u_m. As for the minima v_R of (P_R), the properties of g imply that $G(r,u) \leq cu^2$, whence

$$\int_{B_R} G(v_R) \leq c\|v_R\|_{L^2(B_R)}^2.$$

From

$$\int_{B_R} G(v_R) = \tfrac{1}{2}\|v_R\|_R^2 - J_R(v_R) \geq -J_R(v_R),$$

and using the fact that $\lim_{R\to\infty} J_R(v_R) = -\infty$, see (A4.3), we get

$$c\|v_R\|_{L^2(B_R)}^2 \geq -J_R(v_R) \to +\infty, \qquad R \to \infty.$$

(ii) The existence of a solution of (A4.2) can also be determined when f is increasing and superlinear. In this case the approximating problems still have the MP solution u_R. Moreover, the proof of the a priori estimates of u_R can be obtained in a more direct way.

(iii) Theorem A4.2 holds for $k = 0$, too. In such a case the vortex is a solid ball. Morover, if $g = h$, the Heaviside function, our solution u coincides with the Hill solution, in view of the uniqueness results mentioned before. ∎

A4.3 Other results

Here we briefly outline some further results on vortex theory.

A4.3.1 A result by Fraenkel and Berger

L. E. Fraenkel and M. Berger in a remarkable paper [96], addressed problem (A4.1), but introducing a further vortex strength parameter $\lambda \in \mathbb{R}$. They proved the following theorem.

Theorem A4.4 *Let $f : \mathbb{R}^+ \mapsto \mathbb{R}$ be a locally Lipschitz, nondecreasing function such that $\exists p > 0$, and $c_1, c_2 > 0$, $0 < f(u) \leq c_1 + c_2|u|^p$ for all $u > 0$.*

Then there exist $\lambda > 0$ and $\psi \in C^2(\Pi) \cap C^1(\Pi \setminus \partial A)$ such that

(i) $\psi(r, z) = \psi(r, -z)$, $\partial \psi / \partial z < 0$ for $z > 0$,
(ii) there holds

$$-L\psi = \lambda r^2 g(\psi - \tfrac{1}{2} W r^2 - k), \qquad a.e. \text{ in } \Pi,$$

together with the boundary conditions as in (A4.1),
(iii) the corresponding vortex core A is nonempty and bounded.

The proof is also based on an approximating procedure with problems on balls B_R. Moreover, the fact that g is not continuous is surmounted by approximating g with a continuous piecewise linear function g_δ, with Lipschitz primitive G_δ. Looking for

$$\max\left\{ \int_{B_R} G_\delta(r^2 u - r^2 - k^2) : u \in E_R, \ \|u\|_R^2 = \eta \right\},$$

one finds a $u_{R,\delta}$, where the maximum is achieved, satisfying $-\Delta u = \lambda_{R,\delta} g_\delta(r^2 u - r^2 - k^2)$, for some Lagrange multiplier $\lambda_{R,\delta} \in \mathbb{R}$. Using the fact that $u_{R,\delta}$ has been found as a maximum, one shows that $u_{R,\delta}$ is symmetric and satisfies $\partial u_{R,\delta}/\partial z < 0$ for $z > 0$. Furthermore, these properties and $\|u\|_R^2 = \eta$ allow one to show that the approximated vortex core is uniformly bounded, like in Lemma A4.1(ii). Then one can still pass to the limit as $\delta \to 0$ and $R \to \infty$, yielding a pair (λ, u) satisfying (i)–(iii) of the theorem.

A4.3.2 Bifurcation from the Hill spherical vortex

Another interesting question is to see whether for k small there exists a vortex ring that bifurcates from the Hill spherical one. Here, the constants λ and W are supposed to be fixed and k plays the role of bifurcation parameter. Assuming that $g = h$, the Heaviside function, local bifurcation has been proved for example in [138], while the following global bifurcation result was established in [31]. Let Σ denote the set of $(k, \psi) \in \mathbb{R} \times C^2(\Pi) \cap C^1(\Pi \setminus \partial A)$ such that ψ is a positive solution of (A4.1).

Theorem A4.5 *There exists an unbounded, connected component $\Sigma_0 \subset \Sigma$ bifurcating from the Hill solution: $(0, \psi_H) \in \overline{\Sigma}_0$. Moreover, ψ satisfies (ii) and (iii) of Theorem A4.2.*

The proof relies once more on an approximation of (A4.2) with problems on balls B_R. We put $W = 2$ and $\lambda = 1$. Let H_R denote the completion with respect to the norm induced by

$$\langle u, v \rangle = \frac{1}{2\pi^2} \int_{B_R} \nabla u(x) \cdot \nabla v(x) \, dx,$$

of $u \in C_0^\infty(B_R)$ which are cylindrically symmetric and even in z. Then an $u \in H_R$ satisfying (P_R) is sought as a solution of

$$\langle u, v \rangle = \frac{1}{2\pi^2} \int_{B_R} h(r^2 u - r^2 - k^2) v \, dx, \quad \forall v \in H_R.$$

To transform this integral equation into a fixed point problem in H_R it suffices to define the nonlinear operator $N(R, k; \cdot) : H_R \mapsto H_R$ by setting $N(R, k : u) = w$, where $w \in H_R$ is the unique solution of

$$\langle w, v \rangle = \frac{1}{2\pi^2} \int_{B_R} h(r^2 u - r^2 - k^2) v \, dx, \quad \forall v \in H_R.$$

With this notation, (P_R) is equivalent to the functional equation $u = N(R, k; u)$. As in Theorem A4.4, one needs a further approximation of h with smooth nonlinearities h_δ, yielding an equation of the form $u - N(R, k, \delta; u) = 0$, with N compact and of class C^1. Taking advantage of the fact that the nonlinearity is exactly the Heaviside function, one finds that for $k = 0$, the equation $u - N(R, 0, \delta; u) = 0$ has a solution $u_{R,\delta}$ such that its local degree is -1. Then the global properties of the topological degree yield, like in the Rabinowitz global bifurcation theorem, a global branch $\Sigma_{R,\delta}$ of nontrivial solutions of $u = N(R, k, \delta; u)$ emanating from $k = 0$ and $u = u_{R,\delta}$. Finally, one can pass to the limit by using, as in Section A3.1, the Whyburn topological Lemma A3.1. This limiting procedure can be carried out because one proves that the solutions on the approximating branches $\Sigma_{R,\delta}$ possess properties similar to those established in Lemma A4.1.

Remark A4.6 Although the branch Σ_0 is proved to be unbounded, it is not known what is its behaviour. It is conjectured that Σ_0 is unbounded in the k direction so that one could find a solution of (A4.1) for all $k \geq 0$. Another open problem is to extend Theorem A4.5 to any nondecreasing nonlinearity g, possibly different from the Heaviside function. ∎

Appendix 5

Perturbation methods

In this appendix we discuss a general perturbation method which is useful for treating problems of different natures. This relies on a combination of the Lyapunov–Schmidt reduction together with variational techniques. In order to keep the appendix short, we sketch only the main ideas of the construction, and treat only a few examples. We refer the interested reader to the recent monograph by the authors [19], where the subject is investigated in more detail and several references are given.

A5.1 An abstract result

Given a Hilbert space E, we consider a class of functionals of the type

$$I_\varepsilon(u) = I_0(u) + \varepsilon G(u), \qquad (A5.1)$$

where $I_0 \in C^2(E, \mathbb{R})$ is considered the *unperturbed functional* and $G \in C^2(E, \mathbb{R})$ the *perturbation*. Typically, we assume that the critical points of I_0 correspond to solutions of some autonomous problem in $\mathbb{R}^n$ which possesses some group of invariance, for example given by the translations in space or sometimes also by dilation. For this reason critical points of I_0 usually arise in *manifolds*, and the present goal is to understand the effect of the perturbation εG on the structure of such manifolds, and in particular the persistence of some critical points.

We assume that there exists a d dimensional smooth manifold Z, with $0 < d = \dim(Z) < \infty$, such that every $z \in Z$ is a critical point of I_0. The set Z is called a *critical manifold* for I_0.

Let $T_z Z$ denote the tangent space to Z at z. If Z is a critical manifold for I_0, then for every $z \in Z$ one has $I_0'(z) = 0$. Differentiating this identity along Z, we also get

$$(I_0''(z)[v]|\phi) = 0, \quad \forall v \in T_z Z, \ \forall \phi \in E,$$

and this shows immediately that $T_z Z \subseteq \text{Ker}[I_0''(z)]$. In particular, $I_0''(z)$ has a nontrivial kernel (whose dimension is at least d) and hence every $z \in Z$ is a degenerate critical point of I_0. We shall require that the dimension of $\mathrm{Ker}[I_0''(z)]$ is exactly d. Precisely we will assume the following condition

$$T_z Z = \mathrm{Ker}[I_0''(z)], \quad \forall z \in Z. \tag{ND}$$

In addition to (**ND**) we will assume that

$$\text{for all } z \in Z, I_0''(z) \text{ is a Fredholm map}^1 \text{of index } 0. \tag{Fr}$$

Definition A5.1 *A critical manifold Z will be called nondegenerate, ND in short, if* (**ND**) *and* (**Fr**) *hold.*

In the spirit of Chapter 2, and in particular Section 2.2, solutions of $I_\varepsilon' = 0$ will be found through a reduction to a finite dimensional problem. Let us define $W = (T_z Z)^\perp$ and let $(q_i)_{1 \le i \le d}$ be an orthonormal set which spans $T_z Z$. Below, we will assume understood that Z admits a local C^2 parametrization $z = z_\xi$, $\xi \in \mathbb{R}^d$. Furthermore, we also suppose that $q_i = \partial_{\xi_i} z_\xi / \| \partial_{\xi_i} z_\xi \|$.

We look for critical points of I_ε in the form $u = z + w$ with $z \in Z$ and $w \in W$. If $P : E \to W$ denotes the orthogonal projection onto W, the equation $I_\varepsilon'(z + w) = 0$ is equivalent to the following system

$$\begin{cases} P I_\varepsilon'(z + w) = 0 & \text{(auxiliary equation)} \\ Q I_\varepsilon'(z + w) = 0 & \text{(bifurcation equation)} \end{cases} \tag{A5.2}$$

where $Q = (\mathrm{Id} - P)$. We show next that the auxiliary equation is solvable under the above assumption.

Lemma A5.2 *Let* (**ND**) *and* (**Fr**) *hold. Given any compact subset Z_c of Z there exists $\varepsilon_0 > 0$ such that: for all $|\varepsilon| < \varepsilon_0$, for all $z \in Z_c$, the auxiliary equation in* (A5.2) *has a unique solution $w = w_\varepsilon(z)$ such that:*

(i) *$w_\varepsilon(z)$ belongs to $W = (T_z Z)^\perp$ and is of class C^1 with respect to $z \in Z_c$ and $w_\varepsilon(z) \to 0$ as $|\varepsilon| \to 0$, uniformly with respect to $z \in Z_c$, together with its derivative with respect to z, w_ε';*

(ii) *more precisley one has that $\|w_\varepsilon(z)\| = O(\varepsilon)$ as $\varepsilon \to 0$, for all $z \in Z_c$.*

Proof. Let $F : \mathbb{R} \times Z \times W \to W$ be defined by setting

$$F(\varepsilon, z, w) = P I_0'(z + w) + \varepsilon P G'(z + w).$$

Clearly F is of class C^1 and one has $F(0, z, 0) = 0$ for every $z \in Z$. Moreover, letting $D_w F(0, z, 0)$ denote the partial derivative with respect to w evaluated at

---

^{1} A linear map $A \in L(\mathcal{H}, \mathcal{H})$ is *Fredholm* if the kernel is finite dimensional and the image is closed and has finite codimension. The *index* of A is $\dim(\mathrm{Ker}[A]) - \mathrm{codim}(\mathrm{Im}[A])$

$(0, z, 0)$, from (**ND**) and (**Fr**) one easily finds that $D_w F(0, z, 0)$ is invertible as a map from W into itself.

Then, by the implicit function theorem for all $z \in Z_c$ we get a solution $w_\varepsilon = w_\varepsilon(z) \in W$ satisfying (i). Let us point out explicitly that w'_ε (the derivative of w_ε with respect to ξ) for $\varepsilon = 0$ is zero. Actually w'_ε satisfies

$$PI_0''(z + w_\varepsilon)[q + w'_\varepsilon] + \varepsilon PG''(z + w_\varepsilon)[q + w'_\varepsilon] = 0,$$

where $q = \sum_{i=1}^d \alpha_i q_i \in T_z Z$. Then for $\varepsilon = 0$ we get $PI_0''(z)[q + w'_0] = 0$. Since $q \in T_z Z \subseteq \mathrm{Ker}[I_0''(z)]$, then $PI_0''(z)[q] = 0$, and this implies $w'_0 = 0$.

Let us now show (ii). Setting $\widetilde{w}_\varepsilon = \varepsilon^{-1} w_\varepsilon(z)$ we have to prove that $\|\widetilde{w}_\varepsilon\| \le$ constant for $|\varepsilon|$ small. Recall that w_ε satisfies $PI_\varepsilon'(z + w_\varepsilon) = 0$; using a Taylor expansion we find

$$I_\varepsilon'(z + w_\varepsilon) = I_0'(z + w_\varepsilon) + \varepsilon G'(z + w_\varepsilon)$$
$$= I_0'(z) + I_0''(z)[w_\varepsilon] + \varepsilon G'(z) + \varepsilon G''(z)[w_\varepsilon] + o(\|w_\varepsilon\|).$$

Since $I_0'(z) = 0$ we get $I_\varepsilon'(z+w_\varepsilon) = I_0''(z)[w_\varepsilon] + \varepsilon G'(z) + \varepsilon G''(z)[w_\varepsilon] + o(\|w_\varepsilon\|)$, and the equation $PI_\varepsilon'(z + w_\varepsilon) = 0$ becomes

$$PI_0''(z)[w_\varepsilon] + \varepsilon PG'(z) + \varepsilon PG''(z)[w_\varepsilon] + o(\|w_\varepsilon\|) = 0. \tag{A5.3}$$

Dividing by ε we infer that $\widetilde{w}_\varepsilon$ verifies $PI_0''(z)[\widetilde{w}_\varepsilon] + PG'(z) + PG''(z)[w_\varepsilon] + \varepsilon^{-1} o(\|w_\varepsilon\|) = 0$. Since $\varepsilon^{-1} o(\|w_\varepsilon\|) = o(\|\widetilde{w}_\varepsilon\|)$ we deduce $PI_0''(z)[\widetilde{w}_\varepsilon] = -PG'(z) - PG''(z)[w_\varepsilon] + o(\|\widetilde{w}_\varepsilon\|)$. Recalling that $w_\varepsilon \to 0$ as $|\varepsilon| \to 0$, we get

$$PI_0''(z)[\widetilde{w}_\varepsilon] \to -PG'(z), \qquad \text{as } \varepsilon \to 0,$$

and this implies (ii). ∎

We next give a criterion for the solvability of the bifurcation equation in (A5.2). If $w_\varepsilon(z)$ is the function constructed in Lemma A5.2, we reduce the existence of critical points of I_ε to a finite dimensional problem by defining the *reduced functional* $\Phi_\varepsilon : Z_c \to \mathbb{R}$ as

$$\Phi_\varepsilon(z) = I_\varepsilon(z + w_\varepsilon(z)). \tag{A5.4}$$

Proposition A5.3 *Let $I_0, G \in C^2(E, \mathbb{R})$ and suppose that I_0 has a smooth critical manifold Z which is nondegenerate, in the sense that (**ND**) and (**Fr**) hold. Given a compact subset Z_c of Z, let us assume that Φ_ε has, for $|\varepsilon|$ sufficiently small, a critical point $z_\varepsilon \in Z_c$. Then $u_\varepsilon = z_\varepsilon + w_\varepsilon(z_\varepsilon)$ is a critical point of $I_\varepsilon = I_0 + \varepsilon G$.*

Proof. We use the preceding notation and, to be short, we write below D_i for D_{ξ_i}, etc. Let ξ_ε be such that $z_\varepsilon = z_{\xi_\varepsilon}$, and set $q_i^\varepsilon = \partial z / \partial \xi_i|_{\xi_\varepsilon}$. Without loss of generality we can assume that $z_\varepsilon \to z^* \in Z_c$ as $\varepsilon \to 0$. From Lemma A5.2 we infer that there exists $\varepsilon_0 > 0$ such that Lemma A5.2 holds. In particular, from statement (i) of Lemma A5.2 and by continuity, one has

$$\lim_{|\varepsilon| \to 0} (D_i w_\varepsilon(z_\varepsilon) \mid q_j^\varepsilon) = 0, \quad i, j = 1, \dots, d.$$

Let us consider the matrix $B^\varepsilon = (b_{ij}^\varepsilon)_{ij}$, where

$$b_{ij}^\varepsilon = (D_i w_\varepsilon(z_\varepsilon) \mid q_j^\varepsilon).$$

From the above arguments we can choose $0 < \varepsilon_1 < \varepsilon_0$, such that

$$|\det(B^\varepsilon)| < 1, \quad \forall \, |\varepsilon| < \varepsilon_1. \tag{A5.5}$$

Fix $\varepsilon > 0$ such that $|\varepsilon| < \min\{\varepsilon_0, \varepsilon_1\}$. Since z_ε is a critical point of Φ_ε we get

$$\left(I_\varepsilon'(z_\varepsilon + w_\varepsilon(z_\varepsilon)) \mid q_i^\varepsilon + D_i w_\varepsilon(z_\varepsilon)\right) = 0, \quad i = 1, \dots, d.$$

From $PI_\varepsilon'(z + w_\varepsilon(z_\varepsilon)) = 0$ we deduce that $I_\varepsilon'(z_\varepsilon + w_\varepsilon(z_\varepsilon)) = \sum A_{i,\varepsilon} q_i^\varepsilon$, where

$$A_{i,\varepsilon} = (I_\varepsilon'(z_\varepsilon + w_\varepsilon(z_\varepsilon)) \mid q_i^\varepsilon).$$

Then we find

$$\left(\sum_j A_{j,\varepsilon} \, q_j^\varepsilon \mid q_i^\varepsilon + D_i w_\varepsilon(z_\varepsilon)\right) = 0, \quad i = 1, \dots, d,$$

namely

$$A_{i,\varepsilon} + \sum_j A_{j,\varepsilon}(q_j^\varepsilon \mid D_i w_\varepsilon(z_\varepsilon)) = A_{i,\varepsilon} + \sum_j A_{j,\varepsilon} b_{ij}^\varepsilon = 0, \quad i = 1, \dots, d. \tag{A5.6}$$

Equation (A5.6) is a $(d \times d)$ linear system whose coefficient matrix $\mathrm{Id}_{\mathbb{R}^d} + B^\varepsilon$ has entries $\delta_{ij} + b_{ij}^\varepsilon$, where δ_{ij} is the Kronecker symbol and b_{ij}^ε are defined above and satisfy (A5.5). Then, for $|\varepsilon| < \varepsilon_1$, the matrix $\mathrm{Id}_{\mathbb{R}^d} + B^\varepsilon$ is invertible. Thus (A5.6) has the trivial solution only: $A_{i,\varepsilon} = 0$ for all $i = 1, \dots, d$. Since the $A_{i,\varepsilon}$ are the components of $\Phi_\varepsilon(z_\varepsilon)$, the conclusion follows. ∎

Some sufficient conditions for finding critical points of Φ_ε are given in the next proposition.

Proposition A5.4 *Suppose I_ε is as in (A5.1), and that I_0, G, Z, Z_c, $w_\varepsilon(z)$ are as in Proposition A5.3. Define the functional $\Gamma : Z \to \mathbb{R}$ as $\Gamma(z) = G(z)$ for every $z \in Z$. Suppose $\bar{z}$ is a local strict maximum or minimum in the interior of Z_c, or that there exists an open set Ω in the interior of Z_c such that $\deg(\Gamma', \Omega, 0) \neq 0$. Then Φ_ε possesses a critical point in Z_c.*

Proof. We have clearly $\Phi_\varepsilon(z) = I_0(z + w_\varepsilon(z)) + \varepsilon G(z + w_\varepsilon(z))$: evaluating each term separately, for the first we have

$$I_0(z + w_\varepsilon(z)) = I_0(z) + (I_0'(z) \mid w_\varepsilon(z)) + o(\|w_\varepsilon(z)\|).$$

Since $I_0'(z) = 0$ we get

$$I_0(z + w_\varepsilon(z)) = c_0 + o(\|w_\varepsilon(z)\|), \tag{A5.7}$$

where $c_0 = I_0|_Z$. Similarly, one has

$$G(z + w_\varepsilon(z)) = G(z) + (G'(z) \mid w_\varepsilon(z)) + o(\|w_\varepsilon(z)\|)$$

$$= G(z) + O(\|w_\varepsilon(z)\|). \tag{A5.8}$$

Putting together (A5.7) and (A5.8) we infer that

$$\Phi_\varepsilon(z) = c_0 + \varepsilon\,[G(z) + O(\|w_\varepsilon(z)\|)] + o(\|w_\varepsilon(z)\|).$$

Since $\|w_\varepsilon(z)\| = O(\varepsilon)$, see Lemma A5.2(ii), we deduce that

$$\Phi_\varepsilon(z) = c_0 + \varepsilon G(z) + o(\varepsilon), \quad \text{where } c_0 = I_0(z). \tag{A5.9}$$

Reasoning in a similar way, one can also prove that

$$\nabla_\xi \Phi_\varepsilon(z) = \varepsilon \nabla_\xi G(z) + o(\varepsilon), \tag{A5.10}$$

where, we recall, the variables ξ are a local parameterization of Z near z. Then the conclusion of the lemma follows immediately from (A5.9) and (A5.10), by the stability properties of strict local maxima (or minima) and of the degree. ∎

Remark A5.5

(a) Suppose that the functional Γ possesses multiple strict maxima or minima $\bar{z}_i$, or that there are different sets Ω_i for which the degrees $\deg(\Gamma, \Omega_i, 0) \neq 0$. Then Φ_ε, and hence I_ε, possesses multiple critical points, localized near the $\bar{z}_i$ or near the Ω_i.

(b) If the critical point $\bar{z}$ is an isolated local extremum of Γ, or if the index $i(\Gamma, \bar{z}) \neq 0$, since we can use formulas (A5.9), (A5.10) on arbitrarily small neighbourhoods of $\bar{z}$, the corresponding critical points of Φ_ε, or of I_ε, converge to $\bar{z}$ as ε tends to zero. ∎

In the next sections we consider applications of the abstract method to some elliptic problems in $\mathbb{R}^n$ or in a bounded domain with a nonlinear term which is a subcritical power. These are only some examples of the possible applications of this method, which is very versatile and allows us to treat a wide class of problems, including cases where a dilation-invariance is present, for example when the critical exponent appears or when the equations arise in a geometric context.

A5.2 Elliptic equations on $\mathbb{R}^n$

We apply in this section the abstract method just derived to study some semi-linear elliptic equations on $\mathbb{R}^n$. As we shall see, these problems are of the form of those considered in Appendix 2. Although the method applies only to some special cases (namely those for which the corresponding functionals are as in (A5.1)), the arguments are quite simple and there is no need of the Palais–Smale condition. Furthermore, one can use a unified approach to find critical points which might have different Morse indices, which is in striking contrast with the methods discussed in the above appendix.

We will consider the elliptic problem

$$\begin{cases} -\Delta u + u = (1 + \varepsilon h(x))u^p, \\ u \in W^{1,2}(\mathbb{R}^n), \qquad\qquad u > 0, \end{cases} \qquad (P_\varepsilon)$$

where $n \geq 3$, $1 < p < (n+2)/(n-2)$ and h is a continuous function on $\mathbb{R}^n$ tending to zero at infinity. Let $E = W^{1,2}(\mathbb{R}^n)$: solutions of (P_ε) are the critical points of the Euler functional $I_\varepsilon : E \mapsto \mathbb{R}$, $I_\varepsilon = I_0 + \varepsilon G$, where

$$I_0(u) = \frac{1}{2}\|u\|^2 - \frac{1}{p+1}\int_{\mathbb{R}^n} u_+^{p+1}\,dx, \qquad G(u) = -\frac{1}{p+1}\int_{\mathbb{R}^n} h(x)\,u_+^{p+1}\,dx.$$

The unperturbed problem $I_0'(u) = 0$ is equivalent to the elliptic equation

$$-\Delta u + u = u^p, \qquad u \in E, \qquad (A5.11)$$

which admits a positive radial solution u_0, see Chapter 11. Since (A5.11) is translation invariant, it follows that, for any $\xi \in \mathbb{R}^n$, $z_\xi(x) := u_0(x - \xi)$ is also a solution of (A5.11). In other words, I_0 has a (noncompact) critical manifold given by

$$Z = \{z_\xi(x) : \xi \in \mathbb{R}^n\} \simeq \mathbb{R}^n.$$

It can be shown, see for example Chapter 4 in [19], that the properties (Fr) and (ND) hold in this particular case, so in order to find critical points of I_ε we can apply Propositions A5.3 and A5.4. Therefore, we are reduced to studying the properties of the functional Γ, which is given by

$$\Gamma(\xi) = G(z_\xi) = -\frac{1}{p+1}\int_{\mathbb{R}^n} h(x)u_0^{p+1}(x - \xi)\,dx, \qquad \xi \in \mathbb{R}^n.$$

We first show the following result.

Lemma A5.6 *Assume h is continuous on $\mathbb{R}^n$ and that it tends to zero at infinity. Then* $\lim_{|\xi|\to\infty} \Gamma(\xi) = 0$.

Proof. Given $\rho > 0$ we set

$$\Gamma_\rho(\xi) := \int_{|x|<\rho} h(x)u_0^{p+1}(x-\xi)\,dx, \quad \Gamma_\rho^*(\xi) = \int_{|x|>\rho} h(x)u_0^{p+1}(x-\xi)\,dx,$$

in such a way that $\Gamma(\xi) = -1/(p+1)[\Gamma_\rho(\xi) + \Gamma_\rho^*(\xi)]$. Since u_0 tends to zero at infinity, it follows immediately that $\Gamma_\rho(\xi)$ tends to zero as $|\xi|$ tends to infinity. Furthermore, since also h tends to zero at infinity, we have that $\Gamma_\rho^*(\xi) = o_\rho(1)$, where $o_\rho(1)$ tends to zero as ρ tends to infinity. By the arbitrarity of ρ we obtain immediately the conclusion.   ∎

The previous lemma allows us to prove the existence of solutions of $(\boldsymbol{P}_\varepsilon)$, provided $\Gamma \not\equiv 0$.

Theorem A5.7 *Let $h \in C_0(\mathbb{R}^n)$, and suppose that $\int_{\mathbb{R}^n} h(x)u_0(x-\xi)^{p+1}(x) \neq 0$ for some $\xi \in \mathbb{R}^n$. Then $(\boldsymbol{P}_\varepsilon)$ has a solution for $|\varepsilon|$ small enough.*

Proof. From Lemma A5.6 it follows immediately that Γ possesses either a global maximum or minimum on $\mathbb{R}^n$, so the conclusion follows immediately from Proposition A5.4.   ∎

Remark A5.8

(a) A condition which implies $\Gamma \not\equiv 0$ is that h has constant sign in $\mathbb{R}^n$, in accordance with Theorems A2.2 and A2.3.

(b) There are cases in which $(\boldsymbol{P}_\varepsilon)$ has multiple solutions. For example, if

$$\int_{\mathbb{R}^n} h(x)u_0^{p+1}(x) = 0, \quad \int_{\mathbb{R}^n} D_ih(x)u_0^{p+1}(x) \neq 0, \quad \text{for some } i = 1, 2, \dots, n,$$

then $\Gamma(0) = 0$ while $D_i\Gamma(0) \neq 0$. Thus Γ possesses both a positive maximum and a negative minimum, which give rise to a pair of distinct solutions of $(\boldsymbol{P}_\varepsilon)$, for $|\varepsilon|$ small, see Remark A5.5(a).   ∎

We state next a generalization of Theorem A5.7, where indeed no assumptions on Γ are required.

Theorem A5.9 *Let $h \in C_0(\mathbb{R}^n)$. Then $(\boldsymbol{P}_\varepsilon)$ has a solution for $|\varepsilon|$ small enough.*

We do not discuss here the proof of this result, which requires some technicalities, but we limit ourselves to discussing the main ideas. First of all, in the proof of Lemma A5.2 one can substitute the implicit function theorem by a fixed point argument. The advantage is that under the assumption of Theorem A5.9 this can be worked out uniformly in $\xi \in \mathbb{R}^n$ (for ε small), so we do not need to restrict our attention to a fixed compact set of Z. Furthermore, this method also gives some *quantitative* estimates on $w_\varepsilon(z)$, which now is globally defined on Z. It is possible to prove that, for a fixed small ε, the norm of $w_\varepsilon(z_\xi)$ tends to zero uniformly for $|\xi| \to +\infty$. As a consequence, one can prove that $\Phi_\varepsilon(\xi) \to c_0$ as $|\xi| \to +\infty$, where $c_0 = I_0|_Z$. This implies that Φ_ε is either constant on $\mathbb{R}^n$, or it must possess a global maximum or minimum, so we can apply Proposition A5.3 to deduce the existence of a critical point of I_ε. The peculiarity of this approach is that one uses the full strength of Proposition A5.3, bypassing the expansions of Φ_ε in terms of Γ.

A5.3 Semiclassical states of nonlinear Schrödinger equations

We consider now the following equation

$$\begin{cases} -\varepsilon^2 \Delta u + V(x)u = u^p & \text{in } \mathbb{R}^n \\ u > 0 & u \in W^{1,2}(\mathbb{R}^n), \end{cases} \tag{A5.12}$$

where $p > 1$ is subcritical and V is a smooth bounded function. Problem (A5.12) arises in the study of the nonlinear Schrödinger equation

$$i\hbar \frac{\partial \psi}{\partial t} = -\hbar^2 \Delta \psi + \tilde{V}(x)\psi - |\psi|^{p-1}\psi \quad \text{in } \mathbb{R}^n,$$

where $\psi : \mathbb{R} \times \mathbb{R}^n \to \mathbb{C}$ is the *wave function*, $\tilde{V} : \mathbb{R}^n \to \mathbb{R}$ is the potential and $\hbar$ is the *Planck constant*. Looking for *standing wave solutions*, namely solutions of the form $\psi(t, x) = e^{-i\omega t/\hbar} u(x)$, the function u is easily seen to satisfy (A5.12), with $V = \tilde{V} - \omega$ and $\varepsilon = \hbar$. Since $\varepsilon = \hbar$ is very small, one is interested is the asymptotic behaviour of solutions in the limit $\varepsilon \to 0$, the so-called *semiclassical limit*. Typically, in this limit solutions concentrate at some points of $\mathbb{R}^n$, which turn out to be stationary for V, mimicking the behaviour of classical particles at equilibria of the potential. Below, we assume the following conditions on the potential V:

(V1) $V \in C^2(\mathbb{R}^n)$, and $\|V\|_{C^2(\mathbb{R}^n)} < +\infty$;
(V2) $\lambda_0^2 = \inf_{\mathbb{R}^n} V > 0$.

We say that a solution v_ε of (A5.12) *concentrates* at x_0 (as $\varepsilon \to 0$) provided

$$\forall \delta > 0, \quad \exists \varepsilon_0 > 0, \; R > 0 : v_\varepsilon(x) \le \delta, \; \forall |x - x_0| \ge \varepsilon R, \; \varepsilon < \varepsilon_0. \tag{A5.13}$$

The main purpose of this section is to describe the following result. Again, we only sketch the main ideas, referring to Chapter 8 in [19] for details and further results.

Theorem A5.10 *Let* (V1) *and* (V2) *hold, and suppose* x_0 *is a nondegenerate critical point of* V, *namely for which* $V''(x_0)$ *is nonsingular. Then there exists a solution* $\bar{v}_\varepsilon$ *of* (A5.12) *which concentrates at* x_0 *as* $\varepsilon \to 0$.

To simplify the notation (and without losing generality) we will suppose that $x_0 = 0$ and that $V(0) = 1$. To frame (A5.12) in the abstract setting, we first make the change of variable $x \mapsto \varepsilon x$ and rewrite equation (A5.12) as

$$\begin{cases} -\Delta u + V(\varepsilon x)u = u^p & \text{in } \mathbb{R}^n \\ u > 0 & u \in W^{1,2}(\mathbb{R}^n). \end{cases} \tag{A5.14}$$

If $u_\varepsilon(x)$ is a solution of (A5.14) then $v_\varepsilon(x) := u_\varepsilon(x/\varepsilon)$ solves (A5.14). We set again $E = W^{1,2}(\mathbb{R}^n)$ and consider the functional $I_\varepsilon \in C^2(E, \mathbb{R})$,

$$I_\varepsilon(u) = \frac{1}{2} \int_{\mathbb{R}^n} (|\nabla u|^2 + V(\varepsilon x)u^2) - \frac{1}{p+1} \int_{\mathbb{R}^n} |u|^{p+1}. \tag{A5.15}$$

As usual, we endow E with the norm

$$\|u\|^2 = \int_{\mathbb{R}^n} (|\nabla u|^2 + V(\varepsilon x)u^2) \, dx.$$

With this notation, for $\varepsilon = 0$ the functional I_ε takes the form

$$I_0(u) = \frac{1}{2} \|u\|^2 - \frac{1}{p+1} \int_{\mathbb{R}^n} |u|^{p+1}.$$

Let us highlight that I_0 plays the role of the unperturbed functional by writing

$$I_\varepsilon(u) = I_0(u) + \frac{1}{2} \int_{\mathbb{R}^n} (V(\varepsilon x) - 1)u^2 \, dx \equiv I_0(u) + G(\varepsilon, u).$$

Obviously, for any fixed $u \in E$, we have $G(\varepsilon, u) \to 0$ as $\varepsilon \to 0$ and hence we can still view I_ε as a perturbation of I_0.

We define u_0 and Z as in the previous section. However, the abstract method described before cannot be applied in a straightforward way. Notice that, in general, $G''(\varepsilon, u)$ does not tend to zero as $\varepsilon \to 0$. To see this, let us consider a sequence $v_j \in E$ with compact support contained in $\{x \in \mathbb{R}^n : |x| > 1/j\}$. If, for example, the potential V is such that $V(x) - 1 \equiv c > 0$ for all $|x| \geq 1$, then evaluating $G''(\varepsilon, u)[v_j]^2$ for $\varepsilon = 1/j$ we find

$$G''(\varepsilon, u)[v_j]^2 = \int_{\mathbb{R}^n} (V(\varepsilon x) - 1)v_j^2 \, dx = c\|v_j\|^2.$$

However, the first part of the abstract procedure can still be carried out. Denoting by P the orthogonal projection onto $W = (T_z Z)^\perp$, we look for solutions $u = z_\xi + w$, with $z_\xi \in Z$ and $w \in W$, of the system

$$\begin{cases} PI_\varepsilon'(z_\xi + w) = 0, \\ (I - P)I_\varepsilon'(z_\xi + w) = 0, \end{cases}$$

which is clearly equivalent to $I'_\varepsilon(z_\xi + w) = 0$. At this point, instead of invoking the implicit function theorem, we write $PI'_\varepsilon(z_\xi + w) = PI'_\varepsilon(z_\xi) + PD^2 I_\varepsilon(z_\xi)[w] + R(z_\xi, w)$, where $R(z_\xi, w) = o(\|w\|)$, uniformly with respect to $z_\xi \in Z$ for bounded $|\xi|$. Next, using $(\textbf{Fr})$ and $(\textbf{ND})$ one can show that $PI''_\varepsilon(z_\xi)$ is uniformly invertible for ξ belonging to a fixed bounded set of $\mathbb{R}^n$. Setting $A_{\varepsilon,\xi} = -(PI''_\varepsilon(z_\xi))^{-1}$, the equation $PI'_\varepsilon(z_\xi + w) = 0$ can be written in the form

$$w = A_{\varepsilon,\xi}(PI'_\varepsilon(z_\xi) + R(z_\xi, w)) := N_{\varepsilon,\xi}(w).$$

It is also possible to show that $N_{\varepsilon,\xi}$ is a contraction in some ball of W provided ε is sufficiently small. This allows us to solve the auxiliary equation finding a solution $w_\varepsilon(z_\xi)$ which is of class C^1 with respect to ξ. Furthermore, since $V'(0) = 0$, one finds that $w_\varepsilon(z_\xi) = O(\varepsilon^2)$, uniformly with respect to ξ in a bounded set. At this point we can repeat the expansion of Φ_ε obtaining again

$$\Phi_\varepsilon(\xi) = c_0 + \varepsilon^2 \Gamma(\xi) + o(\varepsilon^2),$$

where $c_0 = I_0(U)$ and

$$\Gamma(\xi) = \frac{1}{2} \int_{\mathbb{R}^n} \langle V''(0)x, x\rangle U^2(x - \xi)\, dx.$$

A straight calculation yields

$$\Gamma(\xi) = \frac{1}{2} \int_{\mathbb{R}^n} \langle V''(0)(y + \xi), (y + \xi)\rangle U^2(y)\, dy$$

$$= \frac{1}{2} \int_{\mathbb{R}^n} \langle V''(0)y, y\rangle U^2(y)\, dy + \frac{1}{2} \int_{\mathbb{R}^n} \langle V''(0)\xi, \xi\rangle U^2(y)\, dy$$

$$= c_1 + c_2 \langle V''(0)\xi, \xi\rangle,$$

where

$$c_1 = \frac{1}{2} \int_{\mathbb{R}^n} \langle V''(0)y, y\rangle U^2(y)\, dy, \qquad c_2 = \frac{1}{2} \int_{\mathbb{R}^n} U^2(x)\, dx.$$

Then $\xi = 0$ is a nondegenerate critical point of Γ and therefore, from the general theory, it follows that for $\varepsilon \ll 1$, I_ε has a critical point $u_\varepsilon = z_{\xi_\varepsilon} + w_\varepsilon(z_{\xi_\varepsilon})$, with $\xi_\varepsilon \to 0$ as $\varepsilon \to 0$. In conclusion, coming back to the solutions v_ε of $(A5.12)$, we find that this equation has a solution $\bar{v}_\varepsilon(x) \sim U((x - \xi_\varepsilon)/\varepsilon)$ that concentrates at $x = 0$, proving Theorem A5.10.

Theorem A5.10 admits several extensions. For example, with a generalization of the previous abstract method, one can construct solutions concentrating at multiple points of $\mathbb{R}^n$. We do not give a precise definition concerning concentration at multiple points, which can be stated with obvious modifications of (A5.13).

Theorem A5.11 [139] *Let* (V1) *and* (V2) *hold, and suppose V possesses k distinct nondegenerate critical points* $x_1, \ldots, x_k$. *Then there exists a solution* $\bar{v}_\varepsilon$ *of* (A5.12) *which concentrates at* $\{x_1\} \cup \cdots \cup \{x_k\}$ *when* $\varepsilon \to 0$.

We conclude this section by mentioning a short list of other results in this direction, which can be proved with suitable adaptations of the above method. For example, one can treat the case of degenerate critical points of the potential, or even when the critical points of V arise in manifolds, see for example [26]. Working in a class of weighed spaces, it is indeed also possible to treat the case of potentials tending to zero at infinity: we refer for example to [30] (see also [29]). Other results include more general nonlinearities $f(u)$, for which condition (**ND**) might fail. In this case, under some mild assumptions on f, one can use penalization techniques and find solutions concentrating at local minima of V, see [88]. Recently, new types of solutions have been produced, which concentrate at sets of positive dimensions, like spheres or curves. When some symmetry is present, the abstract method can still be adapted to this case, see for example [27, 35, 46, 50], but in general totally different techniques are required. We refer the interested reader to the paper [89] and to the comments at the end of the next section.

A5.4 Singularly perturbed Neumann problems

In this section we consider the following singularly perturbed Neumann problem

$$\begin{cases} -\varepsilon^2 \Delta u + u = u^p & \text{in } \Omega \\ \dfrac{\partial u}{\partial \nu} = 0 & \text{on } \partial\Omega \\ u > 0 & \text{in } \Omega, \end{cases} \qquad (N_\varepsilon)$$

where Ω is a smooth bounded domain of $\mathbb{R}^n$, $p > 1$ is subcritical and ν denotes the outer unit normal at $\partial\Omega$. A problem like (N_ε) arises in the study of some reaction-diffusion systems with chemical or biological motivation. Referring to for example [136] for more details we simply mention that, according to the so-called *Turing instability*, systems with different diffusivities may produce stable nontrivial patterns. For example, the stationary *Gierer–Meinhardt* system (consisting of two coupled equations) can be reduced in some circumstances to (N_ε) when one of the diffusivity coefficients is very large and the other very small.

Problems (N_ε) and (A5.12) share some common features, in the sense that also (N_ε) admits solutions, called *spike layers*, concentrating at one or multiple

points of the closure of the domain. We want to describe below the proof of the
following result.

Theorem A5.12 *Suppose $\Omega \subseteq \mathbb{R}^n$, $n \geq 2$, is a smooth bounded domain, and
that $1 < p < (n+2)/(n-2)$ $(1 < p < +\infty$ if $n = 2)$. Suppose $x_0 \in \partial\Omega$ is
a local strict maximum or minimum, or a nondegenerate critical point of the
mean curvature H of $\partial\Omega$. Then for $\varepsilon > 0$ sufficiently small problem (N_ε) admits
a solution concentrating at x_0.*

Using a change of variables, problem (N_ε) can be reduced to the following

$$\begin{cases} -\Delta u + u = u^p & \text{in } \Omega_\varepsilon \\ \dfrac{\partial u}{\partial \nu} = 0 & \text{on } \partial\Omega_\varepsilon \\ u > 0 & \text{in } \Omega_\varepsilon, \end{cases} \qquad (\tilde{N}_\varepsilon)$$

where $\Omega_\varepsilon = 1/\varepsilon\Omega$, or equivalently to finding the existence of critical points of
the functional

$$I_\varepsilon(u) = \frac{1}{2} \int_{\Omega_\varepsilon} (|\nabla u|^2 + u^2) - \frac{1}{p+1} \int_{\Omega_\varepsilon} |u|^{p+1}$$

defined on $E = W^{1,2}(\Omega_\varepsilon)$. The functional I_ε is not of the form (A5.1), and in
particular there is not a standard critical manifold Z. However, it is possible
to modify the above abstract approach in the following way. For ε small, one
defines the set

$$Z_\varepsilon = \{z_x := u_0(\cdot - x) \ : \ x \in \partial\Omega_\varepsilon\},$$

which is a manifold in E diffeomorphic to $\partial\Omega_\varepsilon$. Z_ε turns out to be a *pseudocrit-
ical manifold* for I_ε, in the sense that $\|I_\varepsilon'(z)\|$ is small for every $z \in Z_\varepsilon$. Indeed,
all elements of Z_ε satisfy the first equation in $(\tilde{N}_\varepsilon)$, but not the boundary condition, and hence some term in the expression of $I_\varepsilon'(z)$ appears when integrating
by parts. One has indeed the following result.

Lemma A5.13 *For ε sufficiently small there exists a constant $C > 0$ such that
$\|I_\varepsilon'(z)\| \leq C\varepsilon$ for every $z \in Z_\varepsilon$.*

As in the proof of Lemma A5.2 (but using the contraction mapping theorem
instead of the implicit function theorem), denoting $W_z = (T_z Z_\varepsilon)^\perp$, one can
show that $PI_\varepsilon''(z) : W_z \to W_z$ is uniformly invertible, yielding the existence of
a function $w_\varepsilon(z)$ such that $I_\varepsilon'(z + w_\varepsilon(z)) \in T_z Z_\varepsilon$. Therefore one also finds the
following result.

Lemma A5.14 *For ε sufficiently small, let $\Phi_\varepsilon : Z_\varepsilon \to \mathbb{R}$ be defined by $\Phi_\varepsilon(z) =
I_\varepsilon(z + w_\varepsilon(z))$. Then if z_ε is critical for Φ_ε, $z_\varepsilon + w_\varepsilon(z_\varepsilon)$ is a stationary point of I_ε.*

The behaviour of Φ_ε is determined by the geometry of Ω: in fact, one can prove the following expansion

$$\Phi_\varepsilon(z_x) = \tfrac{1}{2}c_0 - c_1 \varepsilon H(\varepsilon x) + o(\varepsilon), \qquad (A5.16)$$

where c_0 is as in Section A5.2, c_1 is a positive constant and $H(y)$ stands for the mean curvature of $\partial\Omega$ at the point y. Then Theorem A5.12 follows reasoning as in the proof of Proposition A5.4.

As for the nonlinear Shrödinger equation, there exist solutions of types different from those discussed in Theorem A5.12. For example there are solutions which concentrate at the interior of Ω, and their location is determined by the distance function from the boundary. It is well known that problem (N_ε) also admits *multipeak* solutions, which concentrate at multiple points of $\overline{\Omega}$. For example, in [103] the authors prove the existence of solutions which have an arbitrarily large number of peaks, both at the interior and at the boundary of the domain.

Solutions concentrating at higher dimensional sets exist for problem (N_ε) as well, and some of them are known to concentrate at the whole $\partial\Omega$ or at some minimal k dimensional submanifold of $\partial\Omega$, see the papers [120, 123–125]. When some symmetry is present, it is possible to construct solutions which have the profile of interior spikes, and which concentrate at some k dimensional manifolds approaching the boundary when ε tends to zero. For these and related questions (like the analogous problem with Dirichlet boundary conditions) see for example the papers [28, 86, 126, 133]. These results suggest the possible presence of these (and maybe other) types of solutions also in nonsymmetric contexts.

A5.5 Perturbation of even functionals

The case of the perturbation of even functionals does not fit into the preceding set-up but requires a different approach. In this section we will outline some results on this interesting question.

In Theorem 10.12 we have shown that any even functional J satisfying (a) $J \in C^1(E, \mathbb{R})$, $J(0) = 0$ and $J(u) < 0$, for all $u \neq 0$, (b) J is weakly continuous and J' is compact, and (c) $J'(u) \neq 0$ for all $u \neq 0$, possesses infinitely many critical points z_k on unit sphere S of a separable Hilbert space E. A natural problem is to see what happens if J is perturbed by a functional which is not even. The first answer to this question has been given by Krasnoselski.

Theorem A5.15 [110], Theorem 4.6, Chapter VI. *Suppose that (a)–(c) hold and let $J_1 \in C^1(E, \mathbb{R})$ be such that*

$$\sup_{\|u\|\le 1} |J(u)| + \sup_{\|u\|\le 1} \|J'(u)\| \le \text{constant}.$$

Then for any $m \in \mathbb{N}$ there exists $\varepsilon_m > 0$ such that for $|\varepsilon| \leq \varepsilon_m$, the pertubed functional $J_\varepsilon = J + \varepsilon J_1$ has at least m critical points on S.

The proof of Theorem A5.15 relies on the construction of critical levels of J on S which are suitable for perturbation. Roughly, we consider the class

$$\tilde{\mathcal{A}}_m = \{A \subset S : A = \phi(S^m), \ \phi \in C(S^m, S), \ \text{odd}\},$$

where $S^m \subset S$ denotes the m dimensional sphere, and define

$$\tilde{\sigma}_m = \inf_{A \in \tilde{\mathcal{A}}_m} \max_{u \in A} J(u).$$

It is easy to check that $\tilde{\sigma}_m$ is a critical level for J on S. Moreover, since $\gamma(A) \geq \gamma(S^m) = m + 1$ for all $A \in \tilde{\mathcal{A}}_m$, it follows that $\tilde{\sigma}_m \geq \sigma_{m+1}$ (the levels σ_m have been defined in (10.1)). Therefore, using the fact that $\sigma_m \uparrow 0$, we infer that $\tilde{\sigma}_m \uparrow 0$ as well. Next, given $a < 0$, there exists $m = m(a) \in \mathbb{N}$ such that $\tilde{\sigma}_{m(a)} \leq a < \tilde{\sigma}_{m(a)+1}$. Let

$$\mathcal{M}_a = \{T \subset M : T = \chi(A), \ A \in \tilde{\mathcal{A}}_{m(a)}, \ \chi \in C(A, S), \ \sup_{\chi(A)} J \leq a\},$$

and define a new critical level of J on S by setting

$$d(a) = \inf_{T \in \mathcal{M}_a} \max_{u \in T} J(u).$$

It is worth pointing out that the new feature here is that we do not require that the maps χ in the definition of $\mathcal{M}_a$ are odd. For this reason, each level $d(a)$ can be perturbed. Moreover, one can prove that $d(a) \uparrow 0$ as $a \uparrow 0$ and this allows us to find m critical points of J_ε, provided $|\varepsilon| \leq \varepsilon_m$.

Theorem A5.15 applies to nonlinear eigenvalue problems like

$$\begin{cases} -\lambda \Delta u = |u|^{p-1} u + \varepsilon h(x, u) & x \in \Omega \\ u = 0 & x \in \partial\Omega. \end{cases} \tag{A5.17}$$

A result similar in nature to Theorem A5.15 can be proved for a functional like the one studied in Section 7.6. Moreover, Morse theory has been used in [130] to prove perturbation results in the above spirit. As a remarkable application, it has been proved in [130] that the eigenvalue problem

$$\begin{cases} -\lambda \Delta^2 u = u + h(x, u) & x \in \Omega \subset \mathbb{R}^2 \\ u = 0 & x \in \partial\Omega \\ \dfrac{\partial u}{\partial \nu} = 0 & x \in \partial\Omega, \end{cases}$$

has infinitely many solutions in $H^2(\Omega)$ with unit norm, provided h is smooth and such that

$$|h(x, u)| \leq k|u|^\gamma, \qquad \gamma > 3, \ k > 0.$$

Here, one works on the unit sphere of H^2 and the role of the *unperturbed* functional is played by the quadratic functional $\int u^2$. Its critical levels are nothing but the (linear) eigenvalues λ_m of $-\lambda \Delta^2 u = u$, $u = 0$, $\partial u / \partial v = 0$ on $\partial \Omega$ and it is known that $\lambda_m \sim m^{-2}$. This asymptotic behaviour and the growth restriction on h allow us to show that near a sequence of those λ_m, $m \gg 1$, there are critical levels of the functional $\int_\Omega u^2 + \int_\Omega dx \int_0^u h(x, s) \, ds$. This is the reason why the application deals with the bi-Laplacian Δ^2 in $\Omega \subset \mathbb{R}^2$. To find infinitely many solutions of a more general class of nonlinear eigenvalue problems is an open problem.

The interest of the preceding multiplicity result also relies on the fact that there are examples of perturbed functionals J_ε which have only a finite number of critical points on the unit sphere. The following example is due to Krasnoselski. Let

$$J(u) = -\sum_{m=1}^{\infty} \frac{1}{m^2} (u \mid e_m)^2,$$

where e_m is an orthonormal system in E, whose critical levels on S are given by $-1/m^2$. For any $\varepsilon > 0$ we can take $k(\varepsilon) \in \mathbb{N}$ such that the functional

$$J_{k(\varepsilon)}(u) = -\sum_{m=1}^{k} \frac{1}{m^2} (u \mid e_m)^2,$$

satisfies $\sup_{\|u\| \leq 1} |J(u) - J_{k(\varepsilon)}(u)| + \sup_{\|u\| \leq 1} \|J'(u) - J'_{k(\varepsilon)}(u)\| \leq \varepsilon$. Obviously, the *perturbed* functional $J_{k(\varepsilon)}$ has a only a finite number of critical points.

There are other specific, but important, classes of perturbed functionals which possess infinitely many critical points. The model problem is given by the functional $J : H_0^1(\Omega) \mapsto \mathbb{R}$,

$$J(u) = \frac{1}{2} \|u\|^2 - \frac{1}{p+1} \int_\Omega |u|^{p+1} \, dx - \int_\Omega h(x) u \, dx, \qquad h \in L^2(\Omega),$$

whose critical points are solutions of the superlinear Dirichlet BVP

$$\begin{cases} -\Delta u = |u|^{p-1} u + h(x) & x \in \Omega \\ u = 0 & x \in \partial \Omega. \end{cases} \tag{A5.18}$$

The following result is due to A. Bahri [37].

Theorem A5.16 *Let* $1 < p < (n+2)/(n-2)$. *Then* (A5.18) *has infinitely many solutions for a residual set of h in $H^{-1}(\Omega)$.*

On the other hand, if one wants to have a result which is not *generic*, a restriction on p is in order.

Theorem A5.17 *If* $1 < p < n/(n-2)$ *then* (A5.18) *has infinitely many solutions.*

Theorem A5.17 is a particular case of a more general one due to Bahri and Lions [43] and improves some previous results [39, 152, 165]. It is not known whether the bound $n/(n-2)$ is optimal. It is worth pointing out that, though the proofs rely on pertubation arguments, h is not required to be *small*. This is due, roughly, to the fact that the critical levels σ_m of the unperturbed functional (namely when $h = 0$) tend to infinity and the gaps between them increase as $m \to \infty$. It is also natural to ask whether some perturbation result can be obtained for coercive problems like

$$\begin{cases} -\Delta u = \lambda u - |u|^{p-1}u + \varepsilon h(x) & x \in \Omega \\ u = 0 & x \in \partial\Omega, \end{cases} \qquad (D^-_{\lambda,\varepsilon})$$

where $1 < p < (n+2)/(n-2)$. When $\varepsilon = 0$ the nonlinearity is odd and Theorem 10.22 yields the existence of at least k pairs of nontrivial solutions to $(D^-_{\lambda,0})$, provided $\lambda > \lambda_k$ (λ_k denotes the kth eigenvalue of $-\Delta$ on $H^1_0(\Omega)$). The question is whether these solutions persist under perturbation. The situation here is very different from the superlinear problems (A5.18) or the nonlinear eigenvalue problems (A5.17), and none of the preceding perturbation arguments can be used to handle $(D^-_{\lambda,\varepsilon})$. The difference relies on the fact that the Euler functional related to $(D^-_{\lambda,0})$ has a finite number of critical levels and two or more of them could also coincide. This degeneracy allows Dancer [85] to show that the result valid for $\varepsilon = 0$ cannot be extended in such a generality to the case $\varepsilon \neq 0$. A possible conjecture is to show that, given any $k \in \mathbb{N}$, there exists $\Lambda(k) > 0$ and $\varepsilon(k) > 0$ such that $(D^-_{\lambda,\varepsilon})$ has at least $2k$ solutions provided $\lambda > \Lambda(k)$ and $|\varepsilon| < \varepsilon(k)$.

Appendix 6

Some problems arising in differential geometry

We treat here some problems motivated by differential geometry, which amount to solving nonlinear elliptic equations involving the critical Sobolev exponent. This is deeply related to one of the main features of these equations, namely the lack of compactness. The latter is typical of geometric problems, since they are usually characterized by some (at least asymptotically) scaling invariance.

A6.1 The Yamabe problem

We begin this section with a short list of notions in Riemannian geometry, referring for example to [34] for detailed derivations of the geometric quantities, their motivation and (more) applications.

Let (M, g) be a compact n dimensional manifold, endowed with a metric g. Let (U, η), $U \subseteq M$, $\eta : U \to \mathbb{R}^n$, be a local coordinate system and let g_{ij} denote the components of the metric g. We also denote with g^{ij} the elements of the inverse matrix $(g^{-1})_{ij}$, and with dV_g the volume element, which is given by

$$dV_g = \sqrt{\det g}\, dx. \tag{A6.1}$$

The *Christoffel symbols* are defined by $\Gamma_{ij}^l = \frac{1}{2}[D_i g_{kj} + D_j g_{ki} - D_k g_{ij}]g^{kl}$ (D_i stands for the derivative with respect to x_i) while the *Riemann curvature tensor*, the *Ricci tensor* and the *Scalar curvature* are given respectively by

$$R_{kij}^l = D_i \Gamma_{jk}^l - D_j \Gamma_{ik}^l + \Gamma_{im}^l \Gamma_{jk}^m - \Gamma_{jm}^l \Gamma_{ik}^m, \qquad R_{kj} = R_{klj}^l, \qquad R_g = R_{kj} g^{kj}. \tag{A6.2}$$

Hereafter, we use the standard convention that repeated (upper and lower) indices are summed over all their range (usually between 1 and n). For $n \geq 3$, the *Weyl tensor* W_{ijkl} is then defined as

$$W_{ijkl} = R_{ijkl} - \frac{1}{n-2}(R_{ik}g_{jl} - R_{il}g_{jk} + R_{jl}g_{ik} - R_{jk}g_{il})$$

$$+ \frac{R}{(n-1)(n-2)}(g_{jl}g_{ik} - g_{jk}g_{il}).$$

For a smooth function u the components of its gradient $\nabla_g u$ are

$$(\nabla_g u)^i = g^{ij} D_j u. \tag{A6.3}$$

The Laplace–Beltrami operator, applied to a C^2 function $u : M \to \mathbb{R}$, is given by the following expression

$$\Delta_g u = g^{ij} \left(D_{ij}^2 u - \Gamma_{ij}^k D_k u \right) = \frac{1}{|dV_g|} D_m \left(|dV_g| g^{mk} D_k u \right). \tag{A6.4}$$

We say that two metrics g and $\tilde{g}$ on M are *conformally equivalent* if there is a smooth function $\rho(x) > 0$ such that $\tilde{g} = \rho\, g$. If $n \geq 3$, using the (convenient) notation $\tilde{g} = u^{4/(n-2)}\, g$, the scalar curvature $R_{\tilde{g}}$ of $(M, \tilde{g})$ is related to R_g by the following formula

$$-2c_n \Delta_g u + R_g u = R_{\tilde{g}} u^{(n+2)/(n-2)}, \qquad c_n = 2\frac{(n-1)}{(n-2)}. \tag{A6.5}$$

For $n = 2$, the scalar curvature coincides with the Gauss curvature K_g, and if one sets $\tilde{g} = e^{2u}\, g$, we have the analogous transformation rule

$$-\Delta_g u + K_g = K_{\tilde{g}}\, e^{2u}. \tag{A6.6}$$

In the spirit of the *uniformization theorem* for the two dimensional case (a classical result of Poincaré in which conformal metrics of constant Gauss curvature are found), for $n \geq 3$ the *Yamabe problem* consists in finding a conformal metric $\tilde{g}$ with constant scalar curvature $\overline{R}$ on M. By (A6.5), the problem amounts to looking for positive solutions of

$$-2c_n \Delta_g u + R_g u = \overline{R} u^{(n+2)/(n-2)} \qquad \text{on } M. \tag{A6.7}$$

The structure of equation (A6.5) is variational, and the presence of the exponent $(n+2)/(n-2)$ makes the study of (A6.7) a noncompact problem. This implies in particular that the associated Palais–Smale sequences do not converge in general, so the analytic study of (A6.5) is rather difficult.

One can try to find solutions of (A6.7) as minima of the Sobolev-type quotient

$$Q_{M,g}(u) := \frac{\int_M (|\nabla_g u|^2 + (1/2c_n) R_g\, u^2)\, dV_g}{\left(\int_M |u|^{2^*} \right)^{2/2^*}}, \qquad u \in W^{1,2}(M) \setminus \{0\}. \tag{A6.8}$$

Defining $\mu_{M,g}$ to be the infimum of the above quotient over all (nonzero) functions u, one can check from (A6.5) that $\mu_{M,g}$ is a conformal invariant called the *Yamabe invariant*, and is usually denoted by $Y(M, [g])$, where $[g]$ stands for the conformal class of g. A compact manifold is called of positive (respectively of null, or of negative) type depending on whether $Y(M, [g])$ is positive (respectively null, or negative). One can easily check that if the Yamabe problem is solvable, then the constant $\overline{R}$ in (A6.7) necessarily has the same sign as $Y(M, [g])$.

Trudinger [169] showed that one always has $Y(M, [g]) \leq Y(S^n, [g_0])$, where g_0 stands for the standard metric of S^n. Indeed, since the sphere (with one point removed) is conformally equivalent to $\mathbb{R}^n$ through the *stereographic projection*, it is possible to prove that $Y(S^n, [g_0])$ coincides with the best Sobolev constant S for the embedding of $W^{1,2}(\mathbb{R}^n)$ into $L^{2^*}(\mathbb{R}^n)$. Furthermore, if $Y(M, [g]) < S$ the infimum is attained, similarly to arguments of Section 11.2. In particular, when $Y(M, [g]) \leq 0$, solutions can be found rather easily.

A first result concerning manifolds of positive type was given by Th. Aubin [33] for the case of dimension n greater than or equal to 6 and when M is not locally conformally flat. In these dimensions (actually for $n \geq 4$), for the latter condition to hold it is necessary and sufficient that $W_g \neq 0$. To verify the condition $Y(M, [g]) < S$, Aubin used test functions u_ε highly concentrated near a point x_0 of M where $W_g(x_0) \neq 0$, and which are similar to those used in Section 11.2. He proved the following estimates

$$Q_{M,g}(u_\varepsilon) = \begin{cases} S - a_n \varepsilon^4 |W_g(x_0)|^2 + o(\varepsilon^4) & \text{for } n > 6 \\ S - a_n \varepsilon^4 |\log \varepsilon| |W_g(x_0)|^2 + o(\varepsilon^4 \log \varepsilon) & \text{for } n = 6, \end{cases} \quad \text{(A6.9)}$$

where a_n is a dimensional constant. From the last formula, taking ε sufficiently small, it follows immediately that $Y(M, [g]) < S$ and therefore the minimizing sequences for $Q_{M,g}$ stay compact.

The complementary cases in the positive Yamabe class were considered by R. Schoen in [155]. The proof is in the same spirit, except that the test functions $\tilde{u}_\varepsilon$ have to take care of the global geometry of the manifold, and not only of its local nature near the concentration point x_0. Schoen's test functions glue together U_ε (the notation is still from Section 11.2) and the Green function of the *conformal Laplacian*, which is defined by $-2c_n \Delta_g + R_g$. Choosing some (in fact, special) coordinates near x_0, the Green function $G_{M,g}(y, x_0)$ with pole x_0 writes as

$$G_{M,g}(y, x_0) = \frac{1}{(n-2)\omega_{n-1}} \frac{1}{|y - x_0|^{n-2}} + A_{x_0} + O(|y - x_0|), \quad y \neq x_0$$

where $\omega_{n-1} = \mathrm{Vol}(S^{n-1})$, and A_{x_0} is a suitable constant. It is a consequence of the *positive mass theorem* by Schoen and Yau, see [157], that $A_{x_0} > 0$ if $Y(M, [g]) > 0$. Using this result, Schoen proved that

$$Q_{M,g}(u_\varepsilon) = S - a\varepsilon^{n-2} + o(\varepsilon^{n-2}), \quad \text{(A6.10)}$$

where a is a positive constant, obtaining again that $Y(M, [g]) < S$, and hence existence of a solution to the Yamabe problem.

We also mention that in dimensions 3, 4 and 5 there is an alternative proof of existence due to Bahri and Brezis [40] which bypasses the positive mass theorem and uses instead the theory of *critical points at infinity* developed by Bahri and Coron. Solutions found with this method in general are not minimizers for the Sobolev type quotient. Furthermore, some multiplicity results are also available, see for example [19, 55, 144, 156].

A6.2 The scalar curvature problem

The prescribed scalar curvature problem consists in deforming conformally the standard metric g_0 on the sphere S^n so that the new scalar curvature becomes a given function f. By equation (A6.5), since the scalar curvature of the standard sphere is $n(n-1)$, the problem amounts to finding a positive solution to the equation

$$-2c_n\Delta_{g_0}u + n(n-1)u = fu^{(n+2)/(n-2)} \qquad \text{on } S^n. \qquad (A6.11)$$

As one can easily see using integration by parts, a necessary condition for the solvability is that f should be positive somewhere. But a more subtle obstruction is present. Indeed, with a slightly more involved integration by parts (which is similar to that in the proof of Theorem 8.30) Kazdan and Warner [107] proved that any solution of (A6.11) satisfies the identity

$$\int_{S^n} \langle \nabla_{g_0}f, \nabla_{g_0}x_i\rangle dV_{g_0} = 0, \qquad i = 1, \dots, n+1, \qquad (KW)$$

where x_i stands for the ith coordinate function of $\mathbb{R}^{n+1}$ restricted to S^n. For example, this rules out the possibility of existence for functions f of the form $f = 2 + x_i$, which are positive everywhere on the sphere.

Concerning existence of solutions there are several results, and due to the substantial dependence of the solvability on the datum f, they all require rather sophisticated techniques. We are going to discuss briefly the following theorem proved by Bahri and Coron [41].

Theorem A6.1 *Suppose $n = 3$, and that $f : S^3 \to \mathbb{R}$ is a positive Morse function for which $\Delta_{g_0}f(p) \neq 0$ at every critical point p of f. Then, if the following formula holds*

$$\sum_{\{p\,:\,\nabla_{g_0}f(p)=0,\Delta_{g_0}f(p)<0\}} (-1)^{\mathrm{ind}(f,p)} \neq -1, \qquad (A6.12)$$

problem (A6.11) is solvable.

We present here a proof by Chang, Gursky and Yang [77], since it is easier for us to describe it using the contents of the previous chapters (and the previous sections in the appendix).

One first constructs a homotopy f_t between the function f and the constant $n(n-1)$, simply defined by

$$f_t(x) = tf(x) + (1-t)n(n-1), \qquad x \in S^n, t \in [0,1],$$

and considers the one-parameter family of problems

$$-2c_n\Delta_{g_0}u + n(n-1)u = f_t u^{(n+2)/(n-2)} \qquad \text{on } S^n. \tag{A6.13}$$

Solutions of (A6.13) can be found as critical points of the functional $I_t : W^{1,2}(S^n) \to \mathbb{R}$ defined by

$$I_t(u) = c_n \int_{S^n} |\nabla_{g_0}u|^2 \, dV_{g_0} + \frac{n(n-1)}{2} \int_{S^n} u^2 \, dV_{g_0} - \frac{1}{2^*} \int_{S^n} |u|^{2^*} \, dV_{g_0},$$

where dV_{g_0} stands for the volume element of S^n, so I_t has the form $I_t = I_0 + tG$, with

$$I_0(u) = c_n \int_{S^n} |\nabla_{g_0}u|^2 \, dV_{g_0} + \frac{n(n-1)}{2} \int_{S^n} u^2 \, dV_{g_0} - \frac{n(n-1)}{2^*} \int_{S^n} |u|^{2^*} \, dV_{g_0},$$

$$G(u) = \frac{1}{2^*} \int_{S^n} [f(x) - n(n-1)]|u|^{2^*} \, dV_{g_0}.$$

When t is small, this functional fits in the abstract framework described in Appendix 5. We will describe next what the unperturbed manifold Z is in this case, and we will evaluate the effect of the perturbation G. To do this, we introduce some preliminary notation.

For $P \in S^n$ and for $t \in [1, +\infty)$ we define the point $p \in B_1^{n+1}$ (the unit ball of $\mathbb{R}^{n+1}$) by $p = ((t-1)/t)P$. We let $\pi_P : S^n \to \mathbb{R}^n$ denote the stereographic projection through the point $-P$, and in stereographic coordinates on S^n (induced by π_P) we define the map

$$\varphi_{P,t}(y) = ty, \qquad y \in \mathbb{R}^n.$$

We then consider the metric $g_{P,t} = (\varphi_{P,t})^*g_0$, where $(\varphi_{P,t})^*$ stands for the *pull-back* of g_0 through $\varphi_{P,t}$ (see for example [34] for the notation). Since $\varphi_{P,t}$ is *conformal*, there exists a positive function $z_{P,t} : S^n \to \mathbb{R}$ such that $z_{P,t}^{4/(n-2)} g_0 = (\varphi_{P,t})^*g_0$. We then define the manifold

$$Z = \left\{ z_{P,t} \, : \, P \in S^n, t \in [0, +\infty) \right\}.$$

Using the (one-to-one) correspondence $(P,t) \mapsto (t - 1/t)P$, we easily see that Z is homeomorphic to B_1^{n+1}, and indeed diffeomorphic. It turns out that the properties (*Fr*) and (*ND*) are satisfied, so we can apply Propositions A5.3 and A5.4.

Therefore, we analyse next the behaviour of $G|_Z$, especially near the boundary of B_1^{n+1} (through the above identification). It can be shown that $\int_{S^n} |z_{P,t}|^{2^*}$ is independent of (P, t) (in particular equal to $|S^n|$ since $z_{P,0} \equiv 1$), and that $|z_{P,t}|^{2^*}$ converges weakly in the sense of measures to $|S^n|$ times the Dirac delta at P when t tends to $+\infty$. As a consequence one finds that

$$G(z_{P,t}) \to \frac{|S^n|}{2^*}(f(P) - n(n-1)) \qquad \text{as } t \to +\infty.$$

A more detailed analysis shows that

$$G(z_{P,t}) = \frac{|S^n|}{2^*}(f(P) - n(n-1)) + \frac{c_1\Delta_{g_0}f(P)}{t^2} + o\left(\frac{1}{t^2}\right) \qquad \text{as } t \to +\infty,$$

where c_1 is a positive constant depending only on n. Using this expansion one can check that, letting $B_s = sB_1^{n+1}$, the gradient of $\Gamma := G|_Z$ is never zero on ∂B_s if $s \in (0, 1)$ is sufficiently close to 1. In fact, the component of $\nabla\Gamma$ tangent to ∂B_s is (asymptotically) proportional to the gradient of f, while the component normal to ∂B_s has the sign of $\Delta_{g_0}(f)$. By the assumptions of Theorem A6.1, $\nabla_{g_0}f$ and $\Delta_{g_0}f$ can never be zero simultaneously. This allows us to define $\deg(\Gamma, B_s, 0)$ for s close to 1, and it is possible to show that

$$\deg(\Gamma, B_s, 0) = \sum_{\{p \,:\, \nabla_{g_0}f(p)=0, \Delta_{g_0}f(p)<0\}} (-1)^{\text{ind}(f,p)} + 1,$$

so under the assumption (A6.12) the degree is nonzero, and we can apply Proposition A5.4 to obtain a solution of (A6.13) for t sufficiently small.

For $\alpha \in (0, 1)$, let now $L_{g_0} : C^{2,\alpha}(S^n) \to C^\alpha(S^n)$ denote the operator $-2c_n\Delta_{g_0} + n(n-1)$. Then solutions of (A6.13) can also be found as fixed points of the operator $u \mapsto L_{g_0}^{-1}(f_t u^{(n+2)/(n-2)})$. By the results in [77] it turns out that for $t = t_0$ sufficiently small there exists a constant $C(t_0)$ such that

$$\deg\left(u - L_{g_0}^{-1}(f_{t_0}u^{(n+2)/(n-2)}), \left\{\frac{1}{C(t_0)} < v < C(t_0)\right\}, 0\right) = -\deg(\Gamma, B_s, 0).$$

Furthermore, under the assumption that f is Morse, there exists another constant, which we still denote by C_{t_0} such that for $t \in [t_0, 1]$ every (positive) solution v of (A6.13) satisfies $1/C_{t_0} < v < C_{t_0}$. It follows that for $t \in [t_0, 1]$, $u - L_{g_0}^{-1}(f_t u^{(n+2)/(n-2)})$ is different from zero on the boundary of $\{1/C_{t_0} < v < C_{t_0}\}$. Hence, from the homotopy property of the degree, from (A6.12) and the last formulas, we have also

$$\deg\left(u - L_{g_0}^{-1}(f_u^{(n+2)/(n-2)}), \left\{\frac{1}{C(t_0)} < v < C(t_0)\right\}, 0\right) \neq 0.$$

As a conclusion, problem (A6.11) is solvable.

Some comments are in order. First of all, Theorem A6.1 has been extended by Schoen and Zhang in [158], where the authors use the Morse inequalities instead of the degree, so that more general conditions for solvability are given.

In general, the requirements for the existence of solutions strongly depend not only on the datum f, but also on the dimension n. For example, still under the assumption that f is Morse, for $n = 4$ there is another index formula of the form (A6.12), but in which each summand involves multiple points at a time, see [47] and the second part of [117]. If one wants to keep a formula similar to (A6.12), the function f needs to be suitably *flat* near its critical points: naively, for any critical point p_i of f, the main term in the expansion $f(p) - f(p_i)$ should be a homogeneous function of order β in $p - p_i$, with $\beta \in (n-2, n)$, see the first part of [117]. We also refer to the papers [74, 75] (which actually came first) for the case $n = 2$, where the Gauss curvature on the sphere is prescribed, solving equation (A6.6).

Also, there are other kinds of results which are based on different methods and ideas. For example, in [56] (see also [57]) a suitable min-max scheme is used, and the assumptions involve only critical points lying in certain levels of f. Other contributions exploit some symmetry of the datum [58, 70, 104], or the perturbative nature of the problem when f is close to a constant, see for example [18, 25, 76].

Finally, we mention that there are several other problems arising from conformal geometry, which involve operators of higher order or fully nonlinear equations. We refer to the recent monograph [73] for further details and recent results.

References

[1] S. Ahmad, A. C. Lazer and J. L. Paul, *Indiana Univ. Math. J.* **25** (1976), 933–944.

[2] S. Alama and G. Tarantello, *Calc. Var. Partial Differential Equations* **1** (1993), 439–475.

[3] A. D. Alexandrov, *Ann. Mat. Pura Appl.* **58** (1962), 303–354.

[4] H. Amann, *Lectures on some Fixed Point Theorems*, Rio de Janeiro: I.M.P.A. 1974.

[5] H. Amann, Fixed point equations and nonlinear eigenvalue problems in ordered Banach space, *SIAM Rev.* **18** (1976), 620–709.

[6] H. Amann, *Proc. Am. Math. Soc.* **85** (1982), 591–595.

[7] H. Amann and S. Weiss, *Math. Z.* **130** (1973), 39–54.

[8] H. Amann, A. Ambrosetti and G. Mancini, *Math. Z.* **158** (1978), 179–194.

[9] A. Ambrosetti, *Atti Accad. Naz. Lincei* **52** (1972), 660–667.

[10] A. Ambrosetti, *J. Math. Phys. Sci.* **18**(1) (1984), 1–12.

[11] A. Ambrosetti, *J. Anal. Math.* **76** (1998), 321–335.

[12] A. Ambrosetti and M. Badiale, *J. Math. Anal. Appl.* **104**(2) (1989), 363–373.

[13] A. Ambrosetti and M. Badiale, *Proc. R. Soc. Edinburgh, Sect.* A **128** (1998), 1131–1161.

[14] A. Ambrosetti and M. Badiale, Homoclinics, *Ann. Inst. Henri Poincaré* **15** (1998), 233–252.

[15] A. Ambrosetti and J. Gamez, *Math. Z.* **224** (1997), 347–362.

[16] A. Ambrosetti and P. Hess, *J. Math. Anal. Appl.* **73**(2) (1980), 411–422.

[17] A. Ambrosetti and D. Lupo, *J. Nonlinear Anal. TMA* **8**(10) (1984), 1145–1150.

[18] A. Ambrosetti and A. Malchiodi, *J. Funct. Anal.* **168** (1999), 529–561.

[19] A. Ambrosetti and A. Malchiodi, *Perturbation Methods and Semilinear Elliptic Problems on* $\mathbb{R}^n$, Progress in Mathematics, Vol. 240, Basel: Birkhauser, 2005.

[20] A. Ambrosetti and G. Prodi, *A Primer of Nonlinear Analysis*, Cambridge: Cambridge University Press, 1993.

[21] A. Ambrosetti and P. H. Rabinowitz, *J. Funct. Anal.* **14** (1973), 349–381.

[22] A. Ambrosetti and M. Struwe, *Arch. Rat. Mech. Anal.* **108**(2) (1989), 97–109.

[23] A. Ambosetti and R. Turner, *Differential Integral Equations* **1**(3) (1988), 341–349.

[24] A. Ambrosetti, H. Brezis and G. Cerami, *J. Funct. Anal.* **122**(2) (1994), 519–543.

[25] A. Ambrosetti, J. Garcia Azorero and I. Peral, *J. Funct. Anal.* **165** (1999), 117–149.

[26] A. Ambrosetti, A. Malchiodi and S. Secchi, *Arch. Rat. Mech. Anal.* **159** (2001), 253–271.

[27] A. Ambrosetti, A. Malchiodi and W.-Mi. Ni, *Commun. Math. Phys.* **235** (2003), 427–466.

[28] A. Ambrosetti, A. Malchiodi and W.-Mi. Ni, *Indiana Univ. Math. J.* **53** (2004), 297–329.

[29] A. Ambrosetti, V. Felli and A. Malchiodi, *J. Eur. Math. Soc.* **7** (2005), 117–144.

[30] A. Ambrosetti, A. Malchiodi and D. Ruiz, *J. Anal. Math.* **98** (2006), 317–348.

[31] C. J. Amick and R. Turner, *J. Reine Angew. Math.* **384** (1988), 1–23.

[32] D. Arcoya and J. Gamez, *Commun. Partial Differential Equations* **26** (9–10) (2001), 1879–1911.

[33] Th. Aubin, *J. Math. Pures Appl.* **55** (1976), 269–296.

[34] Th. Aubin, *Some Nonlinear Problems in Differential Geometry*, Berlin: Springer, 1998.

[35] M. Badiale and T. d'Aprile, *Nonlinear Anal. Ser. A*, **49** (2002), 947–985.

[36] M. Badiale and A. Pomponio, *Proc. R. Soc. Edinburgh, Sect.* A **134** (2004), 11–32.

[37] A. Bahri, *J. Funct. Anal.* **41** (1981), 397–427.

[38] A. Bahri, *Critical Points at Infinity in Some Variational Problems*, Pitman Research Notes in Mathematics Series, Vol 182, New York: Wiley, 1989.

[39] A. Bahri and H. Berestycki, *Trans. Am. Math. Soc.* **267** (1981), 1–32.

[40] A. Bahri and H. Brezis, *Elliptic Differential Equations Involving the Sobolev Critical Exponent on Manifolds*, PNLDE 20, Boston, MA: Birkäuser.

[41] A. Bahri and J. M. Coron, *J. Funct. Anal.* **95** (1991), 106–172.

[42] A. Bahri and Y. Y. Li, *Rev. Mat. Iberoam.* **6** (1990), 1–16.

[43] A. Bahri and P.-L. Lions, *Commun. Pure Appl. Math.* **41** (1988), 1027–1037.

[44] A. Bahri and P.-L. Lions, *Ann. Inst. Hemri Poincaré*, **14** (1997), 365–413.

[45] C. Bandle, *Isoperimetric Inequalities and Applications*, London: Pitman 1980.

[46] T. Bartsch, and S. Peng, Semiclassical symmetric Schrödinger equations: existence of solutions concentrating simultaneously on several spheres, preprint.

[47] M. Ben Ayed, Y. Chen, H. Chtioui and M. Hammami, *Duke Math. J.* **84** (1996), 633–677.

[48] V. Benci, *Trans. Am. Math. Soc.* **274** (1982), 533–572.

[49] V. Benci and G. Cerami, *J. Funct. Anal* **88** (1990), 90–117.

[50] V. Benci and T. d'Aprile, *J. Differential Equations* **184** (2002), 109–138.

[51] V. Benci and D. Fortunato, *Ann. Mat. Pura Appl.* **32** (1982), 215–242.

[52] V. Benci and P. Rabinowitz, *Invent. Math.* **52** (1979), 241–273.

[53] H. Berestycki and P.-L. Lions, *Arch. Rat. Mech. Anal.* **82** (1983), 313–345.

[54] F. A. Berezin and M. A. Shubin, *The Schrödinger Equation*, Dordrecht: Kluwer, 1991.

[55] M. Berti and A. Malchiodi, *J. Funct. Anal.* **180** (2001), 210–241.

[56] G. Bianchi, *Adv. Differential Equations* **1** (1996), 857–880.

[57] G. Bianchi and H. Egnell, *Arch. Rat. Mech. Anal.* **122** (1993), 159–182.

[58] G. Bianchi, J. Chabrowski and A. Szulkin, *J. Nonlinear Anal. TMA* **25** (1995), 41–59.

[59] R. Böhme, *Math. Z.* **128** (1972), 105–126.

[60] H. Brezis, Analyse Fonctionelle – Théorie et Applications, Paris: Masson, 1983.

[61] H. Brezis and S. Kamin, *Manus. Math.* **74** (1992), 87–106.

[62] H. Brezis and T. Kato, *J. Math. Pures Appl.* **58** (1979), 137–151.

[63] H. Brezis and E. Lieb, *Proc. Am. Math. Soc.* **88** (1983), 486–490.

[64] H. Brezis and L. Nirenberg, *Commun. Pure Appl. Math.* **36** (1983), 437–477.

[65] H. Brezis and L. Nirenberg, *C. R. Acad. Sci. Paris* **317** (1993), 465–472.

[66] H. Brezis and R. E. L. Turner, *Commun. Partial Differential Equations* **2** (1977), 601–614.

[67] F. Browder, *Ann. Math.* **82** (1965), 459–477.

[68] R. F. Brown, *A Topological Introduction to Nonlinear Analysis*, Boston, MA: Birkhäuser, 1993.

[69] D. M. Cao, *J. Nonlinear Anal. TMA* **15** (1990), 1045–1052.

[70] F. Catrina and Z. Q. Wang, *Indiana Univ. Math. J.* **49** (2000), 779–813.

[71] J. Chabrowski, *Variational Methods for Potential Operator Equations. With Applications to Nonlinear Elliptic Equations*, de Gruyter Studies in Mathematics, Vol. 24, 1997.

[72] K. C. Chang, *Infinite-dimensional Morse Theory and Multiple Solution Problems*, PNLDE 6, Boston, MA: Birkhäuser, 1993.

[73] S. A. Chang, *Non-linear Elliptic Equations in Conformal Geometry*, Zurich Lectures in Advanced Mathematics, Zürich: European Mathematical Society, 2004.

[74] S. A. Chang and P. Yang, *Acta Math.* **159** (1987), 215–259.

[75] S. A. Chang and P. Yang, *J. Differential Geom.* **27** (1988), 256–296.

[76] S. A. Chang and P. Yang, *Duke Math. J.* **64** (1991), 27–69.

[77] S. A. Chang, M. J. Gursky and P. Yang, *Calc. Var.* **1** (1993), 205–229.

[78] W. X. Chen and C. Li, *Duke Math. J.* **63** (1991), 615–622.

[79] S. Cingolani and J. L. Gomez, *Adv. Differential Equations* **1**(5) (1996), 773–791.

[80] F. Clarke, *C. R. Acad. Sci. Paris* **287** (1978), 951–952.

[81] M. Conti, S. Terracini and G. Verzini, *Ann. Inst. Henri Poincaré Anal. Non Liniaire*, **19** (2002), 871–888.

[82] R. Courant and D. Hilbert, *Methods of Mathematical Physics*, New York: Interscience, 1962.

[83] M. Crandall and P. H. Rabinowitz, *J. Math. Mech.* **19** (1969/1970) 1083–1102.

[84] B. Dacorogna, *Direct Methods in the Calculus of Variations*, New York: Springer, 1989.

[85] N. Dancer, *Math. Ann.* **272** (1985), 421–440.

[86] E. N. Dancer and S. Yan, A new type of concentration solutions for a singularly perturbed elliptic problem, preprint.

[87] K. Deimling, *Nonlinear Functional Analysis*, Berlin: Springer, 1985.

[88] M. Del Pino and P. Felmer, *Calc. Var.* **4** (1996), 121–137.

[89] M. Del Pino, M. Kowalczyk and J. Wei, Concentration at curves for nonlinear Schrödinger equations, *Commun. Pure Appl. Math.*, to appear.

[90] J. Dugundji, *Pacific J. Math.* **1** (1951), 353–367.

[91] A. L. Edelson and C. Stuart, *J. Differential Equations* **124** (1996), 279–301.

[92] I. Ekeland, *J. Math. Anal. Appl.* **47** (1974), 324–353.

[93] I. Ekeland, *Convexity Methods in Hamiltonian Mechanics*, Berlin: Springer, 1990.

[94] D. de Figueiredo, P.-L. Lions and R. Nussbaum, *J. Math. Pures Appl.* **61** (1982), 41–63.

[95] I. Fonseca and W. Gangbo, *Degree Theory in Analysis and Applications*, Oxford Lecture Series in Mathematics and its Applications, New York: Oxford University Press, 1995.

[96]   L. E. Fraenkel and M. Berger, *Acta Math.* **132** (1974), 13–51.

[97]   N. Ghoussoub, *J. Reine Angew. Math.* **417**, 27–76.

[98]   B. Gidas and J. Spruck, *Commun. Pure Appl. Math.* **34** (1981), 525–598.

[99]   B. Gidas, W. M. Ni and L. Nirenberg, *Commun. Math. Phys.* **68** (1979), 209–243.

[100] B. Gidas, W. M. Ni and L. Nirenberg, *Symmetry of Positive Solutions of Nonlinear Elliptic Equations in R^n*, Mathematical Analysis and Applications A, Adv. in Math. Suppl. Stud. 7a, pp. 369–402, New York: Academic Press, 1981.

[101] D. Gilbarg and N. S. Trudinger, *Elliptic Partial Differential Equations of Second Order*, Berlin: Springer, 1977.

[102] E. Giusti, *Direct Methods in the Calculus of Variations*, Singapore: World Scientific, 2003.

[103] C. Gui and J. Wei, *Can. J. Math.* **52** (2000), 522–538.

[104] E. Hebey, *Bull. Sci. Math.* **114** (1990), 215–242.

[105] E. Heinz, *J. Math. Mech.* **8** (1959), 231–247.

[106] H. Hofer, *Proc. Am. Math. Soc.* **90** (1984), 309–315.

[107] J. L. Kazdan and F. Warner, *Ann. Math.* **101** (1975), 317–331.

[108] H. Kielhofer, *J. Funct. Anal.* **77** (1988), 1–8.

[109] H. Kielöfer, *J. Funct. Anal.* **77** (1988), 1–8.

[110] M. A. Krasnoselski, *Topological Methods in the Theory of Nonlinear Integral Equations*, Oxford: Pergamon Press, 1964.

[111] M. K. Kwong, *Arch. Rat. Mech. Anal.* **105**(3) (1989), 243–266.

[112] E. A. Landesman and A. C. Lazer, *J. Math. Mech.* **19** (1970), 609–623.

[113] S. Lang, *Differential and Riemannian Manifolds*, Graduate Texts in Mathematics, Vol. 160, New York: Springer, 1995.

[114] A. C. Lazer and S. Solimini, *J. Nonlinear Anal.* **10** (1986), 411–413.

[115] J. Leray and J. Schauder, *Ann. Sci. Ec. Norm. Super.* **51** (1934), 45–78.

[116] S. J. Li, *Acta Math. Sci.* **4**(2) (1984), 135–140.

[117] Y. Y. Li, *J. Differential Equations* **120** (1995), 319–410; *Commun. Pure Appl. Math.* **49** (1996), 437–477.

[118] P.-L. Lions, *Ann. Inst. H. Poincaré* **1**(2) (1984), 109–145; **1**(4) (1984), 223–283.

[119] P.-L. Lions, *Rev. Mat. Iberoam.* **1**(1) (1985), 541–597; **1**(2) (1985), 45–121.

[120] F. Mahmoudi and A. Malchiodi, Concentration on minimal submanifolds for a singularly perturbed Neumann problem, *Adv. in Math.*, to appear.

[121] P. Majer, *Topology* **34** (1995), 1–12.

[122] A. Malchiodi, *Calc. Var.* **14** (2002), 429–445.

[123] A. Malchiodi, *G.A.F.A.* **15**(6) (2005), 1162–1222.

[124] A. Malchiodi and M. Montenegro, *Commun. Pure Appl. Math.* **15** (2002), 1507–1568.

[125] A. Malchiodi and M. Montenegro, *Duke Math. J.* **124** (2004), 105–143.

[126] A. Malchiodi, W. M. Ni and J. Wei, *Ann. Inst. Henri Poincaré* **22** (2005), 143–163.

[127] A. Manes and A. M. Micheletti, *Boll. Un. Mat. Ital.* **7**(4) (1973), 285–301.

[128] A. Marino, *Conf. Sem. Mat. Univ. Bari* **132** (1973).

[129] A. Marino and G. Prodi, *Rend. Sem. Mat. Univ. Padova* **41** (1968), 43–68.

[130] A. Marino and G. Prodi, *Boll. Un. Mat. Ital.* **3** (1975), 1–32.

[131] W. Massey, *A Basic Course in Algebraic Topology*, Graduate Texts in Mathematics, Vol 127. New York: Springer, 1991.

[132] J. Milnor, *Morse Theory*, Annals of Mathematics Studies, No. 51, Princeton, NJ: Princeton University Press, 1963.

[133] R. Molle and D. Passaseo, Concentration phenomena for solutions of superlinear elliptic problems, *Ann. Inst. Henri Poincaré* to appear.

[134] M. Morse and G. Van Schaack, *Ann. Math.* **35** (1934), 545–571.

[135] J. Moser, On a nonlinear problem in differential geometry, in *Dynamical Systems* (M. Peixoto ed.), New York: Academic Press, pp. 273–280, 1973.

[136] W.-M. Ni, *J. Anal. Math.* **37** (1980), 208–247.

[137] L. Nirenberg, *Topics in Nonlinear Functional Analysis (with notes by Ralph A. Artino)*, Courant Lecture Notes, Providence, RI: A.M.S., 2001.

[138] J. Norbury, *Proc. Cambridge Philos. Soc.* **72** (1972), 253–284.

[139] Y. G. Oh, *Commun. Math. Phys.* **131** (1990), 223–253.

[140] F. Pacella, *J. Funct.: Anal.* **192** (2002), 271–282.

[141] R. Palais, *Topology* **5** (1966), 115–132.

[142] R. Palais, *Commun. Math. Phys.* 69 (1979). 19–30.

[143] S. Pohozaev, *Sov. Math.* **5** (1965), 1408–1411.

[144] D. Pollack, *Commun. Anal. Geom.* **1** (1993), 347–414.

[145] G. Prodi, *Boll. Un. Mat. Ital.* **22** (1967), 413–433.

[146] P. Pucci and J. Serrin, *Indiana Univ. Math. J.* **35** (1986), 681–703.

[147] P. H. Rabinowitz, *J. Funct. Anal.* **7** (1971), 487–513.

[148] P. H. Rabinowitz, *Théorie du Degreé Topologique et Applications à des Problémes aux Limites Non Linéaires (rédigé par H. Berestycki)*, Notes Université Paris VI, 1973.

[149] P. H. Rabinowitz, *Rocky Mount. J. Math.* **3** (1973), 161–202.

[150] P. H. Rabinowitz, Some minimax theorems and applications to nonlinear partial differential equations, in *Nonlinear Analysis: A Collection of Papers in Honor of Erich Röthe*, New York: Academic Press, pp. 161–177, 1978.

[151] P. H. Rabinowitz, *Ann. Scuola Norm. Sup. Pisa Cl. Sci.* **5** (1978), 215–223.

[152] P. H. Rabinowitz, *Trans. Am. Math. Soc.* **272** (1982), 753–770.

[153] P. H. Rabinowitz, *Minimax Methods in Critical Point Theory with Applications to Differential Equations*, CBMS Reg. Conf. Ser. in Math., Vol. 65, Providence, RI: A.M.S., 1986.

[154] D. H. Sattinger, *Topics in Stability and Bifurcation Theory*, Lecture Notes in Mathematics, Vol. 309, Berilin: Springer, 1973.

[155] R. Schoen, *J. Differential Geom.* **20** (1984), 479–495.

[156] R. Schoen, On the number of constant scalar curvature metrics in a conformal class, in *Differential Geometry: a Symposium in Honor of M. do Carmo*, New York: Wiley, pp. 311–320, 1991.

[157] R. Schoen and S. T. *Yau, Invent. Math.* **92** (1988), 47–71.

[158] R. Schoen and D. Zhang, *Calc. Var.* **4** (1996), 1–25.

[159] J. T. Schwartz, *Commun. Pure Appl. Math.* **17** (1964), 307–315.

[160] J. T. Schwartz, *Nonlinear Functional Analysis*, New York: Gordon and Breach, 1969.

[161] J. Serrin, *Arch. Rat. Mech. Anal*, **43** (1971), 304–318.

[162] E. H. Spanier, *Algebraic Topology*, New York: Springer, 1981.

[163] M. Spivak, *A Comprehensive Introduction to Differential Geometry*, Second edition, Wilmington, DE: Publish or Perish, 1979.

[164] W. Strauss, *Commun. Math. Phys.* **55** (1977), 149–162.

[165] M. Struwe, *Manus. Math.* **32** (1980), 335–364.

[166] M. Struwe, *Variational Methods*, Ergeb. der Math. u. Grenzgeb., Vol. 34, Berlin: Springer, 1996.

[167] C. Stuart, Bifurcation from the essential spectrum, in *Topological Nonlinear Analysis, II* (M. Matzeu and A. Vignoli eds.), PNLDE 27, Boston: MA: Birkhäuser, pp. 397–444, 1997.

[168] A. Szulkin, *Ann. Inst. Henri Poincarè* **5** (1988), 119–13.

[169] N. Trudinger, *Ann. Scuola Norm. Sup.* **22** (1968), 265–274.

[170] G. T. Whyburn, *Topological Analysis*, Princeton, NJ: Princeton University Press, 1955.

[171] M. Willem, *Minimax Theorems*, PNLDE 24, New York: Birkhäuser, 1996.

Index